INTRODUCTION TO THE THEORY OF INFINITESIMALS

Pure and Applied Mathematics

A Series of Monographs and Textbooks

RECENT TITLES

E. R. Kolchin. Differential Algebra and Algebraic Groups

Gerald J. Janusz. Algebraic Number Fields

A. S. B. Holland. Introduction to the Theory of Entire Functions

Wayne Roberts and Dale Varberg. Convex Functions

A. M. Ostrowski. Solution of Equations in Euclidean and Banach Spaces, Third Edition of Solution of Equations and Systems of Equations

H. M. Edwards. Riemann's Zeta Function

Samuel Eilenberg. Automata, Languages, and Machines: Volumes A and B

Morris Hirsch and Stephen Smale. Differential Equations, Dynamical Systems, and Linear Algebra

Wilhelm Magnus. Noneuclidean Tesselations and Their Groups

François Treves. Basic Linear Partial Differential Equations

William M. Boothby. An Introduction to Differentiable Manifolds and Riemannian Geometry

Brayton Gray. Homotopy Theory: An Introduction to Algebraic Topology

Robert A. Adams. Sobolev Spaces

John J. Benedetto. Spectral Synthesis

D. V. Widder. The Heat Equation

Irving Ezra Segal. Mathematical Cosmology and Extragalactic Astronomy

J. Dieudonné. Treatise on Analysis: Volume II, enlarged and corrected printing; Volume IV; Volume V. *In preparation*

Werner Greub, Stephen Halperin, and Ray Vanstone. Connections, Curvature, and Cohomology: Volume III, Cohomology of Principal Bundles and Homogeneous Spaces

I. Martin Isaacs. Character Theory of Finite Groups

James R. Brown. Ergodic Theory and Topological Dynamics

K. D. Stroyan and W. A. J. Luxemburg. Introduction to the Theory of Infinitesimals

In preparation

Clifford A. Truesdell. A First Course in Rational Continuum Mechanics: Volume 1, General Concepts

B. M. Puttaswamaiah and John D. Dixon. Modular Representations of Finite Groups

Melvyn Berger. Nonlinearity and Functional Analysis: Lectures on Nonlinear Problems in Mathematical Analysis

George Gratzer. Lattice Theory

INTRODUCTION TO THE THEORY OF INFINITESIMALS

K. D. Stroyan
Department of Mathematics
The University of Iowa
Iowa City, Iowa

In collaboration with

W. A. J. Luxemburg
Department of Mathematics
California Institute of Technology
Pasadena, California

ACADEMIC PRESS
New York San Francisco London 1976
A Subsidiary of Harcourt Brace Jovanovich, Publishers

ACADEMIC PRESS, INC.
111 Fifth Avenue, New York, New York 10003

United Kingdom Edition published by
ACADEMIC PRESS, INC. (LONDON) LTD.
24/28 Oval Road, London NW1

Library of Congress Cataloging in Publication Data

Stroyan, K D
Introduction to the theory of infinitesimals.

(Pure and applied mathematics, a series of monographs and textbooks ;)
Bibliography: p.
1. Mathematical analysis, Nonstandard. I. Luxemburg, W. A. J., (date) joint author. II. Title. III. Title: Infinitesimals. IV. Series.
QA3.P8 [QA299.82] 510′.8s [515′.33] 76-14344
ISBN 0–12–674150–6
AMS (MOS) 1970 Subject Classifications: 02-00, 02-01, 26A98, 30A76, 46-00, 46-02

PRINTED IN THE UNITED STATES OF AMERICA

To the memory of
ABRAHAM ROBINSON

CONTENTS

Preface, xi
Acknowledgments, xv

PART 1 CLASSICAL INFINITESIMALS

1. INTRODUCTION: WHAT ARE INFINITESIMALS?

2. A FIRST LOOK AT ULTRAPOWERS: A MODEL OF RATIONAL ANALYSIS

2.1 A Free Ultrafilter $\mathscr{U}$ on a Countable Set J, 7
2.2 An Ultrapower of the Rational Numbers, 8
2.3 Some Calculus of Polynomials, 10
2.4 The Exponential Function, 14
2.5 Peano's Existence Theorem, 16
2.6 Summary, 18

3. SUPERSTRUCTURES AND THEIR NONSTANDARD MODELS

3.1 Introduction, 20
3.2 Definition of a Superstructure, 23
3.3 Superstructures Are Big Enough, 25
3.4 Nonstandard Models of Superstructures, 25
3.5 The Formal Language, 30
3.6 Interpretations of the Formal Language, 35
3.7 Models of a Superstructure, 36
3.8 Nonstandard Ultrapower Models, 37
3.9 Bounded Formal Sentences, 42
3.10 Embedding $\mathscr{X}^J$ in Set Theory, 43
3.11 $*$-Transforms of Categories, 46
3.12 Postscript to Chapter 3, 47

4. SOME BASIC FACTS ABOUT HYPERREAL NUMBERS

4.1 Addition, Multiplication, and Order in $^*\mathbf{R}$, 49
4.2 Some Simplifications of the Notation, 51
4.3 $^*\mathbf{R}$ Is Non-Archimedean, 52
4.4 Infinite, Infinitesimal, and Finite Numbers, 53
4.5 Some External Entities, 56
4.6 Further Simplification of Notation and Classical Functions, 58
4.7 Hypercomplex Numbers, 60

APPENDIX A. PRELIMINARY RESULTS ON ORDERED RINGS AND FIELDS

A.1 Terminology, 62

A.2 Ordered Rings and Fields, 64
A.3 Archimedean Totally Ordered Fields, 67

5. FOUNDATIONS OF INFINITESIMAL CALCULUS

5.1 Continuity and Limits, 70
5.2 Uniform Continuity, 77
5.3 Basic Definitions of Calculus, 78
5.4 The Mean Value Theorem, 85
5.5 The Fundamental Theorem of Calculus, 88
5.6 Landau's "Oh-Calculus," 90
5.7 Differential Vector Calculus, 92
5.8 Integral Vector Calculus, 110
5.9 Calculus on Manifolds, 126

6. TOPICS IN INFINITESIMAL CALCULUS

6.1 Peano's Existence Theorem Revisited, 142
6.2 Interchanging Limits, 145
6.3 Euler's Product for the Sine Function, 147
6.4 Robinson's Lemma and Generalized Limits, 150
6.5 Dynamical Systems, 155
6.6 Geometry of the Unit Ball and Boundary Behavior, 159

PART 2 INFINITESIMALS IN FUNCTIONAL ANALYSIS

7. MORE TOOLS FROM MODEL THEORY

7.1 Countable Ultrapowers, 175
7.2 Enlargements, 176
7.3 Comprehensive Models, 180
7.4 Saturated Models, 180
7.5 Ultralimits, 183
7.6 Properties of Polysaturated Models, 187
7.7 The Isomorphism Property of Ultralimits, 188

8. THE GENERAL THEORY OF MONADS AND INFINITESIMALS

8.1 Monads with Respect to a Ring of Sets, 195
8.2 Chromatic Sets, 198
8.3 Topological Aspects of Monad Theory, 199
8.4 Uniform Infinitesimal Relations and Finite Points, 205
8.5 Topological Infinitesimals at Remote Points, 221

9. COMPACTIFICATIONS

9.1 Discrete Čech–Stone Compactification of $\mathbf{N}$, 228
9.2 Measurable Infinitesimals, 231
9.3 The Samuel Compactification of the Hyperbolic Plane, 235
9.4 Normal Meromorphic Functions and Analytic Disks, 238
9.5 The Fatou–Lindelöf Boundary, 242
9.6 Bounded Holomorphic Functions and Gleason Parts, 244
9.7 Fixed Points of Analytic Maps on $\mathcal{M}$, 252
9.8 The Bohr Group, 259

10. LINEAR INFINITESIMALS AND THE LOCALLY CONVEX HULL

10.1 Basic Theory, 268
10.2 Hilbert Spaces, 285
10.3 Banach Spaces, 292
10.4 Distributions, 299
10.5 Mixed Spaces, 306
10.6 (HM)-Spaces, 312
10.7 Postscript to Chapter 10, 314

References, 316

Index, 323

PREFACE

Looking back into the history of mathematics we cannot escape observing that the seventeenth century was a most important epoch in the development of mathematics. During that period two mathematicians of extraordinary power, I. Newton and G. W. Leibniz, appeared on the scene, who, although each from a somewhat different point of view, created a general framework in which the methods of the calculus could be fully developed. It was particularly Leibniz who during his lifetime strongly advocated that a proper way to develop the calculus was within a number system which in addition to the finite numbers would contain infinitely small as well as infinitely large numbers thought of as "ideal" elements. Despite its inherent contradictions Leibniz' view predominated the calculus until the middle of the nineteenth century when it was rejected as unsound and replaced by the ε, δ method of Weierstrass. Infinitesimals survived only as a matter of speech and as an intuitive notion. A mathematical theory of infinitesimals as proposed by Leibniz seemed to be a hopeless quest despite the fact that the problem attracted the attention of some mathematicians of stature.

This situation has changed dramatically during the last decade or so. It is now clear that a consistent and rigorous theory can be developed in a manner almost identical to that suggested by Leibniz some three centuries ago.

In 1958, based on a generalization of Cantor's construction of the reals, C. Schmieden and D. Laugwitz presented such an enlarged number system. Their system, however, is in a certain sense too large in that it contains not only finite, infinitely small, and infinitely large numbers but also numbers of an indeterminate size.

Completely independent of the work of Schmieden and Laugwitz, Abraham Robinson discovered in 1960 an entirely new and different method to attack and completely solve Leibniz' problem of the infinitesimal. Using the general methods of model theory Robinson showed that the structure that formalizes algebraic and order properties of the real numbers can be enlarged to include infinitely small as well as infinitely large numbers and which in a well-defined sense has the same properties as the reals. In due course, Robinson and his followers showed that these methods, nowadays called nonstandard methods, can be successfully applied to branches of mathematics other than analysis, notably algebraic number theory and topology.

This book gives an up-to-date account of these developments. It has its origins in an earlier set of notes on the same topic written by the senior author in 1961. In these notes the author presented a kind of "naïve" approach to Robinson's theory of infinitesimals via the ultrapower construction. The main purpose is to give an elementary exposition of these new developments for the working analyst, and at the same time to point out the connection between on the one hand the Schmieden and Laugwitz approach and on the other the Robinson approach. The authors strongly feel that the present, although far more complete, exposition still pervades strongly the philosophy of the original notes.

The book begins with an introduction to the ultrapower construction applied to the real numbers and treats a little calculus. This is followed by a complete but elementary treatment of the iogicai framework on which the method rests. This part is rounded off by an elementary but careful discussion of the number systems obtained by taking ultrapowers of the real and complex numbers. These chapters are intended for a reader who is not trained in mathematical logic and thus contain many details. Perhaps careful reading of the brief Chapter 2 followed by a casual reading of Chapters 3 and 4 would be sufficient to begin the topics in Chapter 5 or 6. Such a reader could return to the details of Chapters 3 and 4 as needed. It is our hope that most of the first six chapters will be accessible to upper-division undergraduates (and beyond).

Chapter 5, "The Foundations of Infinitesimal Calculus," gives a fairly complete account of the basic theorems of calculus, including differential forms and Lie brackets. It begins with the basic concepts in one variable. A concept which plays a fundamental role is that of "uniform differentiability" in (5.7.5) and (5.7.9) with $k = 1$. In effect one can say directly what it means for $f\colon \mathbf{R} \to \mathbf{R}$ to be continuously differentiable: *if there is an arbitrary (standard) function $f'(x)$ such that whenever $x \in {}^*\mathbf{R}$ and δ is a nonzero infinitesimal, we have*

$$\frac{f(x+\delta) - f(x)}{\delta} \approx f'(x),$$

then $f'(x)$ is continuous and the derivative of f. More importantly, this condition underlies the proofs of many of the basic theorems—recognizing that simplifies them.

Chapter 6 contains a collection of applications of infinitesimals to classical analysis and is essentially independent of Chapter 5. There are many more applications which we could not include, and we hope our readers will develop some of their own; we think that is a lot of fun and sometimes quite illuminating besides.

The latter half of the book is devoted to more advanced topics. The notion of a polysaturated model is introduced, and its existence is shown via the construction of ultralimits. Applications are given to general topology and functional analysis. The last two chapters of the book are devoted to topics such as the Čech compactification of a topological space, the Stone representation spaces, the maximal ideal spaces of H^∞, the Bohr compactification, the theory of topological vector spaces, distributions, and so on. Some of the results of the last three chapters have not been published elsewhere. We hope this part will provide a useful reference for further research as well as an organized account of general infinitesimals. In this regard, polysaturated models seem to be the right framework for function alanalysis; however, other choices might be more appropriate for other applications (for example, $\aleph_1$-saturated nonenlargements or Boolean-valued models). We have made a split—using weaker extensions than Robinson's enlargements for Part 1 and stronger ones for Part 2.

ACKNOWLEDGMENTS

The senior author expresses his thanks to the National Science Foundation for financial support received at various stages during the preparation of this work. He also wishes to thank the junior author who wrote the entire final draft of the manuscript. Without his drive and energies the book would never have been completed. Special thanks are due to Mrs. L. Decker who typed the manuscripts of the earlier versions back in 1961 and 1964.

The following institutions have been instrumental to the junior author in the creation of this book: California Institute of Technology, especially his teachers H. F. Bohnenblust and W. A. J. Luxemburg; the University of Wisconsin, Madison, where an early draft was prepared as lecture notes and where he received stimulation and encouragement from many colleagues and students; Gesamthochschule Paderborn where a second draft was begun for a month of lectures there; Mathematische Forschungsinstitut Oberwolfach where, through the efforts of D. Laugwitz and W. A. J. Luxemburg, two stimulating symposia have taken place; and the University of Iowa where the bulk of the writing was done and where the infinitesimal contributions of day-to-day conversations with several colleagues certainly have added up to a finite amount. Summer fellowships from The University of Iowa supported him during part of the writing of the book.

Detailed suggestions of K. J. Barwise, C. W. Henson, H. J. Keisler, F. Wattenburg, and especially L. C. Moore, Jr. are gratefully acknowledged.

We have dedicated the book to the memory of our teacher and friend Abraham Robinson, whose untimely death has deprived us of his wisdom.

The outstanding craftmanship of A. Burns in typing the final manuscript has made the writing much easier.

Finally, we express our deep appreciation for the assistance and cooperation received from the staff of our publisher, Academic Press, Inc.

W.A.J.L.
K.D.S.

CLASSICAL INFINITESIMALS PART 1

INTRODUCTION: WHAT ARE INFINITESIMALS?

CHAPTER 1

Simply put, infinitesimal analysis is Abraham Robinson's solution to an old problem of Leibniz. Leibniz believed that infinitesimals were ideal numbers, a fiction useful for the art of mathematical invention. Leibniz maintained that a system of numbers including the "real" numbers and the ideal infinite and infinitesimal ones could be devised that was governed by the "same laws" as the ordinary numbers. In 1961, Robinson showed that a nonstandard model of the formal theory of analysis provides such a system of hyperreal numbers and moreover provides rigorous foundations for many of the intuitive correct uses of infinitesimals in mathematics. This book is an up-to-date account of developments surrounding Robinson's splendid application of model theory to analysis.

The history of infinitesimals dates at least as far back as Archimedes. Centuries of development providing determination of special cases of areas, volumes, tangents, centers of gravity, moments of inertia, etc., preceded the development of a general theory of differentiation and integration in the second half of the seventeenth century. Despite earlier controversies it is generally accepted now that Isaac Newton (1642–1727) and Gottfried Wilhelm Leibniz (1646–1716) independently founded the calculus. The controversy is complicated by Newton's secrecy about his methods in his earlier publications. Leibniz' knowledge of Christiaan Huygens' (1629–1695) work on the geometry of the cycloids, tractrix, catenary, and so on, plus his understanding that Newton was in possession of powerful new methods, certainly contributed to his creation of the calculus of differentials. There is some evidence, however, that Newton's fluxion as a general method was created after Leibniz' calculus of differentials appeared as a general method.

Leibniz treated his differentials of various orders like numbers of infinitely small magnitude. That is, he combined them with other numbers through addition and multiplication while considering them smaller than any nonzero ordinary numbers. Reciprocals of differentials were considered to be infinitely large. Leibniz' d-calculus symbolized derivatives by $dy = f(x + dx) - f(x) = f'(x)\,dx$ and the inverse operation integration by $\int f(x)\,dx$. His notation had clear technical advantages and brought about rapid development of the subject on the European continent, where it was accepted.

Despite its power and beauty the foundations of infinitesimal calculus were shaky. The glaring contradictions were apparent to all and the subject of severe criticism. In 1734, Bishop Berkeley gave a clear account of the difficulties in "A Discourse Addressed to an Infidel Mathematician" (see Newman [1956]). The practice of throwing away "higher-order terms" on one side of an equation while maintaining equality led Berkeley to call infinitesimals "... the ghosts of departed quantities." Infinitesimal analysis deals with this in part by introducing a new equivalence relation "is infinitesimally close to" in place of "equals." Very many attempts were made to free foundations of infinitesimals from contradiction both by Leibniz and later. Leibniz thought of his hoped-for ideal numbers in terms like the "imaginary" numbers are treated algebraically. It is not without interest for us to observe that modern logic has shown the reals to be ideal in several respects (the intuitionists, the constructivists, Cohen's work, and Robinson's). Of course none of these discoveries affect real physical measurement, since neither perfect measurement nor infinitesimal tolerances ever were possible. Perhaps even twentieth century mathematics has not found the best answer to the question of foundations.

Leibniz and his followers were never able to state with sufficient precision just what rules were supposed to govern their new system including infinitely small as well as infinitely large quantities. Leibniz did state the principle that what holds for the finite numbers holds for the extended system, but it is not at all clear in his writings to what sort of laws about numbers his principle was supposed to apply. All this eventually led to the downfall of Leibniz' theory of infinitesimals. Cauchy presented a foundation for the calculus in the form of a theory of limits, which Weierstrass later converted into the now generally accepted epsilon–delta method in analysis. (Cauchy used infinitesimals.)

Through the development in algebra which led to the discovery of non-Archimedean fields, new interest arose in infinitesimals. Using non-Archimedean fields, Hahn [1907] was able to develop a consistent theory of infinitesimals. No serious attempt was made at that time, however, to use these fields in analysis and, in fact, there was a strong belief that this could not be done.

We warn the reader that Robinson's theory of infinite numbers is quite distinct from Cantor's theory of cardinal and ordinal numbers. Cantor's infinite numbers do not submit to the rules of ordinary arithmetic and thus violate Leibniz' principle. No doubt the successes of Cantor's theory, especially in the Lebesgue integral and topology, have contributed to the modern disfavor for infinitesimals. We see no conflict; in fact, we believe important mathematics will continue to develop out of the interaction of the two and that neither theory is "real" since they do not submit to experiments.

More recently, Laugwitz [1959] and Laugwitz and Schmieden [1958] proposed a theory of infinitesimals for the prime purpose of representing the Dirac delta function by a point function whose value is infinitely small except near the origin where it is infinitely large, and whose integral is equal to one. Although this theory unquestionably has merits, it does not give a solution to the problem of developing the calculus with the use of infinitely small and infinitely large quantities in the sense of Leibniz.

A complete and satisfactory solution of Leibniz' problem was presented by Abraham Robinson [1961]. He showed that the ideas and methods of model theory can clarify the notion of "infinitely small" and "infinitely large." More precisely, Robinson showed that there exist proper extensions ***R** of the field of real numbers **R** which in a certain sense have the same formal properties as **R**. It is well known that fields that are proper extensions of **R** are non-Archimedean, and so must contain infinitely small numbers as well as infinitely large numbers. We stated earlier, however, that ***R** in some sense has the same properties as **R**, which seems to be a paradox. There is, however, no paradox. If we examine Robinson's ideas more closely, then we see that the statement "***R** has the same properties as **R**" refers to a specified collection of properties of **R**, including the Archimedean property, that is formulated in a certain formal language. The statements of this language have specific interpretations in **R** as well as in ***R**, and the reinterpretation of higher-order properties like the Archimedean property do not retain their full metamathematical strength. This weak interpretation in the extension gives rise to a preferred class of sets called "internal sets" which the formal language "knows about." These have the "same properties"; the external ones do not. First-order properties really are the same.

Soon after Robinson's discovery it became clear that the method could be applied with equal success to mathematical structures other than the real and complex number systems. This new method is now known as the "theory of infinitesimals."

We believe that this theory provides foundations for modern analysis, topology, and some algebra; ones which are as firm as the foundations that have evolved from Weierstrass. This does not imply a conflict either; we simply have two ways of looking at many things. We hope the infinitesimals will prove to be advantageous in the formulation and solution of open questions and in clear understanding of difficult known work.

Our presentation of the subject is somewhat different from Robinson's book, "Non-Standard Analysis" [1966]. We develop the subject from set theory and a "construction" due to J. Łos [1955], called an "ultrapower." This gives us the model explicitly modulo the existence of a free ultrafilter. This approach was developed by W. A. J. Luxemburg [1962] in the first days of infinitesimal analysis and bears a striking similarity to Laugwitz and Schmieden's earlier theory. Our applications in Part 2 to functional analysis require another important contribution of Luxemburg, the use of saturated models, which we construct by means of direct limits of ultrapowers. This approach gives the model theory a certain amount of concreteness which may be helpful to those readers who are not mathematical logicians.

Finally we urge the reader to read Robinson's fascinating account of the history of calculus presented as one chapter of his book. Moreover, we encourage the reader to consult the freshman calculus text by Keisler [1976], the notes by Luxemburg [1962] and Machover and Hirschfeld [1969], the many applications of Robinson's book which we have not covered, and the proceedings of three symposia on the subject: Luxemburg [1969], Luxemburg and Robinson [1972], and Hurd and Loeb [1974].

A FIRST LOOK AT ULTRAPOWERS: A MODEL OF RATIONAL ANALYSIS

CHAPTER 2

In this chapter we give an explicit nonstandard model of analysis over the rational numbers by means of the ultrapower construction of Łos [1955]. We hope to accomplish several things, so perhaps some explanation is in order. First, by means of these simple applications we hope intuitively to convince the reader that a general transfer principle can be formulated. Second, we hope these simple examples serve to make the abstract setting of the model theory more clear and palatable. Third, we hope to convince the reader that this method simultaneously provides many interesting limit procedures of elementary analysis via infinitesimals. This is one reason why we chose the rationals as our base space—the reader can compare this construction of the reals with the more conventional ones. Another more important reason, however, is that we want the reader to begin to think in terms of a set-theoretical hierarchy with more than one level. Therefore we start out a half-level below the algebra of **R** to force the issue. The infinite sums and products and the infinitesimal partitions live higher up. Formal analysis will have to include all these things.

2.1 A FREE ULTRAFILTER $\mathscr{U}$ ON A COUNTABLE SET J

We introduce the notion of a free ultrafilter by means of the following axioms. The existence of free ultrafilters is nonconstructive and a consequence of the axiom of choice.

(2.1.1) DEFINITION *Let J be a countable set. A nonempty set $\mathscr{U}$ of subsets of J [$\varnothing \subset \mathscr{U} \subset \mathscr{P}(J)$ the set of all subsets of J] is called a free ultrafilter provided:*

(1) $\varnothing \notin \mathscr{U}$ (Proper filter).

(2) *If A and $B \in \mathscr{U}$, then $A \cap B \in \mathscr{U}$ and in particular $A \cap B \neq \varnothing$* (Finite intersection property).

(3) *If $A \in \mathscr{U}$ and $B \in \mathscr{P}(J)$, the power set of J, and if $A \subseteq B$, then $B \in \mathscr{U}$* (Superset property).

(4) *If $B \in \mathscr{P}(J)$, then either $B \in \mathscr{U}$ or $J \backslash B = \{j \in J : j \notin B\} \in \mathscr{U}$* (Maximality).

(5) *No finite subset of J is an element of $\mathscr{U}$* (Freeness).

The reader should observe that the Frechet filter $\mathrm{Fr}(J) = \{A \in \mathscr{P}(J) : (J \setminus A)$ is finite$\}$ satisfies all the axioms except (4). The fixed ultrafilter $\{A \in \mathscr{P}(J) : j_0 \in A\} = \mathscr{U}(j_0)$ satisfies all except (5).

2.2 AN ULTRAPOWER OF THE RATIONAL NUMBERS

In this section we introduce a number system that we will use to develop a little classical analysis without the usual limits. Let $\mathbf{Q}$ be the ordered field of rational numbers. (The usual thing, ratios of integers.)

Now consider the set of all sequences of rational numbers $\mathbf{Q}^J = \{a \mid a: J \to \mathbf{Q}\}$; the value of $a \in \mathbf{Q}^J$ at $j \in J$ is denoted by $a(j)$ $(J = \{0, 1, 2, \ldots\})$. We could introduce algebraic operations pointwise: $a, b, c \in \mathbf{Q}^J$,

$$a + b = c \quad \text{provided} \quad a(j) + b(j) = c(j) \quad \text{for all} \quad j \in J,$$

$$a \cdot b = c \quad \text{provided} \quad a(j) \cdot b(j) = c(j) \quad \text{for all} \quad j \in J,$$

$$a \le b \quad \text{provided} \quad a(j) \le b(j) \quad \text{for all} \quad j \in J.$$

This algebraic system has some nice features: the ordinary rationals are embedded in it as constant sequences and nonconstant sequences could serve as limits. Unfortunately, it has many undesirable properties; for example, it is not totally ordered and it has zero divisors, hence is not a field. We eliminate these problems simply by making identifications modulo a free ultrafilter $\mathscr{U}$. Now we define hyperrationals.

(2.2.1) DEFINITION *We replace equality by the relation*

$$a =_{\mathscr{U}} b \qquad \textit{provided} \qquad \{j : a(j) = b(j)\} \in \mathscr{U},$$

that is, a and b are equivalent mod $\mathscr{U}$ *provided they agree on an element of* $\mathscr{U}$. *Similarly, the algebraic operations are given as follows:*

$$a + b =_{\mathscr{U}} c \qquad \textit{provided} \qquad \{j : a(j) + b(j) = c(j)\} \in \mathscr{U},$$

$$a \cdot b =_{\mathscr{U}} c \qquad \textit{provided} \qquad \{j : a(j) \cdot b(j) = c(j)\} \in \mathscr{U},$$

and

$$a \le_{\mathscr{U}} b \qquad \textit{provided} \qquad \{j : a(j) \le b(j)\} \in \mathscr{U}.$$

The set $\mathbf{Q}^J$ *with equality replaced by* $=_{\mathscr{U}}$ *and the operations as above will be denoted by* $^*\mathbf{Q}$. *We shall denote the set of constant sequences in* $^*\mathbf{Q}$ *by* $^\sigma\mathbf{Q}$ *and use the value of a constant sequence for its name, that is, the sequence* $a(j) = \frac{1}{2}$ *for all j will just be referred to as* $\frac{1}{2}$ *both in* $\mathbf{Q}$ *and* $^\sigma\mathbf{Q}$ *even though it is a sequence in* $^\sigma\mathbf{Q}$. *We shall also denote the set of natural number sequences* *by* ***N** *and the constant natural number sequences by* $^\sigma\mathbf{N} = \{0, 1, 2, \ldots$ (as constant sequences)$\}$.

The prefix σ stands for the embedded *standard* copy.

First, we observe that $=_{\mathscr{U}}$ *is an equivalence relation*

$$a =_{\mathscr{U}} a \qquad \text{because} \qquad \{j : a(j) = a(j)\} = J.$$

By (4) in the definition of $\mathscr{U}$ either J or $\varnothing \in \mathscr{U}$ and by (1) $\varnothing \notin \mathscr{U}$. Clearly, $a =_{\mathscr{U}} b$ if and only if $b =_{\mathscr{U}} a$. Finally, if $a =_{\mathscr{U}} b$ and $b =_{\mathscr{U}} c$, then $\{j : a(j) = c(j)\} \supseteq \{j : a(j) = b(j) = c(j)\}$, and since $\{j : a(j) = b(j)\} \in \mathscr{U}$ and $\{j : b(j) = c(j)\} \in \mathscr{U}$ by assumption, (2) and (3) tell us that $a =_{\mathscr{U}} c$.

The importance of (5) in the definition of $\mathscr{U}$ is that ***Q** is strictly larger than $^\sigma\mathbf{Q}$. For example, since J is countable there is a sequence $\omega(j)$ which is one-to-one and onto **N**. This sequence satisfies $\{j : \omega(j) = n\} = \{j_n\}$, a singleton, for each $n \in \mathbf{N}$. Thus $\omega \neq_{\mathscr{U}} q$ for any $q \in {}^\sigma\mathbf{Q}$. In fact, we can see that ω is positive and *infinite* since the set where it is positive is J and the set where any $q \in {}^\sigma\mathbf{Q}$ exceeds ω is at most finite, therefore not in $\mathscr{U}$. *Since ω exceeds every constant sequence, we say it is infinite.*

Since $\omega \neq 0$, $1/\omega$ is defined simply by $v(j) = 1/\omega(j)$ because $v\omega = \omega v = 1$ (on an element of $\mathscr{U}$ at least). Naturally, $1/\omega$ is *infinitesimal*, meaning that for each $n \in {}^\sigma\mathbf{N}$, $-1/n < 1/\omega < 1/n$. This is because we already know $\omega(j) \geq n$ for j on an element of $\mathscr{U}$, hence $0 < 1/\omega(j) < 1/n$ on an element of $\mathscr{U}$.

We shall denote the *finite numbers* by $\mathcal{O}(\mathbf{Q})$ or $\mathcal{O}$, "big oh." These are the sequences $a \in {}^*\mathbf{Q}$ for which there are constant sequences p and $q \in {}^\sigma\mathbf{Q}$ such that $p \leq a \leq q$. We denote the *infinitesimals* defined above by $o(\mathbf{Q})$ or o, "small oh."

Now we shall summarize some algebraic properties of ***Q**. The details of this theorem are contained in the later chapters on ordered rings and the basic facts about ***R**. We are only interested in a summary since we intend to work with these numbers rather than study them at this point.

(2.2.2) THEOREM ***Q** *is an ordered field,* $\mathcal{O}$ *an ordered ring with* o *a maximal order ideal, and* $\mathcal{O}/o$ *a complete ordered (Archimedean) field, hence isomorphic to the reals* **R**. *(The canonical homomorphism from* $\mathcal{O}$ *to* **R** *is called standard part* $o \to \mathcal{O} \xrightarrow{\text{st}} \mathbf{R}$.*)*

The proof that ***Q** is an ordered field follows from the corresponding properties in **Q** applied to sequences using the axioms for $\mathscr{U}$. They are similar to the discussion above concerning $=_{\mathscr{U}}$ and ω. We mention only the following consequence of (4) since it is typical.

Consider the two sequences of rationals

$$(0, 1, 0, 1, 0, \ldots) \qquad \text{and} \qquad (1, 0, 1, 0, \ldots).$$

Multiplying pointwise we obtain

$$(0 \cdot 1, 1 \cdot 0, 0 \cdot 1, \ldots) = (0, 0, 0, \ldots)$$

and it might seem that we have zero divisors (in fact we do not). Axiom (4) tells us that one of these sequences is equal to zero mod $\mathscr{U}$ and the other is equal to one mod $\mathscr{U}$, depending on the particular ultrafilter.

The point of the second part of the theorem is that the finite hyperrationals are big enough to do analysis with and that is what we shall proceed with, leaving the proof of this theorem to the reader once he has progressed to the point where it is easy.

An infinitesimal ε satisfies $-1/n \leq \varepsilon \leq 1/n$ for every constant sequence $n \in {}^{\sigma}\mathbf{N}$ and if $r = \max(-p, q)$ where p and q are the constant sequence bounds for the finite number x,

$$-r/n \leq \varepsilon x \leq r/n \qquad \text{for every} \quad n.$$

Hence, taking $n = r \cdot m$, $-1/m \leq \varepsilon x \leq 1/m$ for any m and therefore εx is infinitesimal. This is one part of the statement that o is an order ideal in $\mathcal{O}$. *An infinitesimal times a finite number is infinitesimal.*

2.3 SOME CALCULUS OF POLYNOMIALS

The function $f(x) = x^2$ extends naturally to $^*\mathbf{Q}$, that is, $f(x)$ on constant sequences, has the same constant sequence answer and $x \cdot x$ defines a perfectly good function on $^*\mathbf{Q}$. We call the extension $^*f$.

The number 2 is not attained by x^2 over $\mathbf{Q}$ and hence also not over $^*\mathbf{Q}$ (since $a^2(j) \neq 2$ for any j). Drawing the graph of x^2 over a piece of the finite numbers we might be tempted to leave a hole at (?, 2), but without infinitely powerful microscopes this is not necessary. There are plenty of numbers $a \in {}^*\mathbf{Q}$ for which a^2 is infinitesimally close to 2. Precisely, *a is infinitesimally close to b, written $a \approx b$*, means $a - b \in o$, or the difference is between $-1/n$ and $+1/n$ for standard $n \in {}^{\sigma}\mathbf{N}$. One way to exhibit such a number is by applying the classical algorithm for the square root and defining a sequence $a(j)$ as follows (think of $J = \{0, 1, 2, \ldots\}$):

$$a(0) = 0, \qquad a(1) = 1, \qquad a(2) = 1.4, \qquad a(3) = 1.41.$$

In general, $a(j)$ is the partial answer computed at step j:

$a(1) = 1$	1	2.000000
		1
	24	100
$a(2) = 1.4$	4	96
	281	400
$a(3) = 1.41$	1	281
	2824	11900
$a(4) = 1.414$	4	11296

In this case, $a^2 \approx 2$ because the number of terms $(2 - a^2(j))$ in the sequence $2 - a^2$ greater than 10^{-n} is finite for each $n \in \mathbf{N}$. The only terms greater are those that correspond to the number of steps in the computation before the answer is correct to n places.

Other numbers in $^*\mathbf{Q}$ whose square is nearly 2 are given as follows. Define a sequence $b: J \to \mathbf{Q}$ by

$$b(0) = 1,$$
$$b(1) = 1 + \tfrac{1}{2},$$
$$b(2) = 1 + 1/(2 + \tfrac{1}{2}),$$
$$b(3) = 1 + 1/(2 + 1/[2 + \tfrac{1}{2}]),$$

that is, $b(j)$ is the jth step in the continued fraction $\langle 1, 2, 2, 2, \ldots \rangle$. Define another sequence $c: J \to \mathbf{Q}$ by

$$c(0) = 1, \qquad c(j + 1) = \big(c(j) + 2/c(j)\big)/2.$$

In other words, guess and average your guess with the division error.

In the case of a, b, and c it is a perfectly standard undertaking to show that given $n \in \mathbf{N}$, $a(j)$, $b(j)$, and $c(j)$ eventually satisfy $|d^2(j) - 2| < 1/n$ for $d = a, b, c$. This means each is within an infinitesimal of $\sqrt{2}$ by Axiom (5). Notice that we do not claim $a = b = c$, only that $a^2 \approx b^2 \approx c^2 \approx 2$.

Suppose a is finite and ε infinitesimal; then $^*f(a + \varepsilon) = (a + \varepsilon)^2 = a^2 + \varepsilon(2a + \varepsilon) \approx a^2 = {}^*f(a)$, because an infinitesimal times a finite number is infinitesimal. This is the intuitive formulation of **continuity** *at a, an infinitesimal change in x only produces an infinitesimal change in f*. Later we will see that this is equivalent to the epsilon–delta formulation of continuity at (the standard part of) a.

When x is not finite, $f(x)$ need not satisfy this infinitesimal perturbation condition; for example, let $x = \omega$ and take the infinitesimal change $1/\omega$: $(\omega + (1/\omega))^2 = \omega^2 + 2 + (1/\omega^2) \approx \omega^2 + 2$. For now we leave it to the reader to express this discontinuity in terms of epsilons and deltas in the standard model.

The functions $g(x) = x^n$ for $n \in \mathbf{N}$ also extend naturally to $^*\mathbf{Q}$, and since we can form sums and scalar products in $^*\mathbf{Q}$ we actually obtain a whole class of nonstandard polynomials of finite degree. For example, $\omega \cdot x^3 + (x/\omega)$, where ω is the infinite number above, is such a polynomial. Which of these functions are continuous, and where?

The function

$$\operatorname{sgn}(x) = \begin{cases} +1 & \text{if } x > 0 \\ 0 & \text{if } x = 0 \\ -1 & \text{if } x < 0 \end{cases}$$

has the natural extension $^*\mathrm{sgn}$ with the same description. At $x = 0$, when $\varepsilon = 1/\omega$, $^*\mathrm{sgn}(0 + \varepsilon) = +1$, while when $\varepsilon = -1/\omega$, $^*\mathrm{sgn}(0 + \varepsilon) = -1$ and infinitesimal changes in x produce finite changes in the function, so it is *discontinuous*.

Now we turn to the *calculation of the slope of the tangent to* $f(x) = x^2$ *at a finite point a*. Let $\varepsilon \neq 0$ be an infinitesimal. We compute the slope of the line through the two points $(a, f(a))$ and $(a + \varepsilon, f(a + \varepsilon))$ as follows:

$$\frac{\Delta f}{\Delta x}(a) = \frac{f(a + \varepsilon) - f(a)}{\varepsilon} = \frac{(a + \varepsilon)^2 - a^2}{\varepsilon} = 2a + \varepsilon.$$

Now notice that if we pick a different infinitesimal ε the answer is different, but no one would notice the difference unless they were looking through an infinitely powerful microscope. More precisely, *the standard part of* $\Delta f(a)/\Delta x$ *is always the same*, the different lines all go through $(a, f(a))$ and have nearly the same slope. *When this is the case we write* $df(a)/dx$ *for the common standard part* of the ratios of infinitesimal changes in f over infinitesimal changes in x.

The reader may wish to calculate df/dx in this way for some standard polynomials and investigate what happens with nonstandard coefficients and at infinite points. The function $A(x) = |x|$ has the natural extension $^*A(x) = \max(x, -x)$. At $a = 0$ when $\varepsilon = 1/\omega > 0$ we obtain $\Delta A(0)/\Delta x = +1$, and when $\varepsilon = -1/\omega$ we obtain $\Delta A(0)/\Delta x = -1$, so we cannot introduce a single standard symbol for the infinitesimal changes, *or* $dA(0)/dx$ *does not exist.*

We can *compute sequential limits by means of infinitesimals*, for example,

$\lim_{n\to\infty}(1/n) = 0$, simply because when Ω is infinite, $1/\Omega$ is always infinitesimal. *The common standard part is the value to which the sequence tends.*

Suppose $|a| < 1$ and $|a| \not\approx 1$; then $\lim_{n\to\infty} a^n = 0$, but to compute this for infinite values of n we must say what a^ω means.

$*$-finite powers *are described by saying what sequence x^λ is when $x \in {}^*\mathbf{Q}$ and $\lambda \in {}^*\mathbf{N}$*, and again the definition is pointwise $(x^\lambda)(j) =_{\mathscr{U}} x(j)^{\lambda(j)}$. This makes sense since on the right we have only standard rationals to standard powers. Now when ω is the sequence $\omega(j) = j$ we can form $\Omega =_{\mathscr{U}} \omega^\omega$ and $\Omega(j) = j^j$. Also, when $n \in {}^\sigma\mathbf{N}$, $x^n =_{\mathscr{U}} x \cdot x \cdots x$, n times, so this extends the ordinary notion of powers to all ${}^*\mathbf{N}$.

Now let $b = 1/|a|$, $0 \neq |a| < 1$ and $a \not\approx 1$, so $b = 1 + c$, c finite, not infinitesimal, and positive. By the binomial expansion $b^\lambda > \lambda \cdot c$, that is, for all j on a set of $\mathscr{U}$, $b(j)^{\lambda(j)} > \lambda(j) \cdot c(j)$. Therefore, $0 < a^\lambda < 1/(\lambda \cdot c) = (1/\lambda)(1/c)$, an infinitesimal times a finite number, when λ is infinite, so $a^\lambda \approx 0$. The limit is the common standard part of the infinite power values, $\lim_{n\to\infty} a^n = 0$.

Next, consider the case a^Ω when $a = -1$ and Ω infinite. By checking the pointwise definition of nonstandard powers we verify that $a^{m \cdot n} =_{\mathscr{U}} (a^m)^n$ and $a^{m+n} =_{\mathscr{U}} a^m \cdot a^n$. Thus there are two cases for a^Ω. First, $\Omega =_{\mathscr{U}} 2 \cdot v$ for some infinite v and

$$(-1)^\Omega =_{\mathscr{U}} (-1)^{2v} =_{\mathscr{U}} (1)^v =_{\mathscr{U}} 1.$$

Second, $\Omega =_{\mathscr{U}} 2 \cdot v + 1$ for some infinite v and

$$(-1)^\Omega =_{\mathscr{U}} (-1)^{2v}(-1) =_{\mathscr{U}} -1.$$

Since a single standard answer cannot express the tendency at infinite powers we say $\lim_{n\to\infty}(-1)^n$ *does not exist.*

Let $x \in {}^*\mathbf{Q}$ and $v, \lambda \in {}^*\mathbf{N}$ the **$*$-finite sum** $\sum_{n=v}^{\lambda} x^n$ *is defined pointwise as*

$$\left(\sum_{n=v}^{\lambda} x^n\right)(j) =_{\mathscr{U}} \sum_{n=v(j)}^{\lambda(j)} x(j)^n.$$

Notice that when $v, \lambda \in {}^\sigma\mathbf{N}$ this agrees with the ordinary sum, but that λ *can be infinite.*

Now when $|x| < 1$ and $x \not\approx 1$, *the geometric series* $\sum_{n=0}^{\infty} x^n = 1/(1-x)$ since $\sum_{n=0}^{\Omega} x^n =_{\mathscr{U}} (1 - x^{\Omega+1})/(1-x)$ (prove this j pointwise) and we already saw that when Ω is infinite $x^\Omega \approx 0$, so $\sum_{n=0}^{\Omega} x^n \approx 1/(1-x)$ for all infinite Ω.

The infinite harmonic series $\sum_{k=1}^{\infty} 1/k$ *diverges*, because infinite nonstandard sums $\sum_{k=1}^{\omega} 1/k$ and $\sum_{k=1}^{\Omega} 1/k$ are *not* all nearly the same finite number. For example,

$$\sum_{k=1}^{2^{\omega+1}} 1/k - \sum_{k=1}^{2^{\omega}} 1/k =_{\mathcal{U}} \sum_{k=2^{\omega}+1}^{2^{\omega+1}} 1/k =_{\mathcal{U}} \sum_{k=1}^{2^{\omega}} 1/(2^{\omega} + h)$$
$$> \sum_{h=1}^{2^{\omega}} 1/2^{\omega+1} \geq 2^{\omega}/2^{\omega+1} = \tfrac{1}{2}.$$

The reader may be interested in showing that $\sum_{k=1}^{\omega} 1/k$ is always positively infinite. Since this is the case we could say that the harmonic series tends to "$+\infty$" if we introduce that symbol as a new *standard* number.

EXERCISE: Show that the alternating harmonic series tends to a limit, for infinite upper sum index

$$\sum_{k=1}^{\Omega} (-1)^k/k \approx \sum_{k=1}^{\omega} (-1)^k/k.$$

2.4 THE EXPONENTIAL FUNCTION

The exponential function plays a fundamental role in analysis and does not quite fit in ordinary rational analysis. We will show here that it nearly fits in hyperrational analysis by using infinite operations. Naturally, the reader will see similarities between our approach and conventional epsilon–delta methods (also see Sections 6.3 and 6.4).

Two basic properties of the exponential function are the equations

$$d(\exp(x))/dx = \exp(x) \qquad \text{and} \qquad \exp(x + y) = \exp(x)\exp(y).$$

We claim that for finite x one can select

$$\exp(x) \approx \sum_{k=0}^{\omega} x^k/k! \qquad \text{or} \qquad \exp(x) \approx (1 + (x/\omega))^{\omega}$$

and that both choices give infinitesimally nearly the same answer and are nearly independent of the particular infinite ω. The **infinite operations** are defined pointwise as above:

$$\left(\sum_{k=0}^{\omega} \frac{x^k}{k!}\right)(j) =_{\mathcal{U}} \sum_{k=0}^{\omega(j)} \frac{x(j)^k}{k!} \qquad \text{and} \qquad \left(1 + \frac{x}{\omega}\right)^{\omega}(j) =_{\mathcal{U}} \left(1 + \frac{x(j)}{\omega(j)}\right)^{\omega(j)}.$$

For example, our substitute for Euler's e is the sequence

$$\sum_{k=0}^{\omega} 1/k! = (0, 1, 2\tfrac{1}{2}, 2\tfrac{2}{3}, \ldots) \qquad \text{or} \qquad \mathrm{st}\left(\sum_{k=0}^{\omega} 1/k!\right) = e.$$

Let $S_\omega(x) = \sum_{k=0}^{\omega} x^k/k!$. We indicate that this is the "preexponential function" by the properties:

(1) *$S_\omega(x)$ is finite when x is finite.*

(2) *$S_\omega(x) \approx S_\Omega(x)$ for finite x and both ω and Ω infinite.*

(3) *$\Delta S_\omega(x)/\Delta x \approx S_\omega(x)$ for finite x, infinite ω, and infinitesimal Δx, that is, $S_\omega(x)$ nearly satisfies the differential equation for the exponential function.*

(4) *$S_\omega(x) \cdot S_\omega(y) \approx S_\omega(x+y)$, x and y finite, the functional equation for the exponential is nearly satisfied.*

For now we leave it to the reader to examine $(1 + (x/\omega))^\omega$.

We begin with (3) by taking x finite, $y \approx x$, $y \neq_{\mathscr{U}} x$, ω infinite, and writing $(\Delta S_\omega(x)/\Delta x) - S_\omega(x)$. All nonstandard sums are defined pointwise and the steps with $=_{\mathscr{U}}$ are verified by the j pointwise definitions of an element of $\mathscr{U}$. The reader should also verify that $x^\omega/\omega!$ is infinitesimal as an exercise.

Next, we show that

$$\left[\frac{1}{y-x}\left(\sum_{k=0}^{\omega}\frac{y^k}{k!} - \sum_{k=0}^{\omega}\frac{x^k}{k!}\right) - \sum_{k=0}^{\omega}\frac{x^k}{k!}\right]$$

$$=_{\mathscr{U}} \left[\sum_{k=1}^{\omega}\frac{y^k - x^k}{k!(y-x)} - \sum_{k=1}^{\omega+1}\frac{kx^{k-1}}{k(k-1)!}\right]$$

$$=_{\mathscr{U}} \sum_{k=1}^{\omega}\frac{1}{k!}\left[\frac{y^k - x^k}{y-x} - kx^{k-1}\right] + \frac{x^\omega}{\omega!}$$

$$\approx \sum_{k=2}^{\omega}\frac{1}{k!}\left[\sum_{n=0}^{k-1} x^n y^{k-(n+1)} - kx^{k-1}\right]$$

$$=_{\mathscr{U}} \sum_{k=2}^{\omega}\frac{1}{k!}\left[\left(\sum_{n=0}^{(k-1)}(n+1)x^n y^{k-(n+1)} - \sum_{n=0}^{(k-1)} nx^n y^{k-(n+1)}\right) - kx^{k-1}\right]$$

$$=_{\mathscr{U}} \sum_{k=2}^{\omega}\frac{1}{k!}\left[\sum_{n=1}^{k-1} nx^{(n-1)}y^{(k-n)} - \sum_{n=1}^{k-1} nx^n y^{k-(n+1)}\right]$$

$$=_{\mathscr{U}} \sum_{k=2}^{\omega}\frac{1}{k!}\left[\sum_{n=1}^{k-1}(y-x)nx^{(n-1)}y^{k-(n+1)}\right]$$

$$=_{\mathscr{U}} (y-x)\sum_{k=2}^{\omega}\left[\frac{1}{k!}\left(\sum_{n=1}^{k-1} nx^{(n-1)}y^{k-(n+1)}\right)\right].$$

Now x and y are finite so there is a constant sequence r such that $|x| < r$ and $|y| < r$. We estimate the inside sum of the last line as

$$\left|\sum_{n=1}^{k-1} nx^{(n-1)}y^{k-(n+1)}\right| \leq k(k-1)r^{k-2}$$

by applying the triangle inequality, and making the worst choices for $|x|$, $|y|$, and n, there are $k - 1$ terms. Now we have

$$\left|\frac{\Delta S_\omega(x)}{\Delta x} - S_\omega(x)\right| \leq |y - x| \sum_{k=2}^{\omega} \frac{r^{k-2}(k(k-1))}{k!} \leq |y - x| S_\omega(r),$$

and it suffices to show that $S_\omega(r)$ is finite [Property (1)] in order to complete the proof of (3) (infinitesimal times finite is infinitesimal).

EXERCISE: Prove that $S_\omega(x)$ is finite for finite x by obtaining a bound b for the finite sums $\sum_{k=0}^{\omega(j)} x(j)^k/k!$ which holds for all j. For example, you can use comparison with a geometric series. Since the bound holds for all j on an element of $\mathscr{U}$, $|S_\omega(x)| < b$, the constant sequence.

EXERCISE: Show that $S_\omega(x) \approx S_\Omega(x)$ for finite x and infinite ω and Ω by writing the difference, factoring out a term, and estimating the rest.

The proof that $S_\omega(x)$ nearly satisfies the functional equation proceeds as follows. We write $S_\omega(x)S_\omega(y)$ as an iterated sum, with the formulas coming from the finite expressions for $\omega(j)$. We add some terms to write this as "diagonal sums" and find that this gives us $S_{2\omega}(x + y)$ which is nearly $S_\omega(x + y)$ by the exercise above. The infinitesimal difference $[S_{2\omega}(|x + y|) - S_\omega(|x + y|)]$ is more than the terms we added to the sum to write it as diagonals, so $S_\omega(x)S_\omega(y) \approx S_\omega(x + y)$:

$$\begin{aligned} S_\omega(x)S_\omega(y) &= \left(\sum_{k=0}^{\omega} \frac{x^k}{k!}\right)\left(\sum_{k=0}^{\omega} \frac{y^k}{k!}\right) \approx \sum_{n=0}^{2\omega}\left(\sum_{k=0}^{n} \frac{x^k}{k!}\,\frac{y^{n-k}}{(n-k)!}\right) \\ &=_{\mathscr{U}} \sum_{n=0}^{2\omega}\left[\sum_{k=0}^{n}\left(\frac{n!}{k!(n-k)!}\right)\frac{x^k y^{n-k}}{n!}\right] \\ &=_{\mathscr{U}} \sum_{n=0}^{2\omega} \frac{(x + y)^n}{n!} \approx S_\omega(x + y). \end{aligned}$$

Perhaps this is enough of our precalculus look at the exponential function. *The main thing we want to point out is that pointwise (internal) extension of formal sums leaves the infinite formulas looking exactly like ordinary sums, even for infinite operations.* The infinitesimal changes we made of course are different (external).

2.5 PEANO'S EXISTENCE THEOREM

Our last application in this chapter is included to show how an infinitesimal partition of an interval could be introduced in $^*\mathbf{Q}$. We also define an extension of a family of functions in order to obtain a piecewise linear function subordinate to the partition. The construction is naturally suggested in trying to prove

(2.5.1) THEOREM *If* $f : [0, 1] \times \mathbf{R} \to \mathbf{R}$ *is continuous and bounded,* $|f(t, x)| < M$, *and* $U_0 \in \mathbf{R}$ *is given, then there exists a* $u: [0, 1] \to \mathbf{R}$ *with* $u(0) = U_0$ *and* $du(t)/dt = f(t, u(t))$.

To begin with we have one technicality to dispense with concerning real numbers. Since we have not introduced them yet (except as standard parts of finite hyperrationals) we have to say what functions like f and u defined for real numbers mean in our setting.

First, we can let the variable t run over [0, 1] in $^*\mathbf{Q}$, that is, $\{a \in {}^*\mathbf{Q} : 0 \le a \le 1\}$. Then the set of standard parts is [0, 1] in $\mathbf{R}$. The variable x can range over $\mathcal{O}$.

In place of the given function f what we need is a function F defined componentwise on sequences in $[0, 1] \times \mathcal{O}$ such that $\text{st}\, F(t, x) = f(\text{st}(t), \text{st}(x))$. (In terms of later terminology, F must be internal.) That is, for any $(t, x) \in [0, 1] \times \mathcal{O}$ so that $(\text{st}(t), \text{st}(x))$ defines a real number in $[0, 1] \times \mathbf{R}$, $F(t, x)$ is within an infinitesimal of the answer of f. In particular, if $t_1 \approx t_2$ and $x_1 \approx x_2$, then $F(t_1, x_1) \approx F(t_2, x_2) \approx f(\text{st}(t_1), \text{st}(x_1))$, so that f is well defined by F. From here on we will work with such an F.

We begin by taking an arbitrary $n \in \mathbf{N}$ and partitioning [0, 1] (in $^*\mathbf{Q}$) into n equal parts $0 = t_0,\ t_1 = 1/n,\ \ldots,\ t_i = i/n,\ \ldots,\ t_n = 1$ (Fig. 2.5.1). By induction we define a piecewise linear function $v_n(t)$ by the following formula, filling in linearly:

$$\frac{v_n(t_{i+1}) - v_n(t_i)}{t_{i+1} - t_i} = F(t_i, v_n(t_i)), \qquad v_n(t_0) = U_0.$$

We have the telescoping sum

$$v_n(t_\omega) = U_0 + \sum_{i=0}^{\omega-1} [v_n(t_{i+1}) - v_n(t_i)]$$

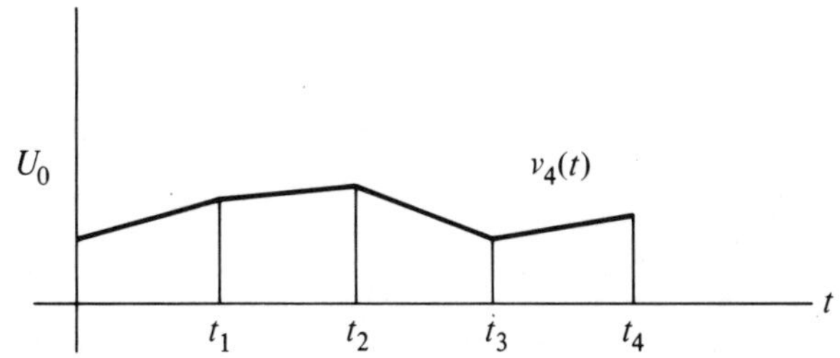

Fig. 2.5.1

for $0 \le \omega \le n$, which we rewrite as

$$v_n(t_\omega) = U_0 + \sum_{i=0}^{\omega-1} \frac{v_n(t_{i+1}) - v_n(t_i)}{t_{i+1} - t_i}(t_{i+1} - t_i).$$

Now take an infinite natural number λ. By letting $(t_i)(j) = t_{i(j)}$ for $0 \le i \le \lambda$ and $[v_\lambda(t)](j) = v_{\lambda(j)}(t(j))$, we have a function that is piecewise linear subordinate to the partition of [0, 1] into λ equal parts. By applying the above formula with components $\lambda(j)$ and $\omega(j)$, $0 \le \omega \le \lambda$, we obtain

$$\text{(I)} \qquad v_\lambda(t_\omega) = U_0 + \sum_{i=0}^{\omega-1} \frac{v_\lambda(t_{i+1}) - v_\lambda(t_i)}{t_{i+1} - t_i}(t_{i+1} - t_i).$$

Now since f is bounded by M, F is bounded by the constant sequence M, so by the definition of v_λ above [applied first to $n \in \mathbf{N}$ and then components $\lambda(j)$]

$$|v_\lambda(t) - v_\lambda(s)| \le M|t - s|.$$

When $t \approx x$, $v_\lambda(t) \approx v_\lambda(s)$ and even more, if we consider the function $\text{st}(v_\lambda(t)) = u(\text{st}(t))$, u is well defined and continuous as a real-valued function. This is the solution. This is at least believable since the fundamental theorem of calculus says

$$u(t) = U_0 + \int_0^t f(s, u(s))\, ds$$

and formula (I) above says the following. Picking t_ω so that $t \approx t_\omega$,

$$v_\lambda(t) \approx v_\lambda(t_\omega) = U_0 + \sum_0^{t_\omega} F(t_i, v_\lambda(t_i))(t_{i+1} - t_i).$$

The sum is a Riemann sum on an infinitesimal partition and to finish the proof of Peano's theorem we have to show that $F(t, v_\lambda(t))$ is at most a uniform infinitesimal from $f(t, u(t))$. In that case the standard part of the Riemann sum is the integral above. To show that F is a uniform infinitesimal from f we could use the fact that u stays in the compact rectangle $[0, 1] \times [U_0 - M, U_0 + M]$ where f is uniformly continuous. Since this is messy in this component-laden context (and actually requires some care with F), and since it follows from purely formal considerations, we leave it as incentive for studying formal analysis in greater detail.

2.6 SUMMARY

We hope this chapter has been detailed enough that the reader realizes that some general formal transfer principle from **Q** to ***Q** must exist. We also hope it has been vague enough that the reader challenged some of our remarks (and found them correct, we trust). We hope this chapter will help the reader through the greater abstraction and formalism of the next. We will return to some of the details that we feel a little guilty about, especially Peano's theorem, since it demonstrates a nontrivial application of the logic. We also remind the reader that via pointwise definitions we gave extensions higher up than the algebra in the theory of analysis, for example, the extension of the family of piecewise linear functions in Peano's theorem.

CHAPTER 3 SUPERSTRUCTURES AND THEIR NONSTANDARD MODELS

The main principles of this chapter are given in Section 3.4; a summary is given in Section 3.12.

3.1 INTRODUCTION

At this stage we wish to abandon the point of view of the preceding chapter as far as doing analysis with rational numbers is concerned. Instead, we want to give a set-theoretical definition of "all of standard analysis." Once we have done this, extensions of sums, factorials, powers, partitions, families of functions, and the like will automatically be included in our ultrapower.

More than saying what "all of standard analysis" is, we want to give an abstract construction which will encompass an infinite ground set X and a mathematical theory based on it. This will allow us to do various kinds of nonstandard algebra, topology, and the like.

The earlier version of infinitesimal analysis (Robinson [1961] and Luxemburg [1962]) rests on the formulation of the properties of $\mathbf{R}$ that can be expressed in a first-order language, which means the quantification in the formal language is permitted only on variables ranging over real numbers. One need not go far in analysis, however, to realize the need for a richer language in which statements such as "For all nonempty sets of natural numbers ..." or "There exists a continuous function ..." can be formulated. In connection with this it is good to observe that even some of the axioms of the real number system are outside the language of the lower predicate calculus. For example, Dedekind's completeness axiom involving quantification with respect to ordered pairs of sets (Dedekind cuts) is such an axiom. The principle of induction is another example of a statement that cannot be expressed in the language of the lower predicate calculus. In order to cope with this problem a version of type theory could be used as a framework (Robinson [1966]).

Although this solution is satisfactory from many points of view, it has the drawback of appearing unnecessarily complicated to most mathematicians. In the article by Robinson and Zakon [1967] it was shown that another way out is a set-theoretical approach. This approach is based on the fact, familiar to every mathematician, that the various branches of mathematics can all be thought of as embedded in set theory, and so the basic concepts of analysis, algebra, and topology can all be defined in terms of the membership relation and sets. The formal language may then be a first-order language whose variables range over sets or points and whose constants denote certain sets or points. Our approach will be based on that idea. The types that Robinson [1966] uses are, in a certain sense, more like intuitive set theory; unfortunately their formal description makes them seem obscure.

Our plan then is as follows. First, we construct a superstructure $\mathscr{X}$ which is a *set* but one big enough to contain "all of standard analysis" in the algebraic theory of $(\mathscr{X}, \in, =)$, where $\in$ and $=$ are the algebraic operations. Second, we give the characteristic list of axioms for a nonstandard extension map from one superstructure into a larger one:

$$(\mathscr{X}, \in, =) \xrightarrow{*} (\mathscr{Y}, \in, =).$$

The ground set of $\mathscr{Y}$ is ${}^*X_0$, the extension of X_0, where X_0 is the ground set of $\mathscr{X}$. The properties of the nonstandard atoms in ${}^*X_0$ are "the same" as in X_0, but the higher-order properties only transfer to a restricted class of sets in $\mathscr{Y}$, the internal sets. The description of those properties could be done ad hoc one at a time from the monomorphism axioms of $*$, and we sketch that approach in passing. The easiest and most powerful way to prove the needed metatheorem (Leibniz' principle) is to introduce a formal language and a systematic method of interpreting formulas in $\mathscr{X}$ and the image ${}^*\mathscr{X} \subset \mathscr{Y}$. This is the third step.

We consider the maps

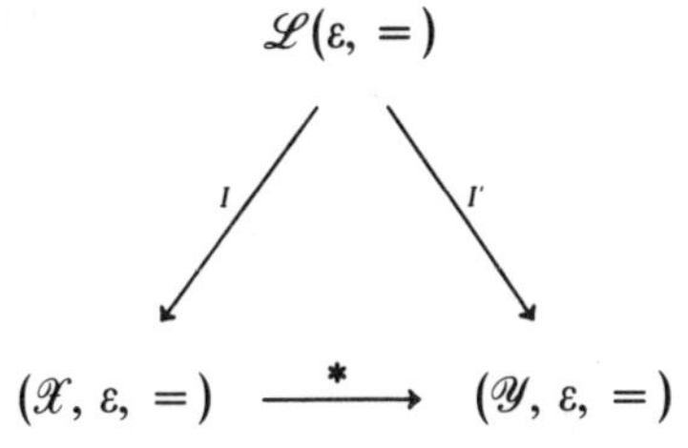

The slippery fact here is that I' is *not onto*, the formal language only has internal sets built in.

The critical reader will ask how we know $*$-maps exist. That is part of the construction of the nonstandard interpretation of $\mathscr{L}$, I', in the third step described above. We actually construct I' and show that $I' \circ I^{-1} = *$ satisfies Leibniz' principle and, in particular, the axioms of a monomorphism.

The construction of I' shows us that every formula of $\mathcal{L}$ whose I-interpretation holds in $\mathcal{X}$ has a true I'-interpretation in the sets of $\mathcal{Y}$. Once we know this we observe that there is a simple method (the $*$-transform) to apply to transfer from $\mathcal{X} \xrightarrow{*} \mathcal{Y}$ that only requires a little care with quantifiers, but essentially no formal logic. The logic tells us that transfer really works, but the goal of the chapter—to learn to transfer properties via Leibniz' principle—really only takes a little practice using the standard and internal definitions of sets. Precisely we have:

(3.1.1) LEIBNIZ' PRINCIPLE *A sentence in $\mathcal{X}$ that has a bounded formalization is true if and only if its $*$-transform is true in $\mathcal{Y}$.*

Roughly speaking, the construction of I' is simply to extend entities by constant sequence extension and to interpret the formal $\in$ and $=$ of $\mathcal{L}$ as $\in_{\mathcal{U}}$ and $=_{\mathcal{U}}$ of the ultrapower as we did in Chapter 2 for various relations, that is,

$$a_j \in_{\mathcal{U}} B_j \qquad \text{if and only if} \quad \{j : a_j \in B_j\} \in \mathcal{U}.$$

Unfortunately, this is only half of the construction of I'. We embed the $\in_{\mathcal{U}}$ and $=_{\mathcal{U}}$ relations in set theory in order to form I' as an interpretation *in set theory*. The sets that arise as embedded set-valued sequences are the internal sets, and the embedding allows us to discuss *all* subsets (external ones). We are forced to take the latter step in order to be able to treat the set of infinitesimals (external) the same way we treat sets like the unit interval *[0, 1] in the extension:

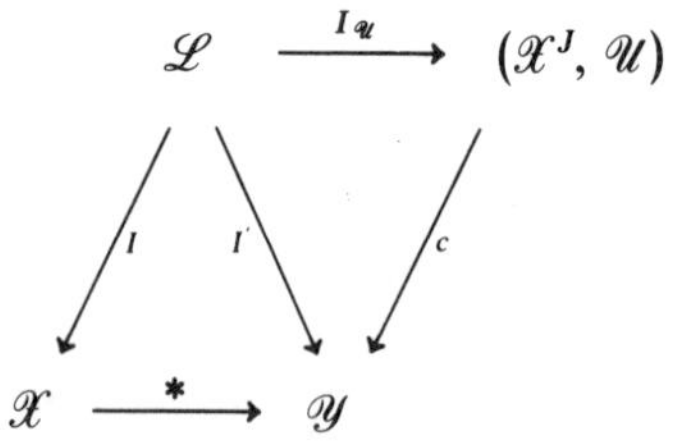

For a proper understanding of the foundations of the method it will certainly be helpful to the reader if he is familiar with some of the results of the predicate calculus. However, the reader who is not familiar with any more than naive set theory should not be discouraged from going on. He simply will have to rely on his informal ideas for the properties of the superstructures to be introduced in the next section.

3.2 DEFINITION OF A SUPERSTRUCTURE

In order to apply the method of infinitesimals to a certain mathematical theory such as real and complex analysis or topology, we first have to construct a superstructure that contains all the mathematical objects under study in the given theory. This can be accomplished by means of the following process.

Let X denote a nonempty set that in the applications may be the set of real numbers or the underlying point set of a topological space, etc. The set X is an infinite set of atoms; in other words, if $a \in X$ we shall not write "$x \in a$."

We define the following sets inductively:

$$X_0 = X, \qquad X_{n+1} = \mathscr{P}\left(\bigcup_{k=0}^{n} X_k\right), \quad n = 0, 1, 2, \ldots,$$

where $\mathscr{P}(Z)$ denotes the power set of Z, that is, the set of all subsets of Z. Now we define a new set $\mathscr{X}$ by means of the following formula, $\mathscr{X} = \bigcup_{n=0}^{\infty} X_n$. We will refer to the *set elements* of $\mathscr{X}$ as the *entities* of $\mathscr{X}$. The elements of $\mathscr{X}$ that are contained in the set $X_0 = X$ will be called the *individuals* of $\mathscr{X}$.

We shall now list some properties of $\mathscr{X}$.

(i) *The empty set is an entity; $\varnothing \in \mathscr{X}$.* Observe that $\varnothing \in \mathscr{P}(X_0) = X_1 \subseteq \mathscr{X}$.

(ii) *For each n, $X_n \in \mathscr{X}$, that is, X_n is an entity of $\mathscr{X}$.* $X_n \in \mathscr{P}(X_n) \subseteq X_{n+1}$.

(iii) *$\mathscr{X}$ is transitive in the sense that if y is an entity and $x \in y$, then $x \in \mathscr{X}$.* If y is an entity, then $y \in X_{n+1}$ for some $n \geq 0$. This means $y \subseteq \bigcup_{k=0}^{n} X_k$ and then $x \in \bigcup_{k=0}^{n} X_k$. Finally, each $X_k \subseteq \mathscr{X}$.

(iv) *If y is an entity and $x \subseteq y$, then x is an entity.* $y \in \mathscr{X}$ *means* $y \in X_{n+1}$ or $\bigcup_{k=0}^{n} X_k \supseteq y \supseteq x$ hence $x \in \mathscr{X}$.

(v) *If x is an entity, then $\mathscr{P}(x)$ is also an entity.* Each entity of $\mathscr{X}$ is a subset of $\mathscr{X}$, since $x \in X_{n+1}$ and $x \subseteq X_0 \cup X_n \subseteq \mathscr{X}$. Hence $\mathscr{P}(x) \subseteq \mathscr{P}(X_0 \cup X_n) = X_{n+1}$ and by (iv) is therefore an entity. We caution the reader that arbitrary subsets of $\mathscr{X}$ are not necessarily entities. The entities are only those sets which are also *elements* of $\mathscr{X}$.

(vi) *If x is a finite subset of $\mathscr{X}$, then $x \in \mathscr{X}$, that is, x is an entity.* Let $x = \{a_1, \ldots, a_p\}$ where $a_i \in \mathscr{X}$. Then all the a_i are in $\bigcup_{k=0}^{n} X_k$ for some n. This means $x \in X_{n+1}$.

(vii) *If x is an entity of $\mathscr{X}$, then $\bigcup x = \bigcup[y : y \in x]$ is also an entity.* Recall that we are working with set theory with atoms so that if $y \in x$

and $y \in X_0$, y has no elements [since we will not allow "$z \in y$"]. In this case there is no contribution to the union. Now $x \in X_{n+1}$ for some n so $\{z : \text{there exists } y \in x \text{ with } z \in y\}$ is a subset of $X_{n-1} \cup X_0$ and so an element of X_n therefore $\mathscr{X}$.

The thing we are leading up to is that the set theory of entities is contained in the entities. Once we have done this we will try to convince the reader that the entities over a set are enough with which to describe many mathematical theories including analysis.

It is customary to think of ordered pairs as

$$(x, y) = \{\{x\}, \{x, y\}\}$$

and to define ordered n-tuples inductively as

$$(x_1, \ldots, x_n) = ((x_1, \ldots, x_{n-1}), x_n).$$

(viii) *If $x_1, \ldots, x_n \in \mathscr{X}$, then $(x_1, \ldots, x_n)$ is an entity. Moreover, if $x_1, \ldots, x_n$ are entities, then the Cartesian product $x_1 \times \cdots \times x_n$ is an entity.* This should be fairly clear: (x, y) is formed from finite sets of entities; $x \times y \subseteq \mathscr{P}(\mathscr{P}(x \cup y))$. Now functions and in general n-ary relations can be viewed as subsets of Cartesian products.

(ix) Hence functions and n-ary relations whose domains and ranges are entities are themselves entities. In general given an n-ary relation $\Phi \in \mathscr{X}$ we will say *Φ is an n-ary relation entity of $\mathscr{X}$ when its domains and range are entities.* Here

$$\begin{aligned}\operatorname{dom}_i \Phi = \{x_i : &\text{there exist } x_1, \ldots, x_{i-1}, x_{i+1}, \ldots, x_n \\ &\text{with } (x_1, \ldots, x_n) \in \Phi\}, \quad 1 \le i \le n,\end{aligned}$$

and

$$\begin{aligned}\operatorname{rng} \Phi = \{x_n : &\text{there exist } x_1, \ldots, x_{n-1} \\ &\text{with } (x_1, \ldots, x_n) \in \Phi\}.\end{aligned}$$

The notion of equality between individuals is assumed to be given. Moreover, no atom equals any entity. Equality between entities is the set-theoretical notion:

$$x = y \qquad \text{if and only if} \quad \begin{cases} z \in x & \text{whenever} \quad z \in y \\ w \in y & \text{whenever} \quad w \in x, \end{cases}$$

that is, when they have the same elements. Also if $x = y$ and $x \in z$, then $y \in z$.

(x) *Assuming the axiom of choice in the larger set theory, the axiom of choice also holds in the entities:* If F is a function entity so that $F(x)$ is a nonempty entity for each $x \in \mathrm{dom}(F) \neq \varnothing$, then there is a function entity f so that $f(x) \in F(x)$ for each $x \in \mathrm{dom}(F)$. We must show $f \in \mathscr{X}$. Each $f(x) \in \mathscr{X}$ by transitivity; $\bigcup[\mathrm{rng}(F)]$ is an entity and rng f is a subset of it. Finally $f \subseteq \mathrm{dom}(f) \times \mathrm{rng}(f)$.

(3.2.1) DEFINITION *A superstructure based on a set of atoms X is the set $\mathscr{X}$ together with the notions of equality and membership on the elements of $\mathscr{X}$.*

3.3 SUPERSTRUCTURES ARE BIG ENOUGH

Let us consider the superstructure based on the natural numbers $\mathbf{N} = \{0, 1, 2, \ldots\}$. First, the basic algebra of $\mathbf{N}$ is part of $\mathscr{N}$ since we could use the following entities for sum, product, and "less than":

$$\mathrm{S} = \{(a, b, c) : a, b, c \in \mathbf{N} \text{ and } a + b = c\},$$
$$\mathrm{P} = \{(a, b, c) : a, b, c \in \mathbf{N} \text{ and } a \cdot b = c\},$$
$$\mathrm{L} = \{(a, b) : a, b \in \mathbf{N} \text{ and } a \leq b\}.$$

One construction of real numbers is to form the integers $\mathbf{Z}$ as ordered pairs of natural numbers, rationals as pairs of integers, and reals as Dedekind cuts. Since this process starts with individuals, and pairs of individuals (which are entities), and pairs of these entities, and so forth, each real number and $\mathbf{R}$ itself are entities in $\mathscr{N}$, as are the complex numbers and set of complex numbers, Euclidean spaces, L^p-spaces, bounded holomorphic functions on the unit disk, and virtually anything that comes up in "classical analysis." For this reason it is safe to call the entities of the superstructure $\mathscr{N}$ a "standard model of analysis."

If we did not wish to include a construction of $\mathbf{R}$ from $\mathbf{N}$ in the discussion, we could take $\mathbf{R}$ as the set of atoms and form the superstructure $\mathscr{R}$ based on $\mathbf{R}$.

If we wish to study an abstract topological space, Y say, we could form the superstructure based on the set of atoms $X = Y \cup \mathbf{N}$. Notice the topology τ of Y is an entity being a set of subsets of Y. This way we have analysis and the topological theory in the entities of $\mathscr{X}$. The set theory (and the category theory) of the entities form a sufficient framework for classical mathematics.

3.4 NONSTANDARD MODELS OF SUPERSTRUCTURES

What we wish to do now is show that given a superstructure $(\mathscr{X}, \in, =)$ there is a larger superstructure $(\mathscr{Y}, \in, =)$ and an embedding $*: \mathscr{X} \to \mathscr{Y}$ which preserves all the mathematical structure of $(\mathscr{X}, \in, =)$.

One way to say what it means to "preserve all the mathematical structure of $(\mathscr{X}, \in, =)$" is to give the axioms of an algebraic injection for the operations $\in$ and $=$. We do this now, but do not show that any extension maps exist until later. Another way to say what it means to "preserve all the mathematical structure" is to introduce all the structure in the form of a formal language. We do that later in the process of constructing an example of a $*$-map. We really transfer "all properties" either way.

Let $(\mathscr{X}, \in, =)$ be a superstructure based on a set of atoms X_0. Let $(\mathscr{Y}, \in, =)$ be a superstructure based on another set of atoms $Y_0 = {}^*X_0$. We discuss a mapping $* : \mathscr{X} \to \mathscr{Y}$. This could be a little confusing at first because of the "levels" built in $\mathscr{X}$, for example, each individual $x \in X_0$ has an extension $^*x$, but so does the entity X and $\{^*x : x \in X_0\} \subset {}^*X_0$ yet $\{^*x : x \in X_0\} \neq {}^*X_0$. We assume ${}^*X_0 = Y_0$. Each subset of $X_0 \supseteq A$ also has an extension $^*A$ and the power set $\mathscr{P}(X_0) = X_1$ also has an extension ${}^*X_1$. Now it is important to learn that ${}^*\mathscr{P}(X_0) \neq \mathscr{P}({}^*X_0)$. This follows from the axioms or the ultrapower construction.

Given an extension $*: \mathscr{X} \to \mathscr{Y}$ we define a map σ on the entities by:

$$^\sigma A = \{^*a : a \in A\}.$$

The map σ gives the embedded *standard* copy of A. Axiom (vii) below says $^*A$ is a proper extension of $^\sigma A$ whenever A is infinite. We shall see later that $^\sigma A \notin {}^*\mathscr{P}(A)$ when A is infinite, that is, the standard points of the extension form an external set. Of course, $^\sigma A \in \mathscr{P}(^*A)$ by (i).

(3.4.1) DEFINITION $*$ *is a superstructure monomorphism of* $\mathscr{X}$ *if it is a one-to-one map defined on* $\mathscr{X}$ *satisfying:*

(i) $*$ *preserves* $\in$: If A is an entity of $\mathscr{X}$ and $a \in A$, then $^*a \in {}^*A$.

(ii) $*$ *preserves equality:* If A is an entity,

$$^*\{(x, x) : x \in A\} = \{(y, y) : y \in {}^*A\}.$$

(iii) $*$ *preserves finite sets:* $^*\{a_1, \ldots, a_n\} = \{^*a_1, \ldots, {}^*a_n\}$ for $a_1, \ldots, a_n \in \mathscr{X}$.

(iv) $*$ *preserves basic set operations:* ${}^*\varnothing = \varnothing$, ${}^*(A \cup B) = {}^*A \cup {}^*B$, ${}^*(A \cap B) = {}^*A \cap {}^*B$, ${}^*(A \backslash B) = {}^*A \backslash {}^*B$, ${}^*(A \times B) = {}^*A \times {}^*B$, where A and B are entities.

(v) $*$ *preserves domains and ranges of n-ary relations and commutes with permutations of the variables:* For example, $\operatorname{dom}({}^*\Phi) = {}^*\operatorname{dom}(\Phi)$, ${}^*\operatorname{rng}(\Phi) = \operatorname{rng}({}^*\Phi)$ and if $(x, y) \in \Psi$ if and only if $(y, x) \in \Phi$, then $(z, w) \in {}^*\Psi$ if and only if $(w, z) \in {}^*\Phi$.

(vi) $*$ *preserves atomic standard definitions of sets:*

$$^*\{(x, y) : x \in y \in A\} = \{(z, w) : z \in w \in {}^*A\}$$

when A is an entity.

(vii) $*$ *produces a proper extension:* $^*A \supseteq {}^\sigma A$ with equality if and only if A is a finite set.

These properties imply that the image $^*\mathscr{X}$ is a nonstandard model in the formal logic sense described below, that is, all the boundedly formalizable properties of $\mathscr{X}$ are preserved by $*$. We will sketch what that means now, but leave the axiomatic proof from (3.4.1) of the three main principles to our reader (see Robinson and Zakon [1967]). When we construct $*$ in Section 3.10 the proof of these principles will be apparent from the fundamental theorem of ultrapowers. Leibniz' principle clearly implies the axioms of (3.4.1).

In Section 3.5 we describe the formal language $\mathscr{L}$ and the precise kinds of properties subject to transfer, the bounded formal sentences. In Section 3.6 we describe how to interpret bounded formal sentences in the standard model $\mathscr{X}$. The interpretation is denoted $^I V$ for the formal sentence V. We denote the set of bounded formal sentences whose interpretation in $\mathscr{X}$ is true by K_0. What we are really interested in is the question: When is a theorem of $\mathscr{X}$ formalizable as a bounded sentence in K_0? Remarkably, the answer is: Any theorem whose quantifiers are specified to run over specific entities. For example, we may say "for every real positive epsilon, there is a real positive delta ...," but not just "for every epsilon there is a delta ..."—remember $\mathscr{X}$ contains much more than points of a ground set.

Since we can transfer "all properties with bounded quantifiers" it is surprising at first that $\mathscr{Y}$ is not isomorphic to $\mathscr{X}$, but the following example illustrates the restrictions of transfer. First the method of transfer:

(3.4.2) DEFINITION *If $V \in K_0$, form $^I V$ and put a $*$ on each interpreted constant, the result is the $*$-transform of $^I V$.*

Every subset of **N** has a first element.

This is not permissible, but the following is:

Every element of $\mathscr{P}(\mathbf{N})$ has a first element.

The $*$-transform is:

Every element of $^*\mathscr{P}(\mathbf{N})$ has a first element.

Notice that the $*$-transform is NOT:

Every element of $\mathscr{P}(^*\mathbf{N})$ has a first element.

The last sentence is false! There is no first infinite $*$-natural number. The $*$-transform of the well order says only that "every *internal* subset of $^*\mathbf{N}$ has a first element."

The following theorem is a consequence of (3.8.4) and the embedding discussed in Section 3.10. It can be proved from the axioms of (3.4.1), but we prefer the explicit ultrapower in the interests of concreteness.

(3.4.3) LEIBNIZ' PRINCIPLE *A sentence in $\mathscr{X}$ that has a bounded formalization in $\mathscr{L}$ is true if and only if its $*$-transform is true.*

As an example, recall the application in Chapter 2 where we indicated that there are numbers in $^*\mathbf{Q}$ whose square is infinitesimally close to 2.

The statement that "for each $n \in \mathbf{N}$ there is a $q \in \mathbf{Q}$ such that $|q^2 - 2| < 1/n$" is true and can be considered as a statement in $\mathscr{N}$ (the superstructure on $\mathbf{N}$) so that $|\ |$, product, minus, 2, less than, division, $\mathbf{Q}$, and $\mathbf{N}$ are all entities. The $*$-transform of the statement in quotes is "for each $n \in {}^*\mathbf{N}$, there is a $q \in {}^*\mathbf{Q}$ such that $^*|q^{*2} \,{}^*\!- {}^*2|^* < 1/n$" therefore, in particular, when n is infinite there is a nonstandard rational whose square is within $1/\omega$ (an infinitesimal) of ^{*2}.

(3.4.4) DEFINITION (Standard, internal, and external entities in $\mathscr{Y}$) Any entity A of $\mathscr{X}$ is called *standard.* $^*A$ is also termed a *standard entity* of $\mathscr{Y}$. Any element of a standard entity in the nonstandard model is called an *internal entity*; if $b \in {}^*B$, then b is internal. Internal entities can also be described as those that arise from mappings in the embedding of Section 3.10. The other entities of $\mathscr{Y}$ are called *external.* Many interesting sets are external—see Section 4.5.

The standard subsets of a standard entity A are the elements of $^\sigma\mathscr{P}(A)$. The internal subsets of A are the elements of $^*\mathscr{P}(A)$. The external subsets are the elements of $\mathscr{P}(^*A)$ that are not internal. Thus

$$^\sigma\mathscr{P}(A) \subset {}^*\mathscr{P}(A) \subset \mathscr{P}(^*A),$$

where the inclusions are strict when A is infinite.

For example, consider the case of a free ultrapower over $J = \{0, 1, 2, \ldots\}$ for the superstructure $\mathscr{N}$. The set of even nonstandard numbers $^*E$ arises via the constant sequence

$$E(j) \equiv E = \{0, 2, 4, \ldots\}$$

in the ultrapower and

$$^*E = \{0, 2, 4, \ldots, 2\omega, \ldots, 2\lambda, \ldots\}.$$

Also, "$n \in {}^*\mathbf{N}$ is in ${}^*E$ if and only if there exists $m \in {}^*\mathbf{N}$ such that $2m = n$" by the $*$-transform of the description of E in $\mathcal{N}$.

Standard sets can also be described by means of

(3.4.5) THE STANDARD DEFINITION PRINCIPLE *A set is standard if and only if it can be described as*

$$\{x : x \in {}^*A \;\&\; P(x)\}$$

where $P(x)$ is a property of x which can be formalized as a bounded predicate with only x free and only involving constants from dom(I), *that is, standard constants.*

This principle is clear logically since $[x \in I^{-1}(B)]$ is such a predicate for each B an entity of $\mathcal{X}$. The point of the principle in effect is that we can do (finite) set theory with standard sets and obtain standard answers. It is not too difficult to prove this theorem from the axioms of (3.4.1).

The algebraic approach to $*$-extensions does make the internal sets seem more subtle. We begin with several views of a simple example, the first being the ultrapower.

The set $\{\omega, \omega + 1, \ldots\}$ of nonstandard numbers bigger than (the equivalence class of) $\omega(j) = j$ is internal and not standard. Let $B(j) = \{j, j + 1, j + 2, \ldots\}$. Then $\omega(j) \in B(j)$ and if $\lambda(j) \geq \omega(j)$, $\lambda(j) \in B(j)$. Also, $B \in {}^*\mathcal{P}(\mathbf{N})$, a standard entity, since $B(j) \in \mathcal{P}(\mathbf{N})$ for all j.

An infinite ω exists by axioms (iii) and (vii), although showing it is infinite takes work.

Let us now show how $\{\lambda : \lambda \in {}^*\mathbf{N} \;\&\; \lambda^* \geq \omega\} = B$ can be proved internal strictly from the axioms of (3.4.1), without recourse to the explicit ultrapower construction. First consider the standard map

$$n \mapsto A_n = \{k : k \in \mathbf{N} \;\&\; k \geq n\}.$$

Let $B = {}^*A_\omega$. By axiom (v), B is defined, $\omega \in \text{dom}({}^*A)$, and $B \in \text{rng}({}^*A)$ and thus is internal. We must show, however, that

$$B = \{k : k \in {}^*\mathbf{N} \;\&\; k^* \geq \omega\}.$$

We know that the standard A in the standard model satisfies "for every n in $\mathbf{N}$ and k in $\mathbf{N}$, $k \in A_n$ if and only if $k \geq n$." Since $\mathbf{N} = \{1, 2, \ldots, n - 1\} \cup A_n$ by axioms (ii)–(iv) we know

$${}^*\mathbf{N} = \{{}^*1, {}^*2, \ldots, {}^*(n - 1)\} \cup {}^*A_n$$

for $n \in {}^\sigma\mathbf{N}$, but that is all we can obtain without an involved application of axiom (vi). The function of axiom (vi) algebraically speaking is to permit a start in the induction used to prove the standard definition principle when quantifiers are present. We essentially need the full strength of that here, so we conclude the sketch by showing that the standard definition principle (which can be proved from the axioms as described above) implies our particular result:

$$\mathbf{N} = \{n : n \in \mathbf{N} \,\&\, (\forall k)[k \in \mathbf{N} \Rightarrow [k \in A_n \Leftrightarrow k \geq n]]\},$$

so

$$^*\mathbf{N} = \{n : n \in {}^*\mathbf{N} \,\&\, (\forall k)[k \in {}^*\mathbf{N} \Rightarrow [k \in {}^*A_n \Leftrightarrow k \geq n]]\};$$

in particular, ω is in the set just described.

EXERCISE: Complete the details of this proof even for the special case of this standard definition. We believe the exercise shows how the logic helps even in the algebraic approach.

The most important description of internal sets is

(3.4.6) THE INTERNAL DEFINITION PRINCIPLE *A set is internal if and only if it can be described as*

$$\{x : x \in B \,\&\, P(x)\},$$

where B is internal and $P(x)$ is a property of x which can be described internally, specifically $P(x)$ can be formalized as a bounded predicate with only x free and so that all the constants have internal interpretations, that is, constants in $\mathrm{dom}(I')$. In particular, finite sets of internal things are internal.

The logical proof of this is clear since an internal set must arise as one of the maps B in $X_n{}^J$ for some n. In that case

$$[x \in (I')^{-1}(B)]$$

is the required predicate. This principle is not completely trivial and we ask the reader to look at the proof of Peano's theorem in Chapter 6 as an example of its use. If not convinced, the reader should then return to Chapter 2 and honestly finish that proof of Peano's theorem in that context.

Our example above is internal by its description:

$$\{k : k \in {}^*\mathbf{N} \,\&\, k \,{}^*\!\geq \omega\}!$$

The constants are $({}^*\geq)$ standard and (ω) internal in $^*\mathbf{N}$.

3.5 THE FORMAL LANGUAGE

The tremendous advantage of introducing a formal language is that we will be able to prove a transfer principle from the small model to the larger one. Moreover, with very little careful practice the reader will be able to apply the principle (see the section on $*$-transforms) without the formal language, so those details should not deter the reader who is unfamiliar with formal logic but who is interested in applications.

In the Introduction we discussed in some length Leibniz' approach to the calculus. In Leibniz' theory, the infinitesimals could be combined with real numbers and with themselves by means of the operations of addition and multiplication. In addition, Leibniz formulated a principle for his extended real number system. In simple terms it states that what holds for the finite numbers, that is, ordinary real numbers, should also hold for the numbers in the extended system. Leibniz was well aware of the contradictory nature of his principle, but he and his followers were never able to free Leibniz' theory of infinitesimals from these contradictions. It is perhaps surprising that a precise formulation of Leibniz' principle did not appear until about three centuries later.

A formal language will provide us with a means to single out for a given mathematical structure a family of statements that hold in the structure and that can be reinterpreted in other structures as true statements. The reinterpretation is the rule of Leibniz. In particular, in the case of the real number system the use of the formal language will enable us to make precise to which properties of the reals Leibniz' principle will apply and in what sense.

We shall now turn to the introduction of a formal language $\mathscr{L}$. With the help of a formal language we can express statements concerning mathematical objects much in the way we do in an ordinary language but with greater precision and more systematically. This will enable us to prove things about "all properties" of a superstructure.

In order to describe a formal language it is first necessary to describe the symbols of the language and then we can describe the process of forming sentences. We shall now present a detailed description of the formal language we shall use.

The *atomic symbols of* $\mathscr{L}$ are

(i) The *logical connectives* $\wedge$ (read "and"), $\vee$ (read "or"), $\Rightarrow$ (read "implies" or "only if"), $\Leftrightarrow$ (read "if and only if"), and $\neg$ (read "not").

(ii) The *variables*, denoted by letters (usually from the end of the alphabet) with or without subscripts. These symbols are supposed to constitute a countable set.

(iii) The *quantifiers*, which are the *universal quantifier* denoted by $\forall$ and the *existential quantifier* denoted by $\exists$.

(iv) The *separation symbols*, denoted by [, the left square bracket and], the right square bracket.

(v) The *basic predicates*, which are $\in$ (read "element of" or "member of") and $=$ (read "is equal to" or "identical with").

(vi) The *extra logical constants* which form a set of symbols that we assume is sufficiently large to be put into one-to-one correspondence with the elements of whatever structures are under consideration. Constants are usually denoted by Roman or Greek letters taken from the beginning of the alphabet with or without subscripts. If an object under consideration has already an accepted name such as for the empty set, or 1, 2, ... for the natural numbers, or cos for the cosine function, etc., we shall adopt the convention that this name is also used as a constant symbol of $\mathscr{L}$.

This completes the list of the so-called *atomic symbols* of the language $\mathscr{L}$.

The formation of formulas of $\mathscr{L}$ is to be carried out according to the following list of rules.

The *atomic formulas* of $\mathscr{L}$ are obtained by combining $\in$ and $=$ in the usual way with two letters which may denote variables or constants. For instance, $x \in a$, $b = y$, $x = z$, etc. *Well-formed formulas*, briefly "wff," are now defined inductively. Atomic formulas bracketed by square brackets are wff. If V is a wff, then $[\neg V]$ is a wff. If V, W are wff, then $[V \wedge W]$, $[V \vee W]$, $[V \Rightarrow W]$, and $[V \Leftrightarrow W]$ are wff. If V is a wff, then $[(\exists x)V]$ and $[(\forall x)V]$ are wff provided V does not already contain one of $(\exists x)$ or $(\forall x)$. Thus $[(\exists x)V(z, y)]$ and $[(\forall x)V(x)]$ are both wff, but $[(\forall x)[(\exists x)V(x)]]$ is not a wff.

In summary, bracketing atomic formulas, combining them with connectives and quantifiers according to the rules described above, and iterating this process a finite number of times we obtain all the formulas of $\mathscr{L}$.

The wff of a language $\mathscr{L} = \mathscr{L}(\in, =)$ are divided into *sentences* (*complete wff*) and *predicates* (*open wff*).

(3.5.1) *The scope of a quantifier, for instance* $(\forall \cdot)$, *within a wff* V *is a wff* W *contained in* V *which starts with the left bracket* [*immediately following* $(\forall \cdot)$ *and ends with the corresponding right bracket.*

For example, in $[(\forall x)[[V(x)] \wedge [W(x)]]]$ the scope of $(\forall x)$ is the wff $[[V(x)] \wedge [W(x)]]$ and in $[[(\forall x)[V(x)]] \wedge [W(x)]]$, the wff $[V(x)]$ is the scope of $(\forall x)$.

It is clear that the scope of a quantifier remains the scope of the same quantifier in any wff which is obtained from the given wff by the further repeated application of the connectives and the quantifiers as described above.

(3.5.2) *A wff* V *is called a sentence when every variable* x *contained in*

V is within the scope of $(\forall x)$ *or* $(\exists x)$, *or in the expression* $(\forall x)$ *or* $(\exists x)$. *A wff which is not a sentence is called a predicate.*

For example, $(\forall x)[V(x)]$ is a sentence, $[(\forall x)[V(x)]] \wedge [W(x)]$ is *not* a sentence; $[(\forall x)[a \in b]]$ is a sentence, and $[(\forall x)[(\forall y)[(\forall z)[[[V(x, y)] \wedge [V(y, z)]] \Rightarrow [V(x, z)]]]]]$ is also a sentence. The reader is advised to determine the scope of the quantifiers in the above sentences.

We shall distinguish between free and bound occurrences of variables in a wff.

(3.5.3) *An occurrence of a variable x is bound in a wff whenever either it is the variable of a quantifier* $(\exists x)$ *or* $(\forall x)$ *in the wff, or it is within the scope of the quantifier* $(\forall x)$ *or* $(\exists x)$ *in the wff. Otherwise, the occurrence is said to be free in the wff.*

For example, consider the following wff's: (i) $[x \in y]$; (ii) $[x \in y] \Rightarrow (\forall x)[x \in a]$; (iii) $(\forall x)[[x \in y] \Rightarrow (\forall x)[x = x]]$; and (iv) $(\forall x)[a \in b]$. In (i), the single occurrence of x is free. In (ii), the first occurrence of x is free, but the second and third occurrences are bound. In (iii) and (iv), all occurrences of x are bound. In all the wff's every occurrence of y is free.

A wff V is said to be in *prenex normal form* if in the formation of V from the atomic formulas, the quantifiers are applied after the connectives, that is, if the connectives are in the scope of all the quantifiers occurring in V. In symbols, $V \equiv (qx_1)\cdots(qx_n)W$, where each $(q_i x_i)$ is a universal or existential quantifier, $x_i \neq x_j$ for $i \neq j$, and W contains no quantifiers.

One of the basic results of the first-order predicate calculus states that *every wff can be transformed by means of an effective procedure into a wff in prenex normal form.*

A formula will be called *bounded* when the quantifiers always appear in the following forms:

$$[(\forall x)[[x \in A] \Rightarrow [V]]] \qquad \text{for the universal quantifier}$$

and

$$[(\exists x)[[x \in B] \wedge [V]]] \qquad \text{for the existential quantifier,}$$

where A and B are constants of the language. For example, a function f that is continuous at a (partially formalized) would read

$$(\forall \varepsilon)(\exists \delta)[\varepsilon \in \mathbf{R}^+ \Rightarrow [\delta \in \mathbf{R}^+ \wedge [\text{“}|x - a| < \delta \Rightarrow |f(x) - f(z)| < \varepsilon\text{”}]]].$$

Set-theoretically this will correspond to saying which entity we are quantifying over. In the case of continuity we do not mean "for every ε," but rather "for every positive real number ε" and *we require this explicitly in bounded formulas.*

We shall conclude this section with comments on the usage of the formal language and the English language. The reader should *not* confuse the language in which the text is written with the formal language described in this section. The language of the text is English to which is added some mathematical jargon and some symbols which also may be symbols of $\mathscr{L}$. In this English language, however, we shall make statements about the language $\mathscr{L}$ and about structures which are entirely outside the realm of the language $\mathscr{L}$. For this reason we shall refer to the language in which the text is written as the *metalanguage*.

Later on there will be exercises on how partially to formalize English sentences so that properties can be transferred. We emphasize now that we view $\mathscr{L}$ itself as a mathematical structure. Our plan is to make a connection between $\mathscr{X}$, $\mathscr{L}$, and $\mathscr{Y}$, all viewed in the metalanguage as mathematical objects:

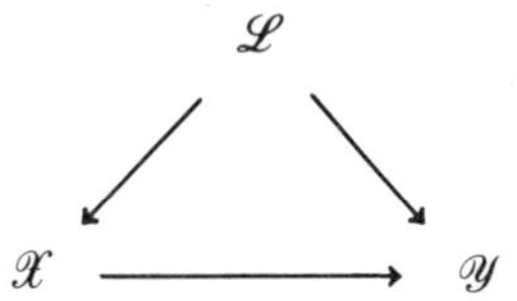

EXERCISES:

1. (i) Let V be a wff in which y does not occur. If y is not free in W, and if $V(x)$ has only free occurrences of x, if any, while $V(y)$ is obtained from $V(x)$ by replacing all occurrences of x by y, show the following equivalences:

$$[(\forall x)V(x)] \Rightarrow W \equiv (\exists y)[V(y) \Rightarrow W];$$

$$[(\exists x)V(x)] \Rightarrow W \equiv (\forall y)[V(y) \Rightarrow W];$$

$$[W \Rightarrow (\forall x)V(x)] \equiv (\forall y)[W \Rightarrow V(y)];$$

$$[W \Rightarrow (\exists x)V(x)] \equiv (\exists y)[W \Rightarrow V(y)].$$

(ii) Show also that the following equivalences hold:

$$\neg(\forall x)V \equiv (\exists x)[\neg V];$$

$$\neg(\exists x)V \equiv (\forall x)[\neg V];$$

$$\neg(\forall x)[x \in a \Rightarrow [V]] \equiv (\exists y)[y \in a \wedge [\neg V]];$$

$$\neg(\exists x)[x \in a \wedge [V]] \equiv (\forall x)[x \in a \Rightarrow [\neg V]].$$

(iii) By using the results given in (i) and (ii) prove the prenex normal form theorem. (*Hint*: Use induction on the total number m of connectives and quantifiers in a wff. For $m = 0$, there is nothing to prove. Assume that we can find a corresponding wff in prenex normal form for all wff's whose total number m of connectives and quantifiers satisfies $m < n$. Assume then that the wff V has a total of n connectives and quantifiers. You have to consider only the following three cases: V is $[\neg V]$; V is $[W_1 \Rightarrow W_2]$; and V is $[(\forall x)W]$.)

2. Find the prenex normal form equivalent to the following wff's:

(i) $(\forall x)[[x \in \alpha] \Rightarrow (\exists y)[[y \in \alpha] \wedge (\forall z)[[z \in \beta] \Rightarrow [u = (y, z)] \wedge [u \in x]]]]$.
(ii) $[[(\forall x)[[x \in \alpha] \Rightarrow [x \in y]]] \Rightarrow [[(\exists y)[y \in \alpha]] \Rightarrow (\exists z)[y \in z]]]$.
(iii) $(\forall x)[[x \in \alpha] \Rightarrow [\neg[(\forall y)[y \in \alpha] \Rightarrow [x = y]]]]$.

3.6 INTERPRETATIONS OF THE FORMAL LANGUAGE

Let $\mathscr{L} = \mathscr{L}(\in, =)$ be a formal language as described in the preceding section. It will now be our task to explain what we mean by the statement that a bounded sentence of $\mathscr{L}$ has an interpretation *in set theory*.

(3.6.1) DEFINITION *A one-to-one mapping I of a subset of the set of all the constants of $\mathscr{L}$,* cns($\mathscr{L}$), *into a superstructure $\mathscr{X}$ will be called an interpretation map of $\mathscr{L}$ in set theory.*

From the interpretation of constants we build up interpretations of bounded sentences $V(\alpha_1, \ldots, \alpha_n; \in, =)$ whose constants $a_1, \ldots, \alpha_n$ are in the domain of I. Atomic formulas $[\alpha = \beta]$ or $[\alpha \in \beta]$ are interpreted in $\mathscr{X}$ as ${}^I[\alpha = \beta]$ "the entity or individual $I(\alpha)$ equals the entity or individual $I(\beta)$" or ${}^I[\alpha \in \beta]$ "the entity or individual $I(\alpha)$ is a member of the entity or individual $I(\beta)$." We interpret $\in$ and $=$ *in the usual set-theoretical way*; this is what we mean by an interpretation in set theory. Notice that being able to interpret *does not* require that the interpretation is true; in particular, when $I(\beta)$ is an individual $[\alpha \in \beta]$ has to have a false interpretation when $\alpha \in \text{dom}(I)$, but has no interpretation when $\alpha \notin \text{dom}(I)$.

The logical connectives are always interpreted as the metamathematical counterpart "and" for $\wedge$, "or" for $\vee$, "not" for $\neg$, "implies" for $\Rightarrow$, "if and only if" for $\Leftrightarrow$.

Now we come to an important point, since a bounded formula has each quantifier specified to run over a constant, the interpretation of $V = (\forall x)[x \in \alpha \Rightarrow W]$ is

$${}^I V = \text{"for all } x \text{ elements of } I(\alpha)\text{, the statement } {}^I W(x)\text{,"}$$

where ${}^I W(x)$ denotes the portion of the formula already interpreted where free occurrences of x are replaced by the elements of $I(\alpha)$. The interpretation of $V = (\exists x)[x \in \alpha \wedge W]$ is

$$^{I}V = \text{“there is an } x \text{ in } I(\alpha) \text{ such that } {}^{I}W(x).\text{”}$$

Building up by the procedure through which we defined formal sentences, we are able to interpret each bounded formal sentence whose constants are in the domain of I.

Let K be a set of bounded sentences of the language $\mathscr{L}(\in, =)$. Then

(3.6.2) DEFINITION *An interpretation I provides a model for a set of bounded sentences K in set theory provided all the constants occurring in sentences of K are in the domain of I,* $\text{dom}(I)$, *and provided the interpretation ^{I}V is true for each V in K.*

We shall not be concerned with how one determines the truth value of an interpreted sentence. We will not prove Riemann's hypothesis. We do assume the interpretation of each sentence is either true or false, whether or not we can prove any particular sentence is true. Observe below that we also are operating on the assumption that set theory is consistent. We are only using the formal language to discuss models.

3.7 MODELS OF A SUPERSTRUCTURE

Let $\mathscr{X} = \mathscr{X}(X, \in, =)$ be a superstructure based on the nonempty set of individuals X, and let $\mathscr{L} = \mathscr{L}(\in, =)$ be a formal language with a set of constants of larger cardinality than the cardinal of $\mathscr{X}$. Assume that I is a standard interpretation map, that is, I is a one-to-one mapping of a subset of the set of constants of $\mathscr{L}$ *onto* $\mathscr{X}$. Then, on the basis of the definitions contained in the preceding section, every bounded sentence V of $\mathscr{L}$ with the property that the constants which occur in V are in the domain of I has an I-interpretation in $\mathscr{X}$. Consider now the set $K_0 = K_0(I)$ *of all the bounded sentences* V of $\mathscr{L}$ such that ^{I}V holds in $\mathscr{X}$. Since we consider all such true statements we conclude that all the statements whose I-interpretation reads $a = a$, $a \in \mathscr{X}$, are contained in K_0. Hence, the set $\text{cns}(K_0)$ of all the constants occurring in the sentences of K_0 is equal to the domain of I.

(3.7.1) DEFINITION *Suppose an interpretation map I' into a superstructure $(\mathscr{Y}, \in, =)$ provides a model for the set of bounded sentences K_0. If $I' \circ I^{-1} = *$ is not onto, we say that $(^{*}\mathscr{X}, \in, =) \subset (\mathscr{Y}, \in, =)$ is a nonstandard set-theoretical model of $(\mathscr{X}, \in, =)$.* All the nonstandard models we shall consider in this book are also proper extensions in the sense of (3.4.1 (vii)).

In algebra or geometry one does not restrict the kind of models permitted; similarly in model theory we want to permit abstract models. In the present setting, let $(\mathscr{A}, M, E)$ be a set $\mathscr{A}$ with two binary relations: M to be interpreted as membership and E to be interpreted as equality. A one-to-one map i from a set of constants of $\mathscr{L}$ into $\mathscr{A}$ induces an interpretation for formulas of $\mathscr{L}$ by interpreting $\in$ in $\mathscr{L}$ as M in $\mathscr{A}$ and $=$ in $\mathscr{L}$ as E in $\mathscr{A}$ as described in Section 3.6.

(3.7.2) DEFINITION *An abstract interpretation i into $(\mathscr{A}, M, E)$ provides an abstract model for a set of sentences of $\mathscr{L}$ if each can be interpreted in terms of M and E, and provided each such sentence is true in $(\mathscr{A}, M, E)$.*

EXERCISE: Let $\mathscr{X} = \mathscr{X}(X, \in, =)$ be a superstructure, and $\mathscr{A} = \mathscr{X}^J$ the set of all maps from J into $\mathscr{X}$. For a fixed $j_0 \in J$ we define

$$(f_j)M(g_j) \qquad \text{if and only if} \qquad f_{j_0} \in g_{j_0}$$

and

$$(f_j)E(g_j) \qquad \text{if and only if} \qquad f_{j_0} = g_{j_0}.$$

Show that there is an interpretation $i: \mathscr{L} \to \mathscr{A}$ which provides a model for $\mathscr{X}$. Would you call the model $(\mathscr{A}, M, E)$ nonstandard?

3.8 NONSTANDARD ULTRAPOWER MODELS

Let $\mathscr{X} = \mathscr{X}(X, \in, =)$ be a superstructure based on the infinite set X, and $\mathscr{L} = \mathscr{L}(\in, =)$ a formal language as above. Assume that I is an interpretation mapping of $\mathscr{L}$ in $\mathscr{X}$ and that $K_1 = K_1(I)$ is the set of all sentences of $\mathscr{L}$ whose I-interpretation holds in $\mathscr{X}$.

It is the purpose of this section to show that in case X is infinite there is a nonstandard model of K_1 with respect to some interpretation mapping $I^{\mathscr{U}}$. Such a nonstandard model will be constructed by means of the ultrapower method, the predicates $\in$, $=$ of $\mathscr{L}$ will be interpreted as $\in_{\mathscr{U}}$ and $=_{\mathscr{U}}$. In Section 3.10 we embed in set theory to obtain a full nonstandard model.

To this end, we shall assume that $\mathscr{U}$ is a δ-incomplete ultrafilter defined over the infinite set J. This means axioms (1)–(4) of Definition (2.7.1) hold. In place of (5), $\{J_n\}$, $n = 1, 2, \ldots$, shall denote a countable partition of J such that $J_n \notin \mathscr{U}$ for all $n = 1, 2, \ldots (\bigcup J_n = J \in \mathscr{U})$. This is equivalent to requiring that there are sets $F_n \in \mathscr{U}$, $n = 1, 2, \ldots$, with $\bigcap F_n \notin \mathscr{U}$.

It is easy to see that if J is countable, then every free ultrafilter over J is δ-incomplete. It is a natural question to ask whether every free ultrafilter is δ-incomplete. The answer to this question is not known. The question itself is known under the name of *Ulam's measure problem*. It was shown, however, that the assumption that every free ultrafilter is δ-incomplete is consistent with the axioms of ZF set theory. Whether the negation of this statement, namely that there exists a δ-complete free ultrafilter, is consistent with the axioms of set theory is not known. It was shown that the cardinal of a set on which a δ-complete free ultrafilter exists must be very large and, in fact, must be an inaccessible cardinal. Since we will not use δ-complete free ultrafilters we shall not discuss this question any further.

Let $\mathscr{A} = \mathscr{X}^J$, the set of all mappings from J into $\mathscr{X}$. We shall now define the binary relations M and E for $\mathscr{A}$ in terms of $\mathscr{U}$. Instead of denoting these binary relations by M and E we shall use the more suggestive notation $\in_{\mathscr{U}}$ and $=_{\mathscr{U}}$, respectively.

(3.8.1) DEFINITION *If $f, g \in \mathscr{X}^J$, then f is called a $\mathscr{U}$-member of g whenever $\{j : f(j) \in g(j)\} \in \mathscr{U}$, and in that case we write $f \in_{\mathscr{U}} g$. Furthermore, f is called $\mathscr{U}$-equal to g whenever $\{j : f(j) = g(j)\} \in \mathscr{U}$, and this will be denoted by $f =_{\mathscr{U}} g$.*

It is our purpose to show that $(\mathscr{X}^J, \in_{\mathscr{U}}, =_{\mathscr{U}})$ is a model of $\mathscr{X}$ with respect to a certain interpretation mapping $I^{\mathscr{U}}$. Such an interpretation mapping needs, in addition to being a one-to-one mapping of a subset of the set of constants cns($\mathscr{L}$) of $\mathscr{L}$ onto $\mathscr{A}$, to have the following properties:

(i) dom $I \subseteq$ dom $I^{\mathscr{U}}$;

(ii) if $\alpha \in$ dom I and $a = I(\alpha)$, then $I^{\mathscr{U}}(\alpha) = f$, where $f(j) = a$ for all $j \in J$ or, in other words, the interpretation mapping $I^{\mathscr{U}}$ is such that its domain includes the domain of the given interpretation mapping I and it induces an embedding of $\mathscr{X}$ into $\mathscr{X}^J$ which assigns to every $a \in \mathscr{X}$ the constant mapping $a(j) = a$ for all $j \in J$;

(iii) the $I^{\mathscr{U}}$-interpretations of the basis predicates $\in$ and $=$ are $\in_{\mathscr{U}}$ and $=_{\mathscr{U}}$, respectively.

In this section we shall prove the central result concerning $(\mathscr{X}^J, \in_{\mathscr{U}}, =_{\mathscr{U}})$ and the interpretation mapping $I^{\mathscr{U}}$. We need a new way of interpreting sentences in prenex normal form which was originally developed by Hilbert and Skolem and later by Herbrand. Before we formulate this method in its generality, we examine a special case.

Consider the sentence $W \equiv (\forall x)(\exists y)V(x, y)$, where V is a wff. The statement that the I-interpretation ${}^IW$ of W holds in $\mathscr{X}$ means that for every constant $\alpha \in$ dom I, there exists a constant $\beta \in$ dom I such that the I-interpretation of $V(\alpha, \beta)$ holds in $\mathscr{X}$. Using the axiom of choice we may conclude that there exists a mapping φ of dom I into dom I such that ${}^IV(\alpha, \varphi(\alpha))$ holds for all $\alpha \in$ dom I. Conversely, if such a mapping φ exists, then ${}^IW$ holds in $\mathscr{X}$. The mapping φ is called a *Herbrand–Skolem functor* or, shortly, a functor. The expression $V(\alpha, \varphi(\alpha))$ is not a wff of the language $\mathscr{L}$. Nevertheless its I-interpretation is a statement in the metalanguage about $\mathscr{X}$. The expression $V(x, \varphi(x))$ is called an *open form sentence.* The open form sentence $V(x, \varphi(x))$ holds in $\mathscr{X}$ if and only if ${}^IV(x, \varphi(x))$ holds in $\mathscr{X}$ for all $x \in \operatorname{dom} I$, which is equivalent to ${}^IW$ holding in $\mathscr{X}$.

Let us now assume that $W = (qx_1)(qx_2)\cdots(qx_n)V$ is given in prenex normal form, where $(q\cdot)$ is either $(\forall\cdot)$ or $(\exists\cdot)$. To obtain an open form sentence from W we delete the prefix $(qx_1)(qx_2)\cdots(qx_n)$, and replace the variables that belong to the existential quantifiers everywhere where they occur in V by distinct symbols $\varphi_1, \ldots, \varphi_{m_n}$, which denote functions of all the variables that belong to the universal quantifiers preceding the existential quantifier in question. If W begins with an existential quantifier, then the corresponding symbol φ is supposed to be a function without variables, that is, a particular constant of the domain of I. The following metamathematical principle should now be evident.

(3.8.2) *Let V be a sentence of $\mathscr{L}$ in prenex normal form which has an I-interpretation ${}^IV$ in $\mathscr{X}$. Then ${}^IV$ holds if and only if there exists an open form sentence of V such that its I-interpretation holds in $\mathscr{X}$.*

This principle is useful in that the Herbrand–Skolem functors provide us with an interpretation of the existential quantifiers in the metalanguage and in set theory.

We shall now formulate and prove the fundamental theorem about ultrapowers.

(3.8.3) FUNDAMENTAL THEOREM *Let $\mathscr{X} = \mathscr{X}(X, \in, =)$ be a superstructure based on an infinite set X and let $K_1 = K_1(I)$ be the set of sentences of $\mathscr{L} = \mathscr{L}(\in, =)$ that hold in $\mathscr{X}$ with respect to an interpretation mapping I of $\mathscr{X}$. Then $(\mathscr{X}^J, \in_{\mathscr{U}}, =_{\mathscr{U}})$ is a nonstandard model of K_1 with respect to the interpretation mapping I' defined above, where $\mathscr{U}$ is a δ-incomplete ultrafilter defined over J.*

PROOF (Luxemburg): We have to show that the $I^{\mathscr{U}}$-interpretation of every sentence $W \in K_1$ holds in $\mathscr{X}^J$. To this end, there is no loss in generality to assume that $W \equiv (qx_1)\cdots(qx_n)V \in K_1$ is in prenex normal form. Since ${}^IW$ holds in $\mathscr{X}$ it follows from (3.8.2) that W admits an open form sentence whose I-interpretation holds in $\mathscr{X}$. This means that there exists a set of Herbrand–Skolem functors $\varphi_1, \ldots, \varphi_v$ for W such that the I-interpretation of the open form sentence $V(y_1, \ldots, y_s, \varphi_1, \ldots, \varphi_v)$ holds in $\mathscr{X}$, where $y_1, \ldots, y_s$ is the set of the different bound variables of V that occur in the universal quantifiers of V, and $v + s = n$. Since V is free of quantifiers it is composed of, say, h atomic formulas $A_1, \ldots, A_h$ which occur in this order in V and which are combined in V with a subset of the set of logical connectives. From the propositional calculus it follows that only certain sets of h-truth values of zeros and ones for $A_1, \ldots, A_h$ will produce the truth value one for V. These sets of truth values do not depend on the individual form of the atomic formulas $A_1, \ldots, A_h$ but only on the family of the logical connectives that combine them in V, and on the order in which they occur in V. A family of h-truth values for $A_1, \ldots, A_h$ can be represented by ordered h-tuples of zeros and ones, where 0 means false and 1 means true. Since the I-interpretation of the open form sentence $V(x_1, \ldots, x_s, \varphi_1, \ldots, \varphi_v)$ holds in $\mathscr{X}$ for all $x_1, \ldots, x_s \in \mathscr{X}$, it follows that among the 2^h truth value h-tuples for $A_1, \ldots, A_h$ there are, say, $n_h \geq 1$ of them which produce the truth value one for V. We shall denote these truth value h-tuples by $t_1, \ldots, t_{n_h}$, $n_h \geq 1$. In other words, for every set of s constants $\alpha_1, \ldots, \alpha_s \in \operatorname{dom} I$, there exists an index i, $1 \leq i \leq n_h$ such that $t_i = (t({}^I A_1), \ldots, t({}^I A_h))$, where $t({}^I A_i)$ is either one or zero according to whether ${}^I A_i$ holds in $\mathscr{X}$ or not. After these preliminaries we can now turn to the proof that the $I^{\mathscr{U}}$-interpretation of W holds in $\mathscr{X}^J$. For this purpose we have to show, according to (3.8.2), that there exists an open form sentence $V(y_1, \ldots, y_s, \overline{\varphi}_1, \ldots, \overline{\varphi}_v)$ of W whose $I^{\mathscr{U}}$-interpretation holds for all sets of s constants $\alpha_1, \ldots, \alpha_s$ of the domain of $I^{\mathscr{U}}$. Such Herbrand–Skolem functors on the domain of $I^{\mathscr{U}}$ can be described in terms of $\varphi_1, \ldots, \varphi_v$ by the following. Let $y_1, \ldots, y_{s_i}$, $1 \leq s_i \leq s$, be the independent variables of φ_i $(i = 1, 2, \ldots, v)$, let $\alpha_1, \ldots, \alpha_{s_i}$ be s_i-constants of dom $I^{\mathscr{U}}$, and let $I^{\mathscr{U}}(\alpha_k) = a_k$ $(k = 1, 2, \ldots, s_i)$. Then $I^{\mathscr{U}}(\overline{\varphi}_i(\alpha_1, \ldots, \alpha_{s_i}))$ is a mapping of J into $\mathscr{X}$ whose value at $j \in J$ is given by the formula $I^{\mathscr{U}}(\overline{\varphi}_i(\alpha_1, \ldots, \alpha_{s_i}))(j) = I(\varphi_i(\alpha_1{}^j, \ldots, \alpha_{s_i}^j))$, where $I(\alpha_k{}^j) = a_k(j)$, $k = 1, 2, \ldots, \alpha_s$. We shall now show that the $I^{\mathscr{U}}$-interpretation of the open form sentence $V(y_1, \ldots, y_s, \overline{\varphi}_1, \ldots, \overline{\varphi}_v)$ holds in $\mathscr{X}^J$ for all $y_1, \ldots, y_s \in \operatorname{dom} I^{\mathscr{U}}$. The $I^{\mathscr{U}}$-interpretation

$$^{I^{\mathscr{U}}}V(y_1, \ldots, y_s, \overline{\varphi}_1, \ldots, \overline{\varphi}_v)$$

of the open form sentence

$$V(y_1, \ldots, y_s, \overline{\varphi}_1, \ldots, \overline{\varphi}_v)$$

is composed of the $I^{\mathscr{U}}$-interpretation ${}^{I^{\mathscr{U}}}A_i$ $(i = 1, 2, \ldots, h)$ of the atomic formulas that make up V. As we stated before, the meaning of the logical connectives remains the same under any interpretation. Since, by the definition of the binary relation $\in_{\mathscr{U}}$ and $=_{\mathscr{U}}$, the truth values of ${}^{I^{\mathscr{U}}}A_i$ $(i = 1, 2, \ldots, h)$ are well determined for any choice of the constants $y_1, \ldots, y_s \in \operatorname{dom} I^{\mathscr{U}}$. We conclude that for any choice of the constants $y_1, \ldots, y_s \in \operatorname{dom} I^{\mathscr{U}}$, we have that ${}^{I^{\mathscr{U}}}V(y_1, \ldots, y_s, \overline{\varphi}_1, \ldots, \overline{\varphi}_v)$ holds in $\mathscr{X}^J$ if and only if there exists an index k, $1 \leq k \leq n_h$ such that $t_k = (t({}^{I^{\mathscr{U}}}A_1), \ldots, t({}^{I^{\mathscr{U}}}A_h))$. Assume that $\alpha_1, \ldots, \alpha_s \in \operatorname{dom} I^{\mathscr{U}}$ and that for every j, $\alpha_i^j, \ldots, \alpha_s^j$ are s-constants contained in the domain of I such that $a_k(j) = (I^{\mathscr{U}}(\alpha_k))(j) = I(\alpha_k^j)$ for all $j \in J$ and $k = 1, 2, \ldots, s$. Then for every $j \in J$, we shall denote by ${}^I A_i(j)$, $i = 1, 2, \ldots, l$, the I-interpretation of A_i upon replacing the variables $y_1, \ldots, y_s$ that may occur in A_i by the constants $\alpha_1^j, \ldots, \alpha_2^j$, respectively. Then, by Definition (3.8.1), we have for all $i = 1, 2, \ldots, s$,

$$t({}^{I^{\mathscr{U}}}A_i) = 1 \qquad \text{if and only if} \qquad \{j : t({}^I A_i(j)) = 1\} \in \mathscr{U},$$

and

$$t({}^{I^{\mathscr{U}}}A_i) = 0 \qquad \text{if and only if} \qquad \{j : t({}^I A_i(j)) = 0\} \in \mathscr{U}.$$

This suggests that we consider the following sets:

$$E_p = \{j : (t({}^I A_1(j)), \ldots, t({}^I A_h(j))) = t_p\},$$

where $p = 1, 2, \ldots, n_h$. Then $E_p \cap E_q = \varnothing$ whenever $p \neq q$ and $\bigcup_{p=1}^{n_h} E_p = J$. The latter statement follows from our assumption that $W \in K_1$, which implies that for every $j \in J$ the I-interpretation of $V(y_1, \ldots, y_s, \varphi_1, \ldots, \varphi_v)$ holds on replacing the variables by $\alpha_1^j, \ldots, \alpha_s^j$, which in symbols means that for every $j \in J$ there is an index $k = 1, 2, \ldots, n_h$ such that $(t({}^I A_i(j)), \ldots, t({}^I A_h(j))) = t_k$. Since $\mathscr{U}$ is an ultrafilter it follows that there exists precisely one index k_0 such that $E_{k_0} \in \mathscr{U}$. Hence $(t({}^{I^{\mathscr{U}}}A_1), \ldots, t({}^{I^{\mathscr{U}}}A_h)) = t_{k_0}$, which is equivalent to saying that $I^{\mathscr{U}}$-interpretation of the open form sentence $V(\alpha_1, \ldots, \alpha_s, \bar{\varphi}_1, \ldots, \bar{\varphi}_v)$ holds in $\mathscr{X}^J$. Using (3.8.2) we conclude that $\mathscr{X}^J$ is a model of K_1.

In order to show that $\mathscr{X}^J$ is a nonstandard model of $\mathscr{X}$ *we shall prove that for every infinite entity* $A \in \mathscr{X}$ *there exists an element* $f(j) \in_{\mathscr{U}} A$ (the constant sequence) *such that* $f(j) \neq_{\mathscr{U}} a$ (the constant sequence) *for all* $a \in A$. Since A is infinite there exists a sequence $\{a_n\}$ of elements of A such that $a_n \neq a_m$ whenever $n \neq m$. Now $\{J_n\}$ $(n = 1, 2, \ldots)$ is a countable partition of J such that $J_n \notin \mathscr{U}$ for all $n = 1, 2, \ldots$, since we have a δ-incomplete ultrafilter $\mathscr{U}$. Define a mapping f of J into $\mathscr{X}$ by means of the definition $f(j) = a_n$ for all $j \in J_n$ $(n = 1, 2, \ldots)$. Then $f \in_{\mathscr{U}} A$ (see (3.4.1)) and $f \neq_{\mathscr{U}} a$ for all $a \in A$. This completes the proof of the fundamental theorem on ultrapowers.

(3.8.4) COROLLARY *Let W be a sentence of $\mathscr{L}$ whose constants are in* dom I. *Then W holds in $\mathscr{X}$ if and only if ${}^{I^{\mathscr{U}}}W$ holds in $\mathscr{X}^J$.*

PROOF: The fundamental theorem implies that if ${}^I V$ holds in $\mathscr{X}$, then ${}^{I^{\mathscr{U}}}V$ holds in $\mathscr{X}^J$. Conversely, assume that ${}^I V$ does not hold in $\mathscr{X}$ and ${}^{I^{\mathscr{U}}}V$ holds in $\mathscr{X}^J$. Then $\neg V \in K_1$, and so ${}^I(\neg V)$ holds in $\mathscr{X}$. Hence, the fundamental theorem implies that then also ${}^{I^{\mathscr{U}}}(\neg V)$ holds in $\mathscr{X}^J$. But ${}^{I^{\mathscr{U}}}(\neg V) \equiv \neg({}^{I^{\mathscr{U}}}V)$ and the assumption ${}^{I^{\mathscr{U}}}V$ holds in $\mathscr{X}^J$ leads to a contradiction.

This corollary proves Leibniz' principle once we embed our ultrapower in set theory.

3.9 BOUNDED FORMAL SENTENCES

The proof of the fundamental theorem depends in an essential way on the notation of an open form sentence. For a sentence $W \in K_1(I)$ the proof of the existence of the corresponding open form sentence depends essentially on the axiom of choice. This in turn makes the fundamental theorem depend on the axiom of choice. Now it is well known from model theory that Gödel's completeness theorem stating that every consistent set of sentences has a model; that is, the compactness principle is logically equivalent to the prime ideal theorem for Boolean algebras. This result of Stone can be deduced from the axiom of choice, but as we have already stated, as an axiom by itself it is weaker than the axiom of choice.

Since the definition of an ultrapower model only involves the prime ideal theorem for Boolean algebras, it seems a natural question to ask whether the validity of the fundamental theorem about ultrapowers depends essentially on the axiom of choice. Although this question seems to be open, we can show that if the set of sentences is restricted to the bounded sentences, then the fundamental theorem can be shown to hold without using the axiom of choice. The prime ideal theorem for Boolean algebras is required so that our δ-incomplete ultrafilter exists (the set of complements of a filter is an ideal).

Let $\mathscr{X} = \mathscr{X}(X, \in, =)$ be a superstructure based in the infinite set X and let $K_1 = K_1(I)$ be the set of all sentences of $\mathscr{L}$ whose I-interpretations hold in $\mathscr{X}$. A sentence $W \in K_1$ is called "bounded" whenever every quantifier occurring in W is of the form $(\forall x)[[x \in a] \Rightarrow \cdots]$ or $(\exists y)[[y \in p] \wedge \cdots]$, that is, if in the I-interpretation the domain of every quantifier is a well defined entity of $\mathscr{X}$. The subset of $K_1(I)$ of the bounded sentences will be denoted by $K_0(I)$. If we restrict the discussion to the bounded sentences, then the fundamental theorem can be shown to hold for K_0 without use of the axiom of choice.

The fact that the proof for K_0 does not depend on the axiom of choice is based on the following simple observation. Let

(F) $$(q_1 x_1) \cdots (q_n x_n) V(x_1, \ldots, x_n) = F$$

be a sentence in prenex normal form. If this is bounded, each quantifier q_i is restricted to run over an entity A_i and the n-ary relation

$$\Phi = \{(x_1, \ldots, x_n) : x_1 \in A_1 \,\&\, \cdots \,\&\, x_n \in A_n \,\&\, {}^I V(x_1, \ldots, x_n)\}$$

is an n-ary relation *entity* in $\mathscr{X}$. Moreover, formula (F) is true in $\mathscr{X}$ if and only if the dom Φ corresponding to the universal quantifiers is all of $A_{i_1} \times \cdots \times A_{i_n}$. Then the constant sequence $\Phi(j) = \Phi$ is given by

$$\Phi(j) = \{(x_1(j), \ldots, x_n(j)) : \\ x_1(j) \in_{\mathscr{U}} A_1 \ \& \cdots \& \ x_n(j) \in_{\mathscr{U}} A_n \ \& \ {}^{I^{\mathscr{U}}}V(x_1(j), \ldots, x_n(j))\}$$

and ${}^{I^{\mathscr{U}}}F$ is true whenever ${}^{I}F$ is.

Robinson [1966, p. 95] makes an interesting observation concerning the axiom of choice in his infinitesimal proof of Tychonoff's theorem. Without the Hausdorff assumption the result is equivalent to the axiom of choice, but with that assumption the axiom is not used in his proof.

For the most part we shall take the attitude that the axiom of choice holds and also (nearly always) only consider bounded sentences.

3.10 EMBEDDING $\mathscr{X}^J$ IN SET THEORY

We now begin the study of the "nonstandardness" of our ultrapower model in greater detail. Recall that the δ-incompleteness of $\mathscr{U}$ tells us that whenever A is an infinite entity of $\mathscr{X}$, there are maps in A^J that are inequivalent mod $\mathscr{U}$ to all of the constant maps in A^J. We proceed by partially embedding $\mathscr{X}^J$ in a superstructure.

This embedding is defined on the finite entities of $\mathscr{X}^J$ and results in the nonstandard I'-interpretation. That is, I' is the set embedding of $I^{\mathscr{U}}$ where the embedding is defined. The reason we embed part of the ultrapower in an ordinary superstructure is that we want to compare sets and functions which arise as sequences in $\mathscr{X}^J$ with arbitrary ones. For example, the set of infinitesimals in ***R** is a collection of all the real-valued sequences $x(j)$ that satisfy $-1/n \le x(j) \le 1/n$ for each constant sequence $1/n$ with $n \in \mathbf{N}$ (Definition (4.4.1)). This is a "real" set, that is, the metamathematical $\in$ rather than $\in_{\mathscr{U}}$ describes its elements. It so happens that the infinitesimals *cannot* be described by $\in_{\mathscr{U}}$—they are an external set—yet we clearly want to be able to discuss them.

The alternative to our approach is to deal with two kinds of set theory (what Machover and Hirschfeld [1969] call "pseudosets"): the internal map sets of $\mathscr{X}^J$ and "real" subsets. We prefer to embed the internal theory in the larger one; unfortunately this is technical, in particular, *not* $\mathscr{X}^J/\mathscr{U}$ in the obvious sense. Despite the technicality, the aim is simple: to make $\in_{\mathscr{U}}$-sets into $\in$-sets. (*Note*: The descending $\in$-chains of $\mathscr{X}^J$ are not included in our embedding—only bounded properties transfer to $\mathscr{Y}$.)

(3.10.1) DEFINITION [0.1] Take $a \in X_0$ and for the constant map $a(j) \equiv a$ associate the equivalence class $*a = [a(j)] = \{b \in \mathscr{X}^J : b(j) =_{\mathscr{U}} a(j)\}$.

[0.2] For $b(j) \in X_0{}^J$, not constant, associate the map with the equivalence class $[b(j)] = b$, by abuse of notation.

[1.1] Let Y_0 be the set of all the equivalence classes of maps $b(j) \in X_0{}^J$. Thus $b \in Y_0$ means $b = [b(j)]$ for some $b(j) \in X_0{}^J$. Define $*X_0 = Y_0$. The remark above about δ-incompleteness means that if we denote the set of classes of constant sequence individual maps by

$$^\sigma X_0 = \{*x : x \in X_0\} \qquad \text{that} \qquad *X_0 \supset {}^\sigma X_0 \quad \text{(strict inclusion)}$$

since X_0 is infinite by assumption (σ stands for *standard* points in the extension).

[1.2] In general, if $A \subseteq X_0$, we define the two sets

$$*A = \{b \in *X_0 : b = [b(j)] \;\&\; b(j) \in_{\mathscr{U}} A(j) \equiv A\} \quad \text{and} \quad {}^\sigma A = \{*a : a \in A\}.$$

It follows then that

$$*A \supseteq {}^\sigma A$$

with equality only if A is finite.

[1.3] When $B(j)$ is a map with values in X_1, that is, $B(j) \subseteq X_0$ for j on $\mathscr{U}$, then we denote the set (abuse of notation)

$$B = \{[b(j)] = b \in Y_0 : b(j) \in_{\mathscr{U}} B(j)\}$$

and then

$$B \subseteq Y_0 = *X_0.$$

The question now arises as to whether we obtain all subsets of $*X_0$ as these mappings, and *a fact of fundamental importance is that we do* **not**. This will be made clear below. This fact makes the construction useful and is the origin of the internal–external terminology.

[2.1] Let $*X_1$ denote the set of all the sets that arise from maps as in [1.3]. Then $*X_1 \subseteq \mathscr{P}(*X_0) = \mathscr{P}(Y_0) = Y_1$.

[2.2] Next take $\mathscr{A} \subseteq X_0 \cup X_1$ in $\mathscr{X}$, that is, $\mathscr{A}$ is a standard entity in X_2. Define

$$*\mathscr{A} = *(\mathscr{A} \cap X_0) \cup \{B \in *X_1 : B \text{ arises from } B(j) \;\&\; B(j) \in_{\mathscr{U}} \mathscr{A}(j) \equiv \mathscr{A}\}.$$

Also, denote the constant sequence elements of $*\mathscr{A}$ by

$$^\sigma\mathscr{A} = \{*A : A \in \mathscr{A}\}.$$

Again,

$$*\mathscr{A} \supseteq {}^\sigma\mathscr{A} \quad \text{with equality only if } \mathscr{A} \text{ is finite.}$$

[2.3] When $\mathscr{B}(J)$ is in $X_2{}^J$, define a set

$$\mathscr{B} = \{B \in {}^*X_0 \cup {}^*X_1 : B \text{ arises from a map } B(j) \ \& \ B(j) \in_{\mathscr{U}} \mathscr{B}(j)\}.$$

[3.1] $^*X_2$ is the set of embeddings of maps $\mathscr{B}(j)$ as in [2.3]. Therefore,

$$^*X_2 \subseteq \mathscr{P}(^*X_0 \cup {}^*X_1) \subseteq \mathscr{P}(Y_0 \cup \mathscr{P}(Y_0)) = Y_2.$$

[3.2] If $\mathscr{A} \subseteq X_0 \cup X_2$, a standard entity,

$$^*\mathscr{A} \text{ is the set of embeddings as above of mappings in } \mathscr{A}^J$$

and

$$^\sigma\mathscr{A} \text{ is the set of embeddings of constant mappings from } \mathscr{A}^J.$$

[3.3] If $\mathscr{B}(j)$ is in $X_3{}^J$, embed a set

$$\mathscr{B} = \{B \in {}^*X_0 \cup {}^*X_2 : B \text{ arises from a map } B(j) \in_{\mathscr{U}} \mathscr{B}(j)\}.$$

[4.1] $^*X_3$ is the set of sets obtained in [3.3], so that

$$^*X_3 \subseteq \mathscr{P}(^*X_0 \cup {}^*X_2) \subseteq \mathscr{P}(Y_0 \cup Y_2) = Y_3.$$

Now proceed inductively defining $*$ and σ for each entity of $\mathscr{X}$ and embedding each map from J into an entity of $\mathscr{X}$ as a set.

We define a superstructure $\mathscr{Y}$ on Y_0 as before, that is,

$$Y_n = \mathscr{P}\left(\bigcup_{k=0}^{n} Y_k\right) \qquad \text{and} \qquad \mathscr{Y} = \bigcup_{n=1}^{n} Y_n.$$

The map $*: \mathscr{X} \to \mathscr{Y}$ takes the individuals into points of Y_0 and entities of $\mathscr{X}$ into elements of $\mathscr{Y}$:

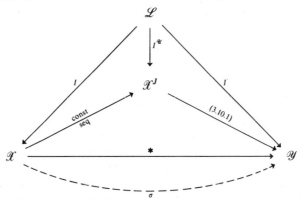

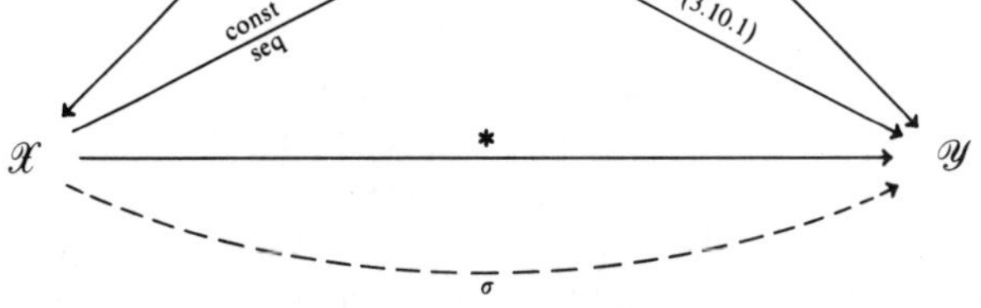

EXERCISE: Show directly that $^\sigma\mathbf{N}$ cannot arise from any set-valued map from J into $\mathscr{P}(\mathbf{N})$. Therefore $^\sigma\mathbf{N} \notin {}^*\mathscr{P}(\mathbf{N})$ [see Theorem (4.5.2)].

Any bounded sentence about $\mathscr{X}$ (that is, whose I-interpretation is in $\mathscr{X}$) can also be interpreted in $(\mathscr{X}^J, \in_{\mathscr{U}}, =_{\mathscr{U}})$ and is true in both or false in both. Now since the above embedding satisfies $b \in B$ if and only if $b(j) \in_{\mathscr{U}} B(j)$ and $A = B$ if and only if $A(j) =_{\mathscr{U}} B(j)$ and since quantifiers of bounded sentences about $\mathscr{X}$ only range over entities of $\mathscr{X}$, such a sentence can be interpreted in $\mathscr{Y}$ where it is true or false, depending on whether it is true or false in $\mathscr{X}$ and $\mathscr{X}^J$. The form of such a sentence interpreted in $\mathscr{Y}$ is called the "$*$-transform" and is particularly simple.

Sections 3.8–3.10 now complete the proofs of Leibniz' principle and the standard and internal definition principles.

3.11 $*$-TRANSFORMS OF CATEGORIES

Part of the formalism of modern algebra is based on category theory. We make a few brief remarks on how we can extend part of this formalism in passing from $\mathscr{X} \xrightarrow{*} \mathscr{Y}$. We define $\mathscr{C}$ to be a *sufficiently small category* provided $\mathrm{Ob}(\mathscr{C}) \subseteq \mathscr{X}$, that is, each object is an entity, and $\mathrm{Mor}_{\mathscr{C}}(A, B)$ is an entity for each $A, B \in \mathrm{Ob}(\mathscr{C})$, in particular, the diagram $A \xrightarrow{f} B$ has only entities written on it. ($\mathrm{Ob}(\mathscr{C})$ itself need not be an entity.) We shall say $\mathscr{C}$ is a "classical category" or *kittygory* when the morphisms are all function entities with the usual composition and $\mathscr{C}$ is sufficiently small as above.

The category of sets restricted to $\mathscr{X}$ becomes the kittygory of entities, which is "classical." Similarly, many other categories can be made "sufficiently small" by restriction to $\mathscr{X}$. Most "classical" mathematics goes on in "classical" categories, so we feel justified in only considering them.

A sufficiently small category extends to $^*\mathscr{C}$ by taking $\mathrm{Ob}(^*\mathscr{C})$ to be the set of internal sets in $\mathscr{Y}$ arising from maps $A\colon J \to \mathrm{Ob}(\mathscr{C})$ under the above embedding. The *standard objects* of $^*\mathscr{C}$ arise from the constant object-valued maps. $\mathrm{Mor}_{\mathscr{C}}(A, B)$ for $A, B \in \mathrm{Ob}(^*\mathscr{C})$ is the set in $\mathscr{Y}$ given by maps $f(j) : J \to \mathrm{Mor}_{\mathscr{C}}(A(j), B(j))$ (that is, $f(j) \in \mathrm{Mor}_{\mathscr{C}}(A(j), B(j))$ on $\mathscr{U}$) under the embedding in $\mathscr{Y}$. *Standard morphisms* arise from constant morphism-valued maps.

$^*\mathscr{C}$ satisfies the associativity and identity laws since the corresponding morphisms $f(j)$, $g(j)$, $h(j)$, $1_{A(j)}$, and $hgf(j)$ do.

The extension of the *category of entities* is the *$*$-category of internal sets and internal maps*. The category of "classical" Banach spaces will be extended to the internal or $*$-Banach spaces. In general, any boundedly formalizable statements about $\mathscr{C}$ carry over to $^*\mathscr{C}$ by **Leibniz' principle**. In other words, the $*$-transform applies to categories in $\mathscr{X}$.

It may also be convenient to have the isomorphic $^{\sigma}\mathscr{C}$ in $\mathscr{Y}$ given by the collection of standard objects and morphisms. Now any finite diagram in $\mathscr{C}$

extends via $*$ to a standard (i.e., in ${}^{\sigma}\mathscr{C}$) diagram in ${}^*\mathscr{C}$ with the same properties (since a finite union of objects is an entity, therefore the properties are admissible). Long diagrams will not be considered here, because they get longer....

Functors and natural transformations between sufficiently small categories extend to the corresponding nonstandard extensions via the pointwise (in j) definitions

$$({}^*\mathscr{F}A)(j) = \mathscr{F}A(j) = {}^*\mathscr{F}A, \qquad A \in \mathrm{Ob}({}^*\mathscr{C}),$$

$$({}^*\mathscr{F}f)(j) = \mathscr{F}f(j) = {}^*\mathscr{F}f, \qquad f \in \mathrm{Mor}({}^*\mathscr{C}),$$

$$\begin{array}{ccc} {}^*\mathscr{F}A & \xrightarrow{{}^*\varphi(A)} & {}^*\mathscr{G}A \\ {\scriptstyle {}^*\mathscr{F}f}\Big\downarrow & & \Big\downarrow{\scriptstyle {}^*\mathscr{G}f} \\ {}^*\mathscr{F}B & \xrightarrow[{}^*\varphi(B)]{} & {}^*\mathscr{G}B \end{array}$$

$$\varphi(A(j)) = {}^*\varphi(A)(j) = {}^*\varphi(A),$$

where the last equality is in the sense of the embedding into $\mathscr{Y}$.

(3.11.1) SUMMARY $*$ *preserves "abstract nonsense,"* that is, we can apply $*$-transform to classical categories and preserve all the diagrams.

A category-theory approach to the basic ideas of this chapter can be found in the work of Koch and Mikkelsen [1974]. They give an interesting description of internal and external.

3.12 POSTSCRIPT TO CHAPTER 3

"You just put $*$'s on everything."

The point of this chapter was to show precisely how naive set theory can be used to construct a nonstandard ultrapower model. From the point of view of applications, if one is willing to believe a map $*$ defined on $\mathscr{X}$ exists and produces a nonstandard model, then one only needs to learn how to form $*$-transforms, distinguish between internal and external sets (that is, sets which arise as mappings $B: J \to X_n$ and arbitrary subsets of ${}^*X_n$), and be able to use the properties in (3.4.1). The internal definition principle from this point of view is more interesting.

A final remark on which sentences in analysis have $*$-transforms is simply that any carefully formulated theorem *whose quantifiers are specified to run over an entity* will have one. Because of this we shall write words like

$*$-continuous, $*$-finite, and so forth, to mean the $*$-transform of the usual notion. We warn the reader that other notions of concepts like continuity in $^*\mathbf{R}$, say, are both possible and often interesting. For example, an internal function f: $^*\mathbf{R} \to {}^*\mathbf{R}$ that is $*$-continuous at $a \in {}^*\mathbf{R}$ reads

$$(\forall \varepsilon \in {}^*\mathbf{R}^+)(\exists \delta \in {}^*\mathbf{R}^+) \text{ such that if } {}^*|x \; {}^*\!- a|\; {}^*\!< \delta,$$
$$\text{then } {}^*|f(x) \; {}^*\!- f(a)|\; {}^*\!< \varepsilon,$$

while another notion of continuity is to measure tolerances with *standard* ε and δ, f is *S-continuous* at a:

$$(\forall \varepsilon \in {}^\sigma\mathbf{R}^+)(\exists \delta \in {}^\sigma\mathbf{R}^+) \text{ such that } {}^*|x \; {}^*\!- a|\; {}^*\!< \delta$$
$$\text{implies } {}^*|f(x) \; {}^*\!- \; f(a)|\; {}^*\!< \varepsilon.$$

The notion of $*$-continuity is internal since it comes from the $*$-transform of "for every real-valued function f on $\mathbf{R}$, f is continuous at a if and only if for every real positive ε, there exists a real positive δ, etc." When expressing this more formally, we need an entity $\mathscr{F}$, the set of all real-valued functions, then $*$-continuity applies to elements of $^*\mathscr{F}$, that is, internal functions. Since $*$-continuity is an internal notion, it is formally the same as continuity in the standard model. As a result, $\sin \omega x$ is $*$-continuous even though it oscillates infinitely fast for infinite $\omega \in ({}^*\mathbf{N} \backslash {}^\sigma\mathbf{N}) = {}^*\mathbf{N}_\infty$. Because of this problem *the external notion* of S-continuity is interesting: $\sin \omega x$ is not S-continuous. (In fact, its standard part may not even be Lebesgue measurable! [see (8.4.45)].) Usually it is more than enough to write a statement carefully and put a $*$ on each constant in order to form $*$-transforms.

Other examples of the care required between internal and external properties are (i) $^*\mathbf{N}$ is $*$-well ordered but not well ordered, and (ii) $^*\mathbf{R}$ is $*$-complete but not complete. (We discuss this below.) The interplay between internal and external notions is at the crux of Robinson's infinitesimal foundations.

This chapter is a formidable list of technicalities for the reader. We hope there is enough detail for easy reading and not so much that the reader is discouraged by it.

SOME BASIC FACTS ABOUT HYPERREAL NUMBERS

CHAPTER 4

In this chapter we begin our study of a nonstandard model of analysis. We shall concern ourselves at first with the algebraic structure of *R as a totally ordered field. Thus we shall need only a few entities of the superstructure, namely, ternary relations for + and · and a binary relation for ≤ as well as the set **R** and the individuals.

Leibniz apparently considered his idealized finite numbers as the ordinary reals with some infinitesimals clustered around zero. As we shall see, the finite hyperreals contain the ordinary reals with new numbers clustered infinitesimally closely around each ordinary real, which is to say that the infinitesimals form an order ideal in the finite numbers such that the quotient ring is isomorphic to **R**. We shall also see that this is all highly external, that is, the finite numbers, the infinitesimals, and the ordinary reals in ***R** are all external sets.

The results below do not depend on the nature of the nonstandard model, but at times it is helpful to think in terms of an ultrapower model. In any event, so far we have only demonstrated existence of nonstandard models by constructing a δ-incomplete ultrapower.

4.1 ADDITION, MULTIPLICATION, AND ORDER IN *R

In order to exploit what we have developed concerning a nonstandard model of a superstructure, let us make it clear how we can view addition, multiplication, and order for **R** within the superstructure $\mathscr{R}$ based on **R**. We take the following sets which are entities of $\mathscr{R}$ since they are sets of n-tuples of real numbers:

$$\mathrm{S} = \{(a, b, c) : a, b, c \in \mathbf{R} \text{ and } a + b = c\},$$

$$\mathrm{P} = \{(a, b, c) : a, b, c \in \mathbf{R} \text{ and } a \cdot b = c\},$$

$$\mathrm{L} = \{(a, b) : a, b \in \mathbf{R} \text{ and } a \leq b\}.$$

Ordinary sentences about +, ·, and ≤ can now be written in our formal language $\mathscr{L}$ using the membership relation and constants corresponding to the entities S, P, and L. For example, "addition is commutative" or "for every x and y in **R**, $x + y = y + x$" can be formalized as follows. Let σ and

ρ be constants of $\mathscr{L}$ such that $I(\sigma) = S$ and $I(\rho) = \mathbf{R}$, where I is the interpretation mapping:

$$\{(\forall x)(\forall y)(\exists z)(\exists w)[[x \in \rho \wedge y \in \rho] \Rightarrow [[z \in \rho \wedge w \in \rho] \wedge [(x, y, z) \in \sigma \wedge (y, x, w) \in \sigma \wedge z = w]]]\}.$$

We also need to express $(x, y, z) \in \sigma$ strictly in the formal language to be completely honest about the formalization, but we leave this to the reader.

Adhering strictly to the formal language can be quite cumbersome. For example, whereas in ordinary practice we write "$x + y = y + x$," in the formal language we are faced with saying what "$x + y$" is for variables x and y. Nonetheless, if the reader will formalize a few sentences completely, he will soon see what is involved and be able to apply model-theoretic arguments to ordinary or partially formalized sentences. For this reason we suggest that he formalize some of the field axioms [(1–9) of Appendix Section A.1].

We now take a nonstandard model $^*\mathscr{R}$ of $\mathscr{R}$, the superstructure based on $\mathbf{R}$. Thus we have the interpretation mappings I and I', the mapping $*$, and the formal language $\mathscr{L}$. We take the entity $^*\mathbf{R}$ of $^*\mathscr{R}$ and define operations as follows: For every $A, B, C \in {}^*\mathbf{R}$

$$A \,{}^*\!\!+ B = C \quad \text{if and only if} \quad (A, B, C) \in {}^*S,$$

$$A \,{}^*\!\cdot B = C \quad \text{if and only if} \quad (A, B, C) \in {}^*P,$$

and

$$A \,{}^*\!\!\leq B \quad \text{if and only if} \quad (A, B) \in {}^*L.$$

We will show that with these operations $^*\mathbf{R}$ is a totally ordered field which contains a natural copy of the ordinary reals. We will first prove this using model-theoretic arguments; later we show how things can be viewed directly as an ultrapower.

The model-theoretic proof proceeds as follows. For each totally ordered field axiom we write a formal sentence whose I-interpretation is that axiom for $\mathbf{R}$. The $*$-transform is then that axiom for $^*\mathbf{R}$ when embedded in set theory. As an example, the $*$-transform of the sentence above for commutativity of addition is "For every x and y in $^*\mathbf{R}$, $x \,{}^*\!\!+ y$ and $y \,{}^*\!\!+ x$ exist in $^*\mathbf{R}$ and are equal." Another example is the property of total order which can be expressed as

$$(\forall x)(\forall y)[[x \in \rho \wedge y \in \rho] \Rightarrow [(x, y) \in \lambda \vee (y, x) \in \lambda]],$$

where $I(\lambda) = L$. The $*$-transform of this sentence is "For every x and y in $^*\mathbf{R}$, either $x \,{}^*\!\!\leq y$ or $y \,{}^*\!\!\leq x$." The remaining axioms are left to the reader. This proves:

(4.1.1) THEOREM ***R** *is a totally ordered field.*

The reader will observe that $*$ is an order isomorphism of **R** onto $^{\sigma}\mathbf{R}$. The zero element and identity element of ***R** are *0 and *1, respectively.

We now look at ***R** viewed as an ultrapower in order to make closer contact with Chapter 2. The set ***R** is the embedded collection of mappings from J into **R** with identification $A = B$ if and only if $\{j : A(j) = B(j)\} \in \mathcal{U}$. The reader can easily verify that the operations given above can be described alternately by

$$A \mathbin{*+} B =_{\mathcal{U}} C \quad \text{if and only if} \quad \{j : C(j) = A(j) + B(j)\} \in \mathcal{U},$$

$$A \mathbin{*\cdot} B =_{\mathcal{U}} C \quad \text{if and only if} \quad \{j : C(j) = A(j) \cdot B(j)\} \in \mathcal{U},$$

and

$$A \mathbin{*\leq} B \quad \text{if and only if} \quad \{j : A(j) \leq B(j)\} \in \mathcal{U}.$$

We can even go one step back in this view of things and consider the ring $\mathbf{R}^J$ of mappings of J into **R** with the operations

$$A \oplus B = C \quad \text{if and only if} \quad C(j) = A(j) + B(j) \quad \text{for all } j \in J,$$

$$A \odot B = C \quad \text{if and only if} \quad C(j) = A(j) \cdot B(j) \quad \text{for all } j \in J,$$

and

$$A \circledleq B \quad \text{if and only if} \quad A(j) \leq B(j) \quad \text{for all } j \in J.$$

This can be compared to the theory of Laugwitz [1959] and Laugwitz and Schmieden [1958], where J is taken to be the natural numbers. This ring $\mathbf{R}^J$ has divisors of zero, although it does contain a copy of **R** as constant mappings. The ring is not totally ordered. The set of elements $\{A \in \mathbf{R}^J : \{j : A(j) = 0\} \in \mathcal{U}\}$ forms a maximal order ideal and the quotient ring is what we described in the preceding paragraph.

In the theory of Laugwitz and Schmieden an equivalence between elements of $\mathbf{R}^J$ is studied which can be viewed as equality relative to the Frechet filter. Two sequences are identified if they agree except at finitely many places.

4.2 SOME SIMPLIFICATIONS OF THE NOTATION

Before going on to the next result we wish to make some simplifications of our notation. Since $^{\sigma}\mathbf{R}$ is an isomorphic copy of **R** in ***R**, no confusion should arise if we no longer place $*$ on *standard individuals* when they occur in ***R**. For example, we will simply write 0 for *0, 1 for *1, 5 for *5, and π for $^*\pi$ in $^\sigma\mathbf{R}$. In the ultrapower context, recall that ^{*0} (say) is the constant map $j \to 0$; it is customary to denote this by 0 anyway. This convention *does not apply to entities* that are not individuals; for sets we continue to write $^\sigma S$ and $^*S$ since, in general, they are quite different.

Since $^*+$, $^*\cdot$, and $^*\leq$ can be viewed as extensions of the operations $+$, $\cdot$, and $\leq$ of $\mathbf{R}$, we also delete the $*$ on them. Thus, if ζ is a positive infinitesimal (say) in $^*\mathbf{R}$ and π has its usual meaning in $\mathbf{R}$, we write $\zeta + \pi$, $0 < \zeta$, and $1 \cdot \zeta$ rather than $\zeta \, ^*\!+ \, ^*\pi$, $^*0 \, ^*\!< \zeta$, and $^*1 \, ^*\!\cdot \, \zeta$.

Absolute value in $^*\mathbf{R}$ can now be obtained in two ways. Since $\leq$ totally orders $^*\mathbf{R}$, for each $A \in {}^*\mathbf{R}$ we can take $|A|$ equal to A, if $-A \leq A$, and equal to $-A$, if $A \leq -A$, i.e., $|A| = \max(-A, A)$. But we also know $|\ |$ is a function from $\mathbf{R}$ to $\mathbf{R}^+$, hence (viewed as a set of ordered pairs) it has a $*$-extension $^*|\ |$, which is a function from $^*\mathbf{R}$ to $^*\mathbf{R}^+$. The functions $\max(\ ,\ )$ and $\min(\ ,\ )$ form $\mathbf{R} \times \mathbf{R}$ into $\mathbf{R}$ also have $*$-extensions, $^*\max(\ ,\ )$ and $^*\min(\ ,\ )$ from $^*\mathbf{R} \times {}^*\mathbf{R}$ into $^*\mathbf{R}$. By writing the appropriate sentences one can show that the two approaches give the same result and that they act on $^\sigma\mathbf{R}$ the same way the original functions act on $\mathbf{R}$. Hence we simply write $|\ |$, $\max(\ ,\ )$, and $\min(\ ,\ )$ for standard as well as nonstandard entries. This liberalization of the notation and some additional conventions later on will help a great deal to simplify the mechanics of the subject and we trust it will not cause confusion.

4.3 *R IS NON-ARCHIMEDEAN

Since $^*\mathbf{R}$ contains a copy of $\mathbf{R}$, if we show that $^*\mathbf{R}$ is not isomorphic to $\mathbf{R}$ we know by Theorem (A.3.2) of Appendix A, that $^*\mathbf{R}$ is non-Archimedean. It is easy to see that the natural copy of the nonnegative integers in $^*\mathbf{R}$ is $^\sigma\mathbf{N}$. Hence, $^*\mathbf{R}$ is non-Archimedean means that there exists $A \in {}^*\mathbf{R}$ such that $|A| > n$ for every $n \in {}^\sigma\mathbf{N}$. [In the next section ((5) in the proof of 4.4.3) we will demonstrate the existence of infinite numbers more directly, but the abstract approach only requires that $^*\mathbf{R}$ is a proper extension rather than an ultrapower.]

Notice that the $*$-transform of the Archimedean property is "For every A in $^*\mathbf{R}$, there exists K in $^*\mathbf{N}$ such that $|A| \leq K$," so that no contradiction arises from the principle that "... insofar as they can be expressed in $\mathscr{L}$, $^*\mathbf{R}$ has the same properties as $\mathbf{R}$." We will see that $^\sigma\mathbf{N}$ is external.

(4.3.1) THEOREM *$^*\mathbf{R}$ is not order isomorphic to $\mathbf{R}$ or any subfield of $\mathbf{R}$, in particular, $^*\mathbf{R}$ is non-Archimedean.*

PROOF: By the remarks above we only need deal with the case in which

$^*\mathbf{R}$ is isomorphic to $\mathbf{R}$. Then there is a one-to-one, onto, order homomorphism of $^*\mathbf{R}$ onto $\mathbf{R}$. The image of $^\sigma\mathbf{R}$ under this map would then be $\mathbf{R}$, and this contradicts the one-to-one assumption. (Recall, $^*\mathbf{R}\backslash{}^\sigma\mathbf{R} \neq \varnothing$.)

4.4 INFINITE, INFINITESIMAL, AND FINITE NUMBERS

(4.4.1) DEFINITION (Infinite, infinitesimal, and finite) *An element A of $^*\mathbf{R}$ is said to be infinite if for every $n \in {}^\sigma\mathbf{N}$,*

$$|A| \geq n;$$

A is said to be finite if for some $m_0 \in {}^\sigma\mathbf{N}$,

$$|A| \leq m_0;$$

and A is said to be infinitesimal if for every $n \in {}^\sigma\mathbf{N}$ $(n \neq 0)$,

$$|A| \leq 1/n.$$

Zero is infinitesimal and it is the only standard infinitesimal since $\leq$ in $^*\mathbf{R}$ agrees with $\leq$ in $\mathbf{R}$ for standard numbers. It follows from the property "$0 < a < b$ implies $1/b < 1/a$," that the reciprocal of an infinite number is infinitesimal and that the reciprocal of a nonzero infinitesimal is infinite. Existence of infinite numbers and nonzero infinitesimals follows from Theorem (4.3.1). It follows from the triangle inequality that if A is finite and B is infinitesimal, $A + B$ is finite. The reader can certainly see other such simple relationships between these categories of numbers.

(4.4.2) DEFINITION ($\mathcal{O}$,o) *The set of finite numbers is denoted by $\mathcal{O}$. The set of infinitesimals is denoted by o.*

This notation is chosen because there is some connection between it and the Landau oh-notations (see Section 5.6). We denote the infinite numbers $^*\mathbf{R}_\infty = {}^*\mathbf{R}\backslash\mathcal{O}$, $^*\mathbf{R}_\infty{}^+ = {}^*\mathbf{R}^+\backslash\mathcal{O}$, and $^*\mathbf{R}_\infty{}^- = {}^*\mathbf{R}^-\backslash\mathcal{O}$.

(4.4.3) THEOREM *$\mathcal{O}$ is a subring of $^*\mathbf{R}$ which is a totally ordered Archimedean integral domain. o is a maximal proper ideal of $\mathcal{O}$ and also an order ideal of $\mathcal{O}$.*

PROOF: We delineate the components of the proof, leaving some as exercises.

(1) $\mathcal{O}$ is a totally ordered subring of $^*\mathbf{R}$. Since the order on $^*\mathbf{R}$ is compatible with the operations and a total order, the restriction will also satisfy these properties. Thus we only need show that for every $A, B \in \mathcal{O}$, $A + B \in \mathcal{O}$, and $A \cdot B \in \mathcal{O}$. We leave this as an exercise.

(2) $\mathcal{O}$ has no zero divisors since *$\mathbf{R}$ is a field.

(3) $\mathcal{O}$ is Archimedean because every element is bounded by a natural number in ${}^\sigma\mathbf{N}$ in order to be in $\mathcal{O}$ in the first place. (${}^\sigma\mathbf{N}$ is the natural copy of $\mathbf{N}$ in $\mathcal{O}$, of course.)

(4) o is an ideal of $\mathcal{O}$, i.e., for every A, $B \in o$ and for any $C \in \mathcal{O}$, $A + B \in o$ and $C \cdot A \in o$. We leave this as an exercise.

(5) o is a maximal proper ideal of $\mathcal{O}$. First, o is proper because $1 \notin o$ and because there is a positive infinitesimal forcing $o \neq \{0\}$. [We can write a positive infinitesimal as follows in our case of a δ-incomplete ultrapower. Of course we have already seen abstractly that *$\mathbf{R}$ is non-Archimedean and that reciprocals take infinite numbers to infinitesimals and conversely. Since $\mathcal{U}$ is a δ-incomplete ultrafilter on J, there exists a partition of J, $(J_n)_{n=1}^{\infty}$, such that $(J_m \cap J_n) = \varnothing$, $\bigcup_{n=1}^{\infty} J_n = J$, and $J_n \notin \mathcal{U}$. We define a map $\zeta: J \to R$ by

$$\zeta(j) = 1/n \qquad \text{for} \quad j \in J_n.$$

Now $\zeta \neq 0$ because

$$\{j : \zeta(j) = 0\} = \varnothing \notin \mathcal{U}.$$

For each $k \in {}^\sigma\mathbf{N}$, $\zeta < 1/k$, because

$$\{j : \zeta(j) \geq 1/k\} = \bigcup_{n=1}^{k} J_n \notin \mathcal{U}.]$$

In order to show that o is maximal we need only observe that o consists of all the nonregular elements of $\mathcal{O}$ (c.f., Appendix Theorem (A.1.3)). Thus if $A \in \mathcal{O} \backslash o$, then there exist k and K in ${}^\sigma\mathbf{N}$ such that

$$1/k \leq |A| < K.$$

In *$\mathbf{R}$ we have then that

$$1/K < |A^{-1}| \leq k \qquad \text{so} \quad A^{-1} \in \mathcal{O}$$

or A is regular. Conversely, since o is proper, it cannot have any regular elements.

(6) o is an order ideal simply because, if $0 < A < B$ in $\mathcal{O}$ and $B \in o$, then $A < 1/k$ for all $k \in {}^\sigma\mathbf{N}$ $(k \neq 0)$, and so $A \in o$. This completes the proof of Theorem 4.3.

(4.4.4) THEOREM *The quotient ring $\mathcal{O}/o$ is order isomorphic to* $\mathbf{R}$.

PROOF: $\mathcal{O}/o$ is a field by Theorem (A.1.2). $\mathcal{O}/o$ is totally ordered by Theorem (A.2.5). $\mathcal{O}/o$ is Archimedean since $\mathcal{O}$ is Archimedean. To prove this, first observe that if $k \neq l$ in ${}^\sigma\mathbf{N}$, then $k + o \neq l + o$. Thus the natural copy of **N** in $\mathcal{O}/o$ is $\{n + o : n \in {}^{\sigma}\mathbf{N}\}$. Now take any $A \in \mathcal{O}$, $|A| \leq m_0$ for some $m_0 \in {}^{\sigma}\mathbf{N}$ so $|A + o| \leq |m_0 + o|$ in the quotient ring since the canonical order homomorphism preserves order.

Since $\mathcal{O}/o$ is a totally ordered Archimedean field, it is isomorphic to a subfield of **R** by Theorem (A.3.2). To show it is actually all of **R** we need only show it contains a natural copy of **R**. To show this, take $a, b \in {}^{\sigma}\mathbf{R} \subset \mathcal{O}$ with $a \neq b$ since $b - a$ is finite $a + o \neq b + o$. Now, however, the canonical homomorphism restricted to ${}^{\sigma}\mathbf{R}$ is one-to-one, so the image in $\mathcal{O}/o$ is a natural copy of **R**. This completes the proof of the theorem.

(4.4.5) DEFINITION (Standard part of a finite number) *The order homomorphism of $\mathcal{O}$ with kernel o onto* **R** *is called the standard part homomorphism and we denote it by* st. (We will use $^\circ$ for discrete standard part, although Robinson [1966] uses it for st.) *We also write* $\mathrm{st}(a) = \hat{a}$ *since this is an example of an infinitesimal hull.*

The existence of the unique map st of $\mathcal{O}$ onto the ordinary copy of the reals is established by the last theorem. It has many important properties which we summarize here for later reference.

(4.4.6) THEOREM *For every* $A, B \in \mathcal{O}$:

(i) $\mathrm{st}(A + B) = \mathrm{st}(A) + \mathrm{st}(B)$,
(ii) $\mathrm{st}(A \cdot B) = \mathrm{st}(A) \cdot \mathrm{st}(B)$,
(iii) $A \leq B$ *implies* $\mathrm{st}(A) \leq \mathrm{st}(B)$,
(iv) $\mathrm{st}(|A|) = |\mathrm{st}(A)|$,
(v) $\mathrm{st}(\max(A, B)) = \max(\mathrm{st}(A), \mathrm{st}(B))$,
(vi) $\mathrm{st}(\min(A, B)) = \min(\mathrm{st}(A), \mathrm{st}(B))$,
(vii) $\mathrm{st}(A) = 0$ *if and only if* $A \in o$,
(viii) $A < B$ *implies* $\mathrm{st}(A) = \mathrm{st}(B)$ *if and only if* $A - B \in o$ *written* $A \approx B$,
(ix) *if* $a \in \mathbf{R}$, *then* $\mathrm{st}(^*a) = a$. *(Here we write* $*$ *on a standard individual for emphasis, otherwise this reads* $\mathrm{st}(a) = a$.)

(4.4.7) DEFINITION (The infinitesimal relation) *For two hyperreals A, B in* ***R** *we shall say A and B are infinitely close to one another, or within an infinitesimal of each other provided* $A - B \in o$. *We also introduce the notation* $A \approx B$ *for this,* i.e., $A \approx B$ *if and only if* $A - B \in o$.

Hence if $A \approx B$ and A and B are in $\mathcal{O}$, then $\mathrm{st}(A) = \mathrm{st}(B)$.

The standard part operation can be defined in other ways for an ultrapower. Dedekind cuts or supremum could be used. (Show how!) The standard part can also be defined as a filter limit as follows. If $A \in \mathcal{O}$, there exists a set $u \in \mathcal{U}$ and a positive standard real number $r > 0$ such that $j \in u$ implies $|A(j)| < r$. Hence the image of the ultrafilter $\mathcal{U}$ under the mapping $j \to A(j)$ of J into $\mathbf{R}$ is the basis for a bounded ultrafilter of subsets of $\mathbf{R}$. Since $\mathbf{R}$ is locally compact, that ultrafilter converges to a unique real number a. A simple observation shows that $\operatorname{st}(A) = a = \lim_{\mathscr{U}} A$.

The last result of this section is:

(4.4.8) THEOREM *A hypernatural number is finite if and only if it is standard, that is,*

$$^*\mathbf{N} \cap \mathcal{O} = {}^\sigma\mathbf{N}.$$

PROOF: It is clear that ${}^\sigma\mathbf{N} \subseteq \mathcal{O}$. *If* $n \in {}^*\mathbf{N}$ is finite, then there exists $m_0 \in {}^\sigma\mathbf{N}$ such that $n \leq m_0$. $K_0(I)$ contains the sentence

$$(\forall x)[[x \in I^{-1}(\mathbf{N})] \Rightarrow [[x \leq I^{-1}(m_0)] \Rightarrow$$
$$[x = I^{-1}(0)] \vee [x = I^{-1}(1)] \vee \cdots \vee [x = I^{-1}(m_0)]]].$$

The I'-interpretation says that n is one of the standard numbers $0, 1, \ldots, m_0$.

4.5 SOME EXTERNAL ENTITIES

We now deal with the question of whether the set ${}^\sigma A$ is internal when A is an infinite set. We know ${}^*A = {}^\sigma A$ if and only if A is finite. We begin with

(4.5.1) THEOREM *The infinite hypernatural numbers*

$$^*\mathbf{N} \backslash {}^\sigma\mathbf{N} = {}^*\mathbf{N}_\infty$$

form an external set.

PROOF: If ${}^*\mathbf{N} \backslash {}^\sigma\mathbf{N}$ is internal, there exists $A \in {}^*\mathscr{P}(\mathbf{N})$ such that $n \in A$ if and only if $n \in {}^*\mathbf{N}$ and $n \notin {}^\sigma\mathbf{N}$. Notice that if $m \in A$, then $m - 1 \in A$ because m must be infinite and in that case $m - 1$ is also.

Now, $\mathbf{N}$ is well ordered, which means a formal sentence corresponding to

$$(\forall X)[[X \in I^{-1}(\mathscr{P}(\mathbf{N}))] \Rightarrow (\exists m)[m \in X \wedge (\forall n)[n \in X \Rightarrow m \leq n]]]$$

is in $K_0(I)$. The I'-interpretation of this sentence is "Every internal subset of ${}^*\mathbf{N}$ has a first element." If A is internal, then A has a first element contrary to the remark in the preceding paragraph. This shows that ${}^*\mathbf{N} \backslash {}^\sigma\mathbf{N}$ is external.

(4.5.2) THEOREM ${}^\sigma\mathbf{N}$ *is external.*

PROOF: If ${}^\sigma\mathbf{N}$ was internal, it follows easily from the internal set theory that ${}^*\mathbf{N} \backslash {}^\sigma\mathbf{N}$ would also be internal.

(4.5.3) THEOREM *Let A be a set in $\mathscr{R}$. Then ${}^\sigma A$ is internal if and only if A is finite.*

PROOF: We only need show that if A is infinite, then ${}^\sigma A$ is external. Since A is assumed infinite there exists a subset B of A and a one-to-one mapping f of B onto $\mathbf{N}$. If ${}^\sigma A$ is internal, then ${}^\sigma A \cap {}^*B = {}^\sigma B$, so ${}^\sigma B$ is internal. Now, ${}^*f({}^\sigma B) = {}^\sigma \mathbf{N}$, but the image of an internal set under an internal map must be internal, contradicting Theorem (4.5.2).

The internal set theory also says ${}^*A \backslash {}^\sigma A$ is external when A is infinite.

(4.5.4) COROLLARY *${}^\sigma\mathbf{R}$, ${}^\sigma\mathbf{Q}$, and ${}^\sigma\mathbf{Z}$ are external.*

(4.5.5) THEOREM *The following sets are external:*

(a) *o, the infinitesimals,*
(b) *$o + a = \{x \approx a\}$, the infinitesimal neighborhood of $a \in {}^*\mathbf{R}$,*
(c) *$\mathcal{O}$, the finite numbers,*
(d) *$\mathcal{O} + a = \{x \sim a\}$, the galaxy of $a \in {}^*\mathbf{R}$,*
(e) *o^+, the positive infinitesimals,*
(f) *o^-, the negative infinitesimals,*
(g) *$\mathcal{O}^+$, the positive finite numbers,*
(h) *$\mathcal{O}^-$, the negative finite numbers,*
(i) *${}^*\mathbf{R}_\infty{}^+$, the positive infinite reals,*
(j) *${}^*\mathbf{R}_\infty{}^-$, the negative infinite reals,*
(k) *${}^*\mathbf{R}_\infty$, the infinite reals.*

PROOF: (1) o is external. Assume it is internal. Since $o \neq \varnothing$ and since o is bounded by 1, it follows that o has a supremum or least upper bound. This is because there is a sentence in $K_0(I)$ corresponding to

$$(\forall x)[[x \in I^{-1}(\mathscr{P}(\mathbf{R})) \wedge \text{“}X \text{ is bounded”}]$$
$$\Rightarrow (\exists z)[z \in I^{-1}(\mathbf{R}) \wedge \text{“}z \text{ is the l.u.b. of } X\text{”}]].$$

The $*$-transform of this sentence is "Every *internal* subset of ${}^*\mathbf{R}$ which is bounded has a least upper bound." Say a_0 is the l.u.b. of o. Since o has positive elements, $0 < a_0$. Furthermore, $a_0 \notin o$ because if it were, $a_0 < 2a_0 \in o$, a contradiction. On the other hand, if $a_0 \notin o$, then $a_0/2$ is an upper bound for o and $a_0/2 < a_0$ so a_0 is not the least upper bound. Since both cases lead to a contradiction, o must be external.

(2) The facts that $\mathcal{O}$ and the infinite numbers are external can be obtained from (1) and the internal definition principle as follows for the latter, the former being left as an exercise: If ${}^*\mathbf{R}\backslash\mathcal{O} = S$ for S in ${}^*\mathscr{P}(\mathbf{R})$, i.e., if ${}^*\mathbf{R}\backslash\mathcal{O} = {}^*\mathbf{R}_\infty$ is internal, then the set

$$\{x : x \in {}^*\mathbf{R} \text{ and } (x = 0 \text{ or } x^{-1} \in S)\}$$

is internal by the **internal definition principle**. This set is easily seen to be o, a contradiction.

REMARK: As the proof that o is external clearly demonstrates, $^*\mathbf{R}$ is not complete in the external sense, that is, there are bounded subsets of $^*\mathbf{R}$ with no least upper bound. The $*$-transform of the formal sentence describing completeness for $\mathbf{R}$ under I does, of course, hold. We could call this property "$*$-completeness" or "internal completeness"; bounded internal sets do have a least upper bound. This shows the limitation of transferring properties from $\mathbf{R}$ to $^*\mathbf{R}$ "... insofar as they can be expressed in $\mathscr{L}$."

If $\varnothing \neq D \subset o$ is internal, the above proof shows that the l.u.b. of D is an infinitesimal. Similarly, the l.u.b. of nonempty, bounded, internal subsets of $\mathcal{O}$ are finite. A nonempty internal subset of $^*\mathbf{R}$ with infinite positive numbers and an infinite bound will have an infinite l.u.b.

REMARK: The standard part operation is *not an internal mapping*. First of all, it maps numbers in $^*\mathscr{R}$ into numbers in $\mathscr{R}$. Even the mapping $a \in \mathcal{O} \to {}^*(\mathrm{st}(a)) \in {}^*\mathbf{R}$ is, however, external since the domain is external. Restriction of the latter mapping to a set will still be external if there are infinitely many points in the range. In that case, the range is of the form ${}^\sigma A$.

4.6 FURTHER SIMPLIFICATION OF NOTATION AND CLASSICAL FUNCTIONS

It has become customary to denote the nonstandard extension of a *standard function* or *standard binary relation* simply by the original name. Thus if one sees $\sin(\zeta)$, where $\zeta \in o$, $\zeta \neq 0$, we understand that $(^*\sin)(\zeta)$ is meant. Several examples of important functions follow. Observe that this is in keeping with our conventions on $\leq$, max, min, and $|\ |$.

EXAMPLES

1. (*The exponential function*) The exponential function e^x (or $\exp(x)$) is a one-to-one map of $\mathbf{R}$ onto $\mathbf{R}^+$. Hence, its nonstandard extension is also a one-to-one map of $^*\mathbf{R}$ onto $^*\mathbf{R}^+$. The extended function satisfies the functional equation

$$(\forall x)(\forall y)[x, y \in {}^*\mathbf{R} \Rightarrow e^{x+y} = e^x e^y].$$

If $h \approx 0$, then $e^h \approx 1$, as one can see from the expansion of Chapter 2. Moreover, the hyperexponential function maps $\mathcal{O}$ into $\mathcal{O}$ and we have

$$\mathrm{st}(e^a) = e^{\mathrm{st}(a)} \qquad \text{for all } a \in \mathcal{O}.$$

This follows from the functional equation since $a = \mathrm{st}(a) + h$ for some $h \in o$,

$$e^a = e^{\mathrm{st}(a)+h} = e^{\mathrm{st}(a)}e^h \approx e^{\mathrm{st}(a)}.$$

(The reader should verify that if $b \approx 1$, then $c \cdot b \approx c$ for $c \in {}^{\sigma}\mathbf{R}$.)

2. (*The logarithmic function*) The hyperlogarithm is a one-to-one map of $^*\mathbf{R}^+$ onto $^*\mathbf{R}$. The functional equation carries over, thus $(\forall x)(\forall y)[x, y \in {}^*\mathbf{R}^+ \Rightarrow \log xy = \log x + \log y]$. We also have $e^{\log x} = x$ for $x \in {}^*\mathbf{R}^+$. If $a \in \mathcal{O}^+ \backslash o$, then $\mathrm{st}(\log(a)) = \log(\mathrm{st(a)})$, because

$$e^{\log(\mathrm{st}(a))} = \mathrm{st}(a) = \mathrm{st}(e^{\log a}) = e^{\mathrm{st}(\log(a))}.$$

From $e^{-\log x} = 1/x$, it follows that $\log h$ is an infinite negative number when $h \in o^+$.

3. (*The trigonometric functions*) The hypersine and hypercosine functions are 2π-periodic functions of $^*\mathbf{R}$ onto $^*[-1, 1]$. For all $x \in {}^*\mathbf{R}$, $\sin^2 x + \cos^2 x = 1$ and the other well-known formulas hold for the extended functions and all hyperreals. If h is infinitesimal, then $\sin h \approx h$. Indeed, from $|\sin x\,| \leq |x|$ it follows that $\sin h$ is infinitesimal and $\sin h - h$ is also infinitesimal. Using the relation $\cos^2 x = 1 - \sin^2 x$, if $h \approx 0$, then $\cos h \approx 1$. If $a \in \mathcal{O}$, then

$$\sin a = \sin(\mathrm{st}(a) + h) = \sin(\mathrm{st}(a)) \cos(h) + \cos(\mathrm{st}(a)) \sin(h)$$

making $\sin a \approx \sin(\mathrm{st}(a))$, or $\mathrm{st}(\sin a) = \sin(\mathrm{st}(a))$. Also, $\mathrm{st}(\cos(a)) = \cos(\mathrm{st}(a))$ for $a \in \mathcal{O}$. If $0 < x < \pi/2$, then

$$\sin x < x < \tan x \qquad \text{or} \qquad \cos x < \sin x/x < 1.$$

Hence, if $h \in o^+$,

$$\cos h < \sin h/h < 1 \qquad \text{so that} \quad \sin h/h \approx 1.$$

For $h \in o$ and $h < 0$, $\sin(-h) = -\sin(h)$ so that for $h \in o$, $h \neq 0$, $\sin h/h \approx 1$.

REMARK: A translation of Euler's proof of his product formula for the sine function is given below with Euler's own use of the words infinite and infinitesimal.

4. (*The greatest integer function*) For $a \in \mathbf{R}$, $[a]$ denotes the greatest integer that is less than or equal to a. The hypergreatest integer function $[\;]$ maps $^*\mathbf{R}$ onto $^*\mathbf{Z}$. If $a \in {}^*\mathbf{R}$ is infinitely large, so is $[a]$.

5. (*The factorial function*) The function $n \to 1, 2, 3, \ldots, n = n!$ from $\mathbf{N}$ into N has an extension to $^*\mathbf{N}$, again properties expressed as sentences of $K_0(I)$ are preserved.

6. ($*$-*Finite sum and product*) In $\mathscr{R}$ if a finite sequence of real numbers $a_0, a_1, \ldots, a_n$ is given, we can form $\sum_{k=0}^{n} a_k$ and $\prod_{k=0}^{n} a_k$. The finite sequence can be thought of as mappings from $\mathbf{N}$ into $\mathbf{R}$ whose domains are of the form $\{k \in N : k \leq m\}$ for some $n \leq N$. Thus we can in turn think of $\sum$ and $\prod$ as mappings from the set of all such finite sequences into $\mathbf{R}$.

The extended mappings $\sum$ and $\prod$ are mappings from *N into *R whose domains are of the form $\{k \in {}^*N : k \leq n\}$, since this can be expressed in $K_0(I)$. When n is infinite we call the sequences $*$-finite sequences and the sums or products $*$-finite sums or $*$-finite products.

4.7 HYPERCOMPLEX NUMBERS

The complex numbers have several important standard descriptions which we investigate here infinitesimally. The points in a Cartesian plane $\mathbf{R}^2 = \{(x, y) : x \in \mathbf{R}\}$ can represent $\mathbf{C}$ by the convention $(x, y) \mapsto x + iy$ where $i = \sqrt{-1}$. In polar coordinates $x + iy = re^{i\theta}$ where $r^2 = x^2 + y^2$ and $\tan\theta = y/x$. By stereographic projection $\mathbf{C}$ can be represented as the points on a 2-sphere less the north pole which is denoted by the love-knot symbol ∞. These are three standard representations we wish to consider.

"Any $z \in \mathbf{C}$ can be decomposed into Cartesian form by taking real and imaginary parts $\mathrm{Re}(z) + i\,\mathrm{Im}(z) = z$." The $*$-transform of this statement (and the converse) give us the Cartesian representation of *$\mathbf{C}$ as *$\mathbf{R}$ × *$\mathbf{R}$. In particular, the real and imaginary part maps extend to *$\mathbf{C}$.

"The modulus or norm of $z \in \mathbf{C}$ is given by $|z| = ((\mathrm{Re}(z))^2 + (\mathrm{Im}(z))^2)^{1/2}$," gives us a notion of length in *$\mathbf{C}$ by applying $*$-transforms.

(4.7.1) DEFINITION *The finite complex numbers $\mathcal{O}(\mathbf{C})$ is the set $\{z \in {}^*\mathbf{C} : |z| \in \mathcal{O}(\mathbf{R})\}$. The infinitesimals $o(\mathbf{C})$ is $\{z \in {}^*\mathbf{C} : |z| \approx 0\}$ and we write $z \approx w$ if $|z - w| \approx 0$.*

EXERCISE: Suppose that $z = x + iy$ for $z \in {}^*\mathbf{C}$, $x, y \in {}^*\mathbf{R}$.

(a) Show that $z \approx 0$ in *$\mathbf{C}$ if and only if $x \approx 0$ and $y \approx 0$ in *$\mathbf{R}$.
(b) Show that z is finite if and only if x and y are finite.
(c) Prove the following theorem.

(4.7.2) THEOREM *$\mathcal{O}/o$ is isomorphic to $\mathbf{C}$ as a field.*

The ring $\mathcal{O}$ has o as a maximal ideal, and the canonical homomorphism $o \to \mathcal{O} \xrightarrow{\mathrm{st}} \mathbf{C}$ is called "*standard part.*" $\mathcal{O}(\mathbf{C}) \cap {}^*\mathbf{R} = \mathcal{O}(\mathbf{R})$, when *$\mathbf{R}$ is embedded in *$\mathbf{C}$ as the "real axis" and embedded, $x \approx y$ in *$\mathbf{C}$ if and only if $x \approx y$ in *$\mathbf{R}$, also embedded standard part on *$\mathbf{R}$ agrees with standard part on *$\mathbf{C}$.

The argument or angle of a nonzero $z \in \mathbf{C}$ is a real number in the interval $(-\pi, \pi]$ so that $z = |z| \exp(i \arg(z))$. The $*$-transform gives the polar decomposition of any $z \in {}^*\mathbf{C}$ as a hyperlength $|z|$ and a hyperangle in $^*(\pi, \pi]$.

Stereographic projection in $\mathbf{R}^3 = \{(x_1, x_2, x_3) : x_j \in \mathbf{R}\}$ of the unit sphere $\mathbf{S} = \{(x_1, x_2, x_3) \in \mathbf{R}^3 : x_1{}^2 + x_2{}^2 + x_3{}^2 = 1\}$ from the point $\infty = (0, 0, 1)$

onto the xy-plane is given by associating $z = x + iy$ with $Z = (x_1, x_2, x_3)$ according to the formulas:

$$z = \frac{x_1 + ix_2}{1 - x_3}, \qquad x_1 = \frac{z + \bar{z}}{1 + |z|^2}, \qquad x_2 = \frac{z - \bar{z}}{i(1 + |z|^2)}, \qquad x_3 = \frac{|z|^2 - 1}{|z|^2 + 1}.$$

The $*$-transform lets us represent $^*\mathbf{C}$ on $^*\mathbf{S}\backslash\{\infty\}$. Notice that *the north pole "∞" is a standard point.*

Geometrically, the points z and Z are the intersection of a line through ∞ with the xy-plane and the sphere, respectively (Fig. 4.7.1). The nonstandard

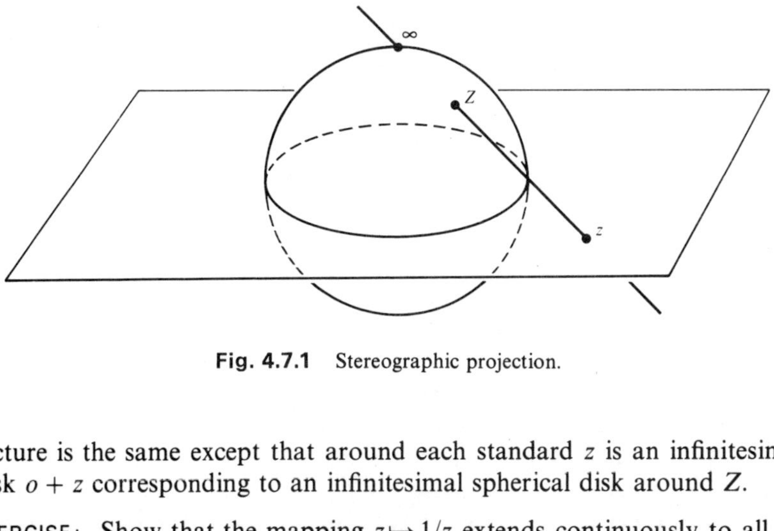

Fig. 4.7.1 Stereographic projection.

picture is the same except that around each standard z is an infinitesimal disk $o + z$ corresponding to an infinitesimal spherical disk around Z.

EXERCISE: Show that the mapping $z \mapsto 1/z$ extends continuously to all $^*\mathbf{S}$ by taking $Z \mapsto$ (stereographic image of $1/z$) when $Z \neq \infty$ and by taking $\infty \mapsto 0$ and $0 \mapsto \infty$. This is simply rotation of $\mathbf{S}$ about the x_1-axis. The straight cord length between Z and W is given by $2|z - w|/((1 + |z|^2)(1 + |w|^2))^{1/2}$.

EXERCISE: Show that the infinite complex numbers $^*\mathbf{C}_\infty$ represented on $^*\mathbf{S}$ are all within an infinitesimal of the standard point ∞ (in the ambient $\mathbf{R}^3$ metric or as measured along the surface).

EXERCISE: A Möbius transformation is a function of the form $z \mapsto (az + b)/(cz + d)$. Show that such a transformation moves no point on the sphere more than an infinitesimal once it moves three points only an infinitesimal. Of course, $a, b, c, z \in {}^*\mathbf{C}$.

APPENDIX A PRELIMINARY RESULTS ON ORDERED RINGS AND FIELDS

In order to describe the algebraic structure of the hyperreals $^*\mathbf{R}$, the ring of finite numbers, and the ring of infinitesimals, we shall give some preliminary results on ordered rings and fields. The first section includes a brief summary of terminology and a few very basic results. The second section contains some results on ordered rings and fields. The final section is concerned with Archimedean totally ordered fields. That section begins with a summary of Dedekind's definition of the real number system.

A.1 TERMINOLOGY

A nonempty set A with two binary operations $+$ and $\cdot$ (mappings of $A \times A$ into A) is called a *ring* provided (1–6) below are satisfied. For emphasis sometimes the triple $\langle A, +, \cdot \rangle$ is referred to as the ring.

(1) For every $a, b, c \in A$ we have $(a + b) + c = a + (b + c)$ [*addition is associative*].

(2) There exists an element $0 \in A$ such that for every $a \in A$, $0 + a = a$ [*zero element or additive identity*].

(3) For every $a \in A$, there exists an element $-a \in A$ such that $a + (-a) = 0$ [*additive inverses*]. (This element is necessarily unique.)

(4) For every $a, b \in A$ we have $a + b = b + a$ [*addition is commutative*].

(5) For every $a, b, c \in A$ we have $a \cdot (b \cdot c) = (a \cdot b) \cdot c$ [*multiplication is associative*].

(6) For every $a, b, c \in A$ we have $a \cdot (b + c) = (a \cdot b) + (a \cdot c)$ and $(b + c) \cdot a = (b \cdot a) + (c \cdot a)$ [*distributive laws*].

If in addition to being a ring A satisfies (7) we call A a *commutative ring*. If A also satisfies (8) we say it has an *identity element* or simply an *identity*. In what follows A shall denote a commutative ring with an identity element, that is, all of (1–8) shall be satisfied.

(7) For every $a, b \in A$, $a \cdot b = b \cdot a$ [*multiplication is commutative*].

(8) There exists an element $1 \in A$, $1 \neq 0$, such that for every $a \in A$, $a \cdot 1 = 1 \cdot a = a$ [*multiplicative identity*].

The reader will observe that the elements 0 and 1 are necessarily unique.

An element $a \in A$ is called *regular* if it has a *multiplicative inverse*, that is, if there exists an element $a^{-1} \in A$ such that $a \cdot a^{-1} = a^{-1} \cdot a = 1$. Such an inverse must also be unique.

Let F, $\langle F, +, \cdot \rangle$, be a commutative ring with identity. Then F is called a *field* if every nonzero element of F is regular, that is, if (9) is satisfied. Thus a field is defined by (1–9).

(9) For every $a \in F$, $a \neq 0$, there exists an element $a^{-1} \in F$ such that $a \cdot a^{-1} = 1$.

We will be primarily interested in the case where A is an *integral domain*, which means A is a commutative ring with identity and has no *divisors of zero*. A *divisor of zero* in A is an element $a \in A$, $a \neq 0$, such that there exists $b \neq 0$ in A for which $a \cdot b = 0$. Observe that a field is an integral domain. Since not all of the preliminary results depend on A being an integral domain we shall mention this additional hypothesis where it is required.

A mapping h from one ring A, $\langle A, +, \cdot \rangle$, into another ring B, $\langle B, \oplus, \odot, > \rangle$, is called a *ring homomorphism*, or simply a *homomorphism*, if for every $a, b \in A$ we have $h(a + b) = h(a) \oplus h(b)$ and $h(a \cdot b) = h(a) \odot h(b)$. If h is one-to-one, h is called an *isomorphism*. If h is one-to-one and onto, the two rings are said to be *isomorphic*; in this case, as far as ring structure is concerned, they are the same. Observe that a homomorphism maps the zero element of A onto the zero element of B. When B is an integral domain (and $h \not\equiv 0$) or when h is onto, h also maps the identity element of A onto the identity element of B. Thus the homomorphic image of a commutative ring with identity is a commutative ring with identity.

The *kernel of a homomorphism* h: $A \mapsto B$ is the name given to $h^{-1}(\bar{0}) = \ker(h)$ ($\bar{0}$ is the zero element of B). Kernels are distinguished by the property of being *ideals* of A. An *ideal* I of A is a subset of A which satisfies

(i) $0 \in I$ and for every $a \in I$, $-a \in I$,
(ii) for every $a, b \in I$, we have $a + b \in I$,
(iii) for every $a \in A$ and for every $b \in I$, we have $a \cdot b \in I$.

If, in addition, $I \neq \{0\}$ and $I \neq A$, we say I is a *proper ideal*. A proper ideal cannot contain any regular elements of A.

Given a ring A and a proper ideal I of A we can construct the *quotient ring* or *residue class ring* A/I as follows. Take for the underlying set the distinct residue classes $a + I = \{a + x \mid x \in I\}$. Define operations $\oplus$ and $\odot$ by $(a + I) \oplus (b + I) = ((a + b) + I)$ and $(a + I) \odot (b + I) = ((a \cdot b) + I)$. Properties (i–iii) make these operations well defined. The reader should verify that this construction yields a commutative ring with identity when A has these properties.

The mapping $h: A \to A/I$ given by $h(a) = a + I$ is called the *canonical homomorphism* of A onto A/I. If $j: A \to B$ is a homomorphism of A onto B such that $\ker(j) = I = \ker(h)$, then we have that B is isomorphic to A/I.

A proper ideal is said to be a *maximal ideal* of A if whenever J is an ideal satisfying $I \subset J \subset A$, then either $I = J$ or $J = A$. In other words, if the set of proper ideals is ordered by inclusion with I as a maximal element, Zorn's lemma gives us

(A.1.1) THEOREM *Every proper ideal is contained in a maximal ideal.*

Maximal ideals can be characterized algebraically as follows:

(A.1.2) THEOREM *A proper ideal I of A is maximal if and only if A/I is a field. (A is a commutative ring with identity.)*

PROOF: Assume that A/I is a field. Then if $a \notin I$ and $b \in A$, there exists an element $c \in A$ such that $b - ac \in I$. Hence, any ideal J containing I as a subset, which also contains a, is equal to A. (This is because $(b + J) = (a + J)(c + J) = (0 + J)(c + J) = J$, forcing b to be an element of J.) This proves that I is maximal.

Conversely, assume that I is maximal and $a \notin I$. Then there exists an element $b \in A$ such that $ab - 1 \in I$, for $aA + I$ is an ideal which contains I forcing $aA + I = A$. Now, however, A/I is a field, for $(a + I)(b + I) = (1 + I)$.

Similarly, a commutative ring with identity which has no proper ideals must be a field. We leave this to the reader for verification.

We conclude the section with a result that will tell us that the infinitesimals form a maximal ideal.

(A.1.3) THEOREM *If the set of nonregular elements of A form an ideal I, then I is maximal.*

PROOF: Suppose that J is an ideal that strictly contains I. Then there exists $a \in J \backslash I$. If $a \notin I$, then a is regular and thus $J = A$ and I is maximal. (Observe I is proper because A has at least one regular element, namely, 1.)

A.2 ORDERED RINGS AND FIELDS

(A.2.1) DEFINITION (Ordered ring) *A ring A is called an ordered ring if A is ordered by $\leq$ and the following two conditions hold:*

(i) *$x \leq y$ implies $x + z \leq y + z$ for all $z \in A$,*
(ii) *for all $x, y \in A$, $0 \leq x$ and $0 \leq y$ implies $0 \leq xy$.*

Thus the order is compatible with the ring operations.

If $\leq$ is a total order on A, then A is called a *totally ordered ring* and then $|a|$ denotes $\max(-a, a)$. We recall that a *total order* on A is a binary relation satisfying the conditions (i) $a \leq a$, for every $a \in A$; (ii) if $a \leq b$ and $b \leq c$, then $a \leq c$; (iii) if $a \leq b$ and $b \leq a$, then $a = b$; and (iv) for every $a, b \in A$ either $a \leq b$ or $b \leq a$. An order (or partial order) satisfies (i–iii).

We now list some properties of a totally ordered integral domain A.

(i) $1 > 0$ *and* $-1 < 0$. Verify by assuming $1 < 0$ and showing then that $1 = (-1)(-1) > 0$, which is a contradiction.

(ii) $x \neq 0$ *implies* $x^2 > 0$. Since A is an integral domain, $x^2 \neq 0$. Since A is totally ordered, either $x > 0$ or $x < 0$ and in either case $x^2 > 0$. *In particular,* -1 *has no square root.*

(iii) *If* $0 \leq x < y$, *then* $x^n < y^n$ *for all* $n \geq 1$.

(iv) *If* $x > 0$, *then* $x + \cdots + x$ (n times) $= nx \neq 0$ *for all* $n \geq 1$.

(v) *A contains a natural copy of the integers* $\mathbf{Z}$.

We prove (iv) and (v) together. We define a mapping from $\mathbf{Z}$ into A as follows: $0 \in \mathbf{Z}$ is mapped to the zero element of A. For positive integers (respectively, negative integers) we proceed by induction: $1 \in \mathbf{Z}$ is mapped to the unit element $1 \in A$; if $n > 0$ in $\mathbf{Z}$ is mapped to $\bar{n} \in A$, then $n + 1$ is mapped to $\bar{n} + 1$ (respectively, $-1 \in \mathbf{Z}$ goes to $-1 \in A$; $m - 1$ to $\overline{m} - 1$). Thus $(n \geq 0) \pm n$ gets mapped to $(\pm 1) + \cdots + (\pm 1)$ (n times).

By (i), $0 < 1$ (respectively, $-1 < 0$) and using (i) of Definition (A.2.1), $0 < 1 < 1 + 1$ (respectively, $(-1) + (-1) < (-1) < 0$) and it follows that the mapping preserves the strict order of $\mathbf{Z}$. In particular, the map is one-to-one.

It follows from preservation of the successor element that our map is an order isomorphism (cf. Definition (A.2.2)).

We could repeat the induction argument $0 < x < x + x < x + x + x < \cdots$ to obtain (iv). We also may view this another way. By the distributive property, for each $x \in A$, $x + \cdots + x$ (m times) equals $\overline{m} \cdot x$, where $\overline{m}$ is as above. Since these elements form an isomorphic copy of $\mathbf{Z}$ *we simply write* $m \cdot x$ for $m \in \mathbf{Z}$. Now (iv) follows from (ii) of Definition (A.2.1) and the assumption that A is an integral domain.

(vi) *If F is a totally ordered field, then F contains a natural copy of the rationals* $\mathbf{Q}$. *A field is necessarily an integral domain and* $\mathbf{Q}$ *is the smallest field containing* $\mathbf{Z}$, *which by* (v) *is in F.*

(A.2.2) DEFINITION (Order homomorphism) *An order homomorphism is a mapping h from one ordered ring A into another ordered ring B which is a ring homomorphism that also satisfies $h(a) \circledS h(b)$ whenever $a \leq b$.*

As in the case of ordinary homomorphisms the kernel of an order homomorphism h is an ideal of A, but h in this case has the additional property that whenever $0 \leq a \leq b$ and $b \in \ker(h)$, then $a \in \ker(h)$. This leads us to

(A.2.3) DEFINITION (Order ideal) *An order ideal of the ordered ring A is a subset I of A which is an ideal and also satisfies the property that whenever $0 \leq a \leq b$ and $b \in I$, then $a \in I$.*

The residue class ring A/I in this case is an ordered ring, and the canonical homomorphism is an order homomorphism. We can say somewhat more, and in doing so clarify the additional property of order ideals.

(A.2.4) THEOREM *Let A be an ordered ring and let I be an ordinary ideal of A. We form A/I and define a relation $\circledleq$ as below. Let h be the canonical homomorphism of A onto A/I. Then*

$$(a + I) \circledleq (b + I) \qquad \text{or, briefly,} \qquad h(a) \circledleq h(b)$$

if and only if there exist x, $y \in A$ such that $x \leq y$ and $h(x) = h(a)$ and $h(y) = h(b)$. With these conventions,

(a) *$h(a) \circledleq h(a)$ for all $a \in A$ [$\circledleq$ is reflexive].*
(b) *$h(a) \circledleq h(b)$ and $h(b) \circledleq h(c)$ implies $h(a) \circledleq h(c)$ [$\circledleq$ is transitive].*
(c) *$\circledleq$ satisfies properties* (i) *and* (ii) *of Definition* (A.2.1).
(d) *$h(a) \circledleq h(b)$ and $h(b) \circledleq h(a)$ implies $h(a) = h(b)$ [$\circledleq$ is antisymmetric]*
if and only if I is an order ideal.

PROOF: (a) Since $\leq$ is an order relation every $a \in A$ satisfies $a \leq a$, thus $h(a) \circledleq h(a)$.

(b) Transitivity implies there exist x, y, $z \in A$ with $x \leq y \leq z$ and $h(a) = h(x)$, $h(b) = h(y)$, and $h(c) = h(z)$. It follows that $x \leq z$ and hence $h(a) \circledleq h(c)$.

(c) $h(a) \circledleq h(b)$ implies that there exist x, $y \in A$ such that $x \leq y$, $h(x) = h(a)$, and $h(y) = h(b)$. Since h is onto we take $c \in A$ and show $(h(a) \oplus h(c)) \circledleq (h(b) \oplus h(c))$ to prove (i). We know $x + c \leq y + c$ so that $h(x + c) \circledleq h(y + c)$. Since h is a homomorphism,

$$h(a) \oplus h(c) = h(x) \oplus h(c) = h(x + c) \circledleq h(y + c) = h(b) \oplus h(c).$$

To show (ii), take $h(a)$ and $h(b) \circledgeq 0$. This implies that there exist x, $y \in A$ with $0 \leq y$, $h(a) = h(x)$, and $h(b) = h(y)$. By (ii) in the ordered ring A, $0 \leq x \cdot y$. Since h is a homomorphism, $h(a) \odot h(b) = h(x) \odot h(y) = h(x \cdot y) \circledgeq 0$. This completes the proof of part (c).

(d) If for every x and y whenever $h(x) \circledleq h(y)$ and $h(y) \circledleq h(x)$ we have

$h(x) = h(y)$, then we must show that I is an order ideal. Take a, $b \in A$ with $0 \leq a \leq b$ and $b \in I$. We have then $0 + I = h(0) \textcircled{\leq} h(a)$ and $h(a) \textcircled{\leq} h(b) = 0 + I$. Thus $h(a) = I$ and $a \in I$ making I an order ideal.

Conversely, let I be an order ideal. If $h(a) \textcircled{\leq} h(b)$ and $h(b) \textcircled{\leq} h(a)$, then it follows that $I \textcircled{\leq} h(a - b)$ and $I \textcircled{\leq} h(b - a)$. (We leave the verification of this last remark to the reader.) Now there exist elements x, $y \in A$ with $0 \leq x$, $0 \leq y$, $h(x) = h(b - a)$, and $h(y) = h(b - a)$. Since $0 \leq x \leq x + y$ and $x + y \in I$, by virtue of the fact that $h(x + y) = h(a + b + b - a)$, we have $x \in I$; thus $h(a - b) = I$ and it follows then that $h(a) = h(b)$. This completes the proof of the theorem.

As an immediate corollary we have

(A.2.5) THEOREM *If A is a totally ordered ring and if I is a proper order ideal, then A/I is a totally ordered ring (with the operations and order given above).*

The reader would do well to observe that the canonical order homomorphism of A onto A/I does not necessarily preserve strict inequalities; even when $a < b$ we may have $h(a) = h(b)$.

The final concept of this section will be studied more closely in the case of fields in the next section:

(A.2.6) DEFINITION (Archimedean totally ordered integral domain) *Let A denote a totally ordered integral domain. We denote the natural copy of the nonnegative integers in A by $\mathbf{N}$. Then A is said to be an Archimedean if for every $a \in A$ there exists an $n \in \mathbf{N}$ so that $|a| < n$.*

In a field the Archimedean property can be stated as "for every $a \in A$, $a > 0$ implies there exists $n \in \mathbf{N}$ such that $n \cdot a > 1$." We shall see that this need not hold in an Archimedean integral domain.

A.3 ARCHIMEDEAN TOTALLY ORDERED FIELDS

The field of rationals and the field of real numbers are examples of totally ordered Archimedean fields. Before we conclude this chapter with a characterization of the Archimedean property for totally ordered fields we shall recall briefly Dedekind's definition of the system of real numbers. Let $\mathbf{Q}$ again denote the field of rational numbers.

(A.3.1) DEFINITION (Dedekind cut) *A subset a of $\mathbf{Q}$ is called a Dedekind cut if* (i) *$a \neq \varnothing$ and $\mathbf{Q} \backslash a \neq \varnothing$,* (ii) *$r \in a$ and $r' \in \mathbf{Q}$ and $r' < r$ imply $r' \in a$,* (iii) *a contains no largest element.*

If a is a Dedekind cut, then $r \notin a$, $r \in \mathbf{Q}$ implies $r' < r$ for all $r' \in a$.

If $r \in \mathbf{Q}$, then the set $a = \{r' : r' \in \mathbf{Q} \text{ and } r' < r\}$ is a Dedekind cut and r is the least upper bound of a in $\mathbf{Q}$. This Dedekind cut is called the *rational cut* defined by r. The Dedekind cut defined by $0 \in \mathbf{Q}$ will be denoted by 0. Thus $0 = \{r : r \in \mathbf{Q} \text{ and } r < 0\}$. Two Dedekind cuts a and b are considered to be equal if they are equal as sets, i.e., if they contain the same rationals.

Let $\mathbf{R}$ be the family of all Dedekind cuts of $\mathbf{Q}$. Then the relation $a \leq b \Leftrightarrow a = b$ or $(\exists r)(r \in \mathbf{Q}$ and $r \in b$ and $r \notin a)$ totally orders R. In particular, for every $a \in \mathbf{R}$ we have either $a < 0$ or $a = 0$ or $a > 0$.

Addition in $\mathbf{R}$ can be introduced as follows: If $a, b \in \mathbf{R}$, then by $a + b$ we denote that subset of $\mathbf{Q}$ which has the following property: $r \in a + b \Leftrightarrow (\exists r')(\exists r'')(r' \in a, r'' \in b$, and $r = r' + r'')$. It is easy to see that $a + b$ is again a Dedekind cut. Then it is easily shown that $\mathbf{R}$ is a totally ordered commutative group, with respect to this definition of addition, which contains the additive group of $\mathbf{Q}$ by means of the rational cuts.

The introduction of multiplication is more complicated: If $a, b \in \mathbf{R}$ and $a \geq 0$, $b \geq 0$, then by $a \cdot b$ or, shortly, ab we denote that subset of $\mathbf{Q}$ that has the following property: $r \in ab \Leftrightarrow r$ is a negative rational or $(\exists r')(\exists r'')(0 \leq r' \in a, 0 \leq r'' \in b$, and $r = r'r'')$. It is easy to see that ab is a Dedekind cut.

Multiplication for general elements of $\mathbf{R}$ is then introduced as follows: If $a, b \in \mathbf{R}$, then $ab = -|a|\,|b|$ whenever $a < 0$ and $b \geq 0$, $ab = -|a|\,|b|$ whenever $a \geq 0$ and $b < 0$, and $ab = |a|\,|b|$ whenever $a < 0$ and $b < 0$.

Then $\mathbf{R}$ turns out to be an Archimedean totally ordered field with the property that every nonempty subset of $\mathbf{R}$ that is bounded above has a least upper bound.

Using Dedekind's definition of the real number system we shall now prove the following theorem which will be essential in the proof of Theorem (4.4.4).

(A.3.2) THEOREM *A totally ordered field is Archimedean if and only if it is isomorphic to a subfield of* $\mathbf{R}$.

PROOF: Since every subfield of $\mathbf{R}$ is Archimedean we have only to show that the condition is necessary. For this purpose assume that F is a totally ordered Archimedean field. If $x, y \in F$ and $x < y$, then there exists an element $n \in \mathbf{N}$ such that $n > 1/(y - x)$. Let m be the smallest integer $> nx$. Then $x < m/n < y$. Hence, $\mathbf{Q}$ is dense in F, and every element x of F is uniquely determined by the Dedekind cut $\{r : r \in \mathbf{Q} \text{ and } r < x\}$. Thus F is embedable in $\mathbf{R}$ in a unique way as a totally ordered subset of $\mathbf{R}$. Furthermore, if $x, y \in F$ and q, r, s, t are rationals such that $q \leq x < r$, $s \leq y < t$, then $s + q \leq x + y < r + t$. Hence, addition in F like addition in

R is uniquely determined by Dedekind cuts of **Q**. This holds also for multiplication as the reader can easily verify himself. This shows that the embedding of F into **R** is an isomorphism, which completes the proof of the theorem.

CHAPTER 5

FOUNDATIONS OF INFINITESIMAL CALCULUS

This chapter is an introduction to calculus using infinitesimals for people who have already seen the "epsilon–delta" introduction. An elementary introduction from scratch can be found in the work of Keisler [1976]. Since we assume familiarity with "epsilon–delta" calculus, we shall make many transitions back and forth between the two formulations. Not all the transitions are necessary; calculus can be rigorously developed to a high level without mention of epsilon–delta.

One place epsilon–delta enters essentially is in numerical approximation. There are even some theoretical considerations where epsilon–delta is advantageous, but we shall emphasize the less familiar infinitesimals. This chapter is a systematic account of the most fundamental results, while the next considers more specialized topics. A reader so inclined could go to Chapter 6 and come back to the necessary results of this chapter when rigor or contact with a familiar formulation is needed.

5.1 CONTINUITY AND LIMITS

The intuitive notion of continuity of a function is that a small change in the independent variable produces a small change in the answer. For example, if $f(x) = x^2$, then $f(x + \varepsilon) = x^2 + \varepsilon(2x + \varepsilon)$ and when x is finite this is nearly x^2, as we saw in Chapter 2.

Standard continuity at a is described by the statement "for every $\varepsilon > 0$, there is a $\delta > 0$, so that if $|x - a| < \delta$, then $|f(x) - f(a)| < \varepsilon$."

We can express *standard continuity at a in the nonstandard model* by the statement

(SC) *For every $\varepsilon \in {}^{\sigma}\mathbf{R}^+$, there is a $\delta \in {}^{\sigma}\mathbf{R}^+$, so that if $|x - a| < \delta$, then $|f(x) - f(a)| < \varepsilon$.*

(5.1.1) THEOREM *Let f be an internal real-valued function defined on the infinitesimal neighborhood of $a \in {}^*\mathbf{R}$. Then the statement* (SC) *is equivalent to*

(IC) *$f(x)$ is infinitesimally close to $f(a)$ whenever x is infinitesimally close to a.*

(5.1.2) DEFINITION *An internal function f satisfying* (IC) *and* (SC) *is said to be S-continuous at $a \in {}^*\mathbf{R}$.*

PROOF OF (5.1.1): Let f satisfy (IC) and let ε be standard positive and fixed. We exhibit a δ for this ε to prove that (IC) implies (SC).

The following set $D(\varepsilon)$ is internal by the internal definition principle, since f is internal:

$$D(\varepsilon) = \{\delta \in {}^*\mathbf{R}^+ : |x - a| < \delta \text{ implies } |f(x) - f(a)| < \varepsilon\}.$$

Also, $D(\varepsilon)$ contains all infinitesimal δ's by the infinitesimal continuity property (IC). Since o is external, $D(\varepsilon)$ contains a standard δ and this establishes (SC).

Conversely, suppose (IC) is false. Then there is an x infinitesimally near a for which $|f(x) - f(a)|$ is not infinitesimal. If we take $\varepsilon = \min(1, |f(x) - f(a)|/2)$, in case $f(x) - f(a)$ is infinite, then we know that for each standard $\delta > 0$, there is a point x within δ of a at which $f(x)$ is farther than ε from $f(a)$. This shows (SC) cannot hold and proves the theorem.

We are using a continuity principle here which we will later generalize: **Cauchy's principle for the infinitesimals around** a:

If $P(x)$ is a boundedly formalizable internal property depending on the free variable x and if $P(m)$ holds for all $m \approx a$, then $P(x)$ holds for all x within some standard δ of a.

The properties in question are:

$P(x) = [x \in \mathbf{R}] \,\&\, [|f(x) - f(a)| < \varepsilon]$, where ε is standard.

$P(x)$ holds to within δ of a means that for every x so that $|x - a| < \delta$ we have $[x \in \mathbf{R}]$ and $[|f(x) - f(a)| < \varepsilon]$.

EXERCISE: Prove this version of Cauchy's principle.

(5.1.3) COROLLARY *Let f be a standard real-valued function defined in a neighborhood of $a \in \mathbf{R}$. Then f is continuous at a in the epsilon–delta sense if and only if $f(x)$ is infinitesimally close to $f(a)$ whenever x is infinitesimally close to a in ${}^*\mathbf{R}$.*

PROOF: By (5.1.1) it suffices to show that ordinary continuity in $\mathscr{R}$ amounts to the external statement (SC) in ${}^*\mathbf{R}$ when f is standard.

Consider the set $D(\varepsilon)$ from the preceding proof. Since f is now standard, $D(\varepsilon)$ is a standard set by the standard definition principle.

Now one argues case by case on ε. If continuity holds in the standard model, $D(\varepsilon) = {}^*\{\delta \in \mathbf{R}^+ : |x - a| < \delta$ implies $|f(x) - f(a)| < \varepsilon\}$ and in particular it is nonempty. A standard $\delta \in D(\varepsilon)$ establishes (SC) for that ε.

Conversely, if (SC) holds, $D(\varepsilon)$ is nonempty in $^*\mathscr{R}$ therefore also in $\mathscr{R}$ by removing $*$'s from its standard definition (Leibniz' principle).

The point is that since $D(\varepsilon)$ is standard we can describe it with or without $*$'s and Leibniz' principle says it is nonempty in one model if and only if it is nonempty in the other. We have simply written a set to reduce "true or false" to "empty or nonempty."

(5.1.4) ***REMARKS:*** See Section (3.12) and Example (8.4.45) where we discuss sinusoids with infinite frequency.

The function

$$\operatorname{sgn}(x) = \begin{cases} -1, & x < 0 \\ 0, & x = 0 \\ +1, & x > 0 \end{cases}$$

is discontinuous at zero since if ε is a positive infinitesimal, $\operatorname{sgn}(\varepsilon) = 1$, $\operatorname{sgn}(-\varepsilon) = -1$, and $\operatorname{sgn}(0) = 0$.

The internal function $f(x) = \operatorname{sgn}(x - \omega)$ is S-continuous everywhere but at ω where it is dis-S-continuous. Now, $f(\omega + \varepsilon) = 1$, $f(\omega - \varepsilon) = -1$, and $f(\omega) = 0$.

The function $f(x) = \sin(1/x)$ has no continuous extension at $x = 0$ since if ω is an infinite odd natural number (that is, $\omega/2 \neq [\omega/2]$), then letting $\varepsilon = 2/(\omega\pi)$ we have $f(\varepsilon) = -f(-\varepsilon)$ and $|f(\varepsilon)| = 1$.

Now we turn to a little of the topology of real numbers from our perspective. Two examples to keep in mind here are the open unit interval (0, 1) and the natural numbers $\mathbf{N} \subseteq \mathbf{R}$. The extension of (0, 1) is $\{x \in {}^*\mathbf{R} : 0 < x < 1\}$, and this contains points near 0 and 1 while not containing 0 and 1 themselves. The extension $^*\mathbf{N}$ contains the standard parts of each of its finite elements, but also contains points that have no standard part. We now give the standard reformulations.

(5.1.5) THEOREM *Let $S \subseteq \mathbf{R}$. Then S is bounded if and only if $^*S$ is contained in the finite points $\mathcal{O}$.*

PROOF: If S is bounded, the statement

$$\text{for all } x \in S, \qquad |x| < b$$

holds with $b \in \mathbf{R}^+$. We apply the $*$-transform to see that for all $x \in {}^*S$, $|x| < b$ and b is finite so $^*S \subseteq \mathcal{O}$.

Conversely, if $^*S \subseteq \mathcal{O}$, the set $B = \{b \in {}^*\mathbf{R}^+ : \text{for all } x \in {}^*S,\ |x| < b\}$ contains $\mathbf{R}_\infty{}^+$, the positive infinite reals. Since B is standard (by the standard definition principle) and nonempty, by removing $*$'s in the definition of B we see that S is bounded (that is, apply the converse of Leibniz' principle. The set B is either empty in both $\mathscr{R}$ and $^*\mathscr{R}$ or in neither.)

(5.1.6) THEOREM *Let $S \subseteq \mathbf{R}$. Then S is closed if and only if $S =$ $\operatorname{st}(^*S \cap \mathcal{O})$, that is, if and only if $^*S$ contains all the standard points near elements of $^*S$.*

In Section 8.3 some general topological theorems about closures and standard parts are given. In Section 8.4 similar results in the uniform category are presented. For these reasons and since "sequentially closed" is perhaps most intuitive in $\mathbf{R}$, we shall take "closed" in this theorem to mean "*each sequence $(s_n) \subseteq S$ for which* $\lim_{n\to\infty} s_n$ *exists* (in the ordinary sense) *has its limit in S.*"

Before proving (5.1.6) we give a reformulation of an epsilon–delta definition of limit. We begin with a discussion of the infinitesimal Cauchy condition.

Suppose $(a_n : n \in \mathbf{N})$ is a standard sequence of reals. We view this as a map $a: \mathbf{N} \to \mathbf{R}$ and still denote $^*a: {}^*\mathbf{N} \to {}^*\mathbf{R}$ by a. For example, a_ω must mean the extension when ω is infinite. Now suppose that whenever ω and Ω are infinite subscripts then $a_\omega \approx a_\Omega$. First, a_ω is finite! To see this, notice that the set of indices m for which "$p, n \geq m$ implies $|a_n - a_p| < 1$" contains all infinite m and therefore a finite m_0, since that set is standard and nonempty. Then all a_k are bounded by $\max[|a_k| : k = 0, 1, \ldots, m_0] + 1$, a standard number. *Thus we can form $b = \operatorname{st}(a_\omega)$ independent of the particular infinite ω. The next result says $b = \lim_{n\to\infty} a_n$.*

EXERCISE: *State and prove a reformulation theorem giving standard and infinitesimal versions of the Cauchy sequence condition* (cf. (8.4.26)). The above discussion should serve as a hint.

(5.1.7) THEOREM *Let $(a_n : n \in \mathbf{N})$ be a standard sequence in $\mathbf{R}$. Then the following are equivalent:*

(SL in $\mathscr{R}$) *For every $\varepsilon \in \mathbf{R}^+$, there is an $M \in \mathbf{N}$ such that if $n > M$, then $|a_n - b| < \varepsilon$.*

(IL in $^*\mathscr{R}$) *a_Ω is infinitesimally close to b whenever Ω is an infinite subscript in $^*\mathbf{N}$.*

NOTE: One can begin the following proof with the weaker hypothesis that $a_\Omega \approx a_\omega$; the discussion above provides b mentioned in (IL).

PROOF: We suppose $b = \operatorname{st}(a_\omega)$ for all infinite ω. Let a standard positive fixed ε be given; we shall produce the M of (SL) for that ε. Consider the set of indices

$$\{m \in {}^*\mathbf{N} : n > m \text{ implies } |a_n - b| < \varepsilon\}$$

which is nonempty since all infinite m's work and is standard by the standard definition principle. By Leibniz' principle the set is nonempty in $\mathscr{R}$ and this demonstrates the existence of M in (SL).

For the converse one could argue on the hypothesis that (IL) is false in the manner of the proof of (5.1.1). Instead, we approach the converse another way. The reader may wish to adopt this approach in another proof of (5.1.1) as an exercise.

Suppose that (SL) holds in $\mathscr{R}$ and that ω in $^*\mathbf{N}$ is infinite. By (SL) there is a standard function $m(\varepsilon)$ so that "if $n \in \mathbf{N}$ and $n > m(\varepsilon)$, then $|a_n - b| < \varepsilon$." To see this, make a choice for M and let $m(\varepsilon)$ equal M for that value of ε.

Now to establish (IL) it is sufficient to show that for each standard positive ε, $|a_\omega - b| < \varepsilon$ for our fixed but arbitrary infinite ω. Let $\varepsilon \in \mathbf{R}^+$ be given. Since ω is infinite and $m(\varepsilon)$ is standard we have $\omega > m(\varepsilon)$ and therefore $|a_\omega - b| < \varepsilon$ by Leibniz' principle applied to the statement in quotes above. This proves the theorem.

PROOF OF (5.1.6): Suppose that $S = \operatorname{st}({}^*S \cap \mathscr{O})$ and $(s_n : n \in \mathbf{N}) \subseteq S$ is convergent. Then $s_\omega \in {}^*S$ by the $*$-transform of "for every $n \in \mathbf{N}$, $s_n \in S$" and $\operatorname{st}(s_\omega) = \lim_{n\to\infty} s_n \in S$.

If t is standard, $t \approx s$ for $s \in {}^*S$ and $t \notin S$, then the set

$$\{s \in {}^*S : |s - t| < 1/n\}$$

is nonempty and standard. By transfer to $\mathscr{R}$ there is a standard point $s_n \in S$ such that $|t - s_n| < 1/n$—such a sequence converges to t outside S and thus S is not closed. This proves Theorem (5.1.6).

EXAMPLES OF SEQUENTIAL LIMITS (Also see Section 2.3):

1. *If* $a \in \mathbf{R}^+$, *then* $\lim_{n\to\infty} a^{1/n} = 1$. Let $b > 0$ be finite and not infinitesimal, for example, $b = {}^*a$. We may assume $b \geq 1$, for otherwise we consider $1/b$. Let $c_n = b^{1/n} - 1$ so that $b = (1 + c_n)^n$ with $c_n \geq 0$. Expanding binomially we see that for each n, $b > nc_n$. Now let ω be an arbitrary infinite number, from which $0 < c_\omega < b/\omega \approx 0$ and, in particular, $a^{1/\omega} \approx 1$.

2. *The* $\lim_{n\to\infty} n^{1/n} = 1$ *also* because if we let $n = (1 + d_n)^n$ so $d_n = n^{1/n} - 1$, then by the binomial expansion,

$$n > \frac{n(n-1)}{2} d_n^{\,2} \qquad \text{for each } n.$$

Therefore when $\omega \in {}^*\mathbf{N}_\infty$, $0 < d_\omega^{\,2} < 2/(\omega - 1) \approx 0$ and $\omega^{1/\omega} \approx 1$.

3. *The infinite series* $\sum_{k=1}^{\infty} 1/k(k+1) = 1$, because when $\omega \in {}^*\mathbf{N}_\infty$,

$$\sum_{k=1}^{\omega} \frac{1}{k(k+1)} = \sum_{k=1}^{\omega} \frac{1}{k} - \sum_{k=1}^{\omega} \frac{1}{(k+1)} = 1 + \sum_{k=1}^{\omega} \frac{1}{k} - \sum_{k=1}^{\omega} \frac{1}{k} - \frac{1}{(\omega+1)} \approx 1.$$

4. *The infinite series* $\sum_{k=1}^{\infty} 1/k^2$ *converges* also since we can compare it term by term with the series above, $0 \le 1/k^2 \le 1/k(k-1)$ for $k \ge 2$.

5. *The infinite series* $\sum_{k=1}^{\infty} 1/k^p$ *converges for* $p > 1$ *and diverges for* $p \le 1$. When $p \le 1$,

$$\sum_{k=1}^{2^{\omega+1}} \frac{1}{k^p} - \sum_{k=1}^{2^\omega} \frac{1}{k^p} = \sum_{k=1}^{2^\omega} \frac{1}{(k+2^\omega)^p} > \frac{2^\omega}{2^{(\omega+1)p}} > \frac{1}{2^p},$$

a finite difference even when ω is infinite. When $p > 1$ and $p \not\approx 1$,

$$\sum_{k=2^n}^{2^{n+1}} \frac{1}{k^p} < \frac{2^n}{(2^n)^p},$$

so given infinite $\omega < \Omega$ we first take the largest λ so that $2^\lambda < \omega$, then the smallest Λ so that $\Omega < 2^\Lambda$ and observe

$$\sum_{k=1}^{\Omega} \frac{1}{k^p} - \sum_{k=1}^{\omega} \frac{1}{k^p} \le \sum_{k=2^\lambda}^{2^\Lambda} \frac{1}{k^p} \le \sum_{k=\lambda}^{\Lambda} (2^{1-p})^k \approx 0$$

by the convergence of the geometric series (see Section 2.3).

6. *The infinite series* $\sum_{k=2}^{\infty} 1/(k \log k)$ *diverges.*

$$\sum_{k=2^n}^{2^{n+1}} \frac{1}{k \log k} > \frac{1}{(2 \log 2)(n+1)}, \qquad \text{so} \qquad \sum_{k=2}^{2^\omega} \frac{1}{k \log k} > \frac{1}{2 \log 2} \sum_{k=1}^{\omega} \frac{1}{k},$$

and this is infinite (see Section 2.3).

7. (Olivier) *If the series* $\sum_{k=1}^{\infty} a_k$ *converges and* $a_k \ge a_{k+1} \ge 0$, *then* $\lim_{k\to\infty} ka_k = 0$. Proof: $0 \approx \sum_{k=[\omega/2]}^{\omega} a_k \ge [\omega/2]a_\omega$, so $\omega a_\omega \approx 0$ for every infinite ω.

(5.1.8) THEOREM *Let* $S \subseteq \mathbf{R}$. *Then* S *is compact if and only if every point of* ${}^*S$ *is infinitesimally near a standard point of* ${}^\sigma S$.

PROOF: We can show that S is closed and bounded if and only if this infinitesimal condition is met by applying (5.1.5) and (5.1.6). A topological version is contained in Section 8.3, a uniform version is in Section 8.4—the connection between near-standard points and compactness is important.

We now consider applications of infinitesimals to the proofs of standard results.

(5.1.9) THEOREM *A continuous function attains its maximum and minimum on a compact set.*

PROOF: Let f be a standard continuous function defined on a compact set K. By (5.1.8) this means $*K$ consists entirely of points near standard points of $^{\sigma}$K. By (5.1.3) at the points $k \in {}^{\sigma}K$ we know $h \approx k$ implies $f(h) \approx f(k)$.

We now show that $f(K)$ is bounded and closed. First, $*f(K)$ contains only finite points. (Note $*f(K) = f(*K) = \{y \in *\mathbf{R} : y = f(k)$ for some $k \in *K\}$.) This is because every point h of $*K$ satisfies $h \approx k \in {}^{\sigma}K$ and $f(h) \approx f(k)$ and $f(k)$ is standard. Second, $f(K)$ is closed because if a standard $g \approx f(k)$ for some $k \in *K$, then $g \approx f(k) \approx f(\text{st}(k))$ forcing $g = f(\text{st}(k))$. Since $f(K)$ is itself compact f attains its maximum and minimum over K.

Our next application has a proof which first appeared in Luxemburg [1962].

(5.1.10) INTERMEDIATE VALUE THEOREM *Let f be a continuous real-valued function defined on the compact interval $[s, t]$. If $f(s) \neq f(t)$, then f attains all the values between $f(s)$ and $f(t)$ on the open interval (s, t).*

PROOF: Let c be a number between $f(s)$ and $f(t)$. We want to show that there is an x, $s < x < t$ so that $f(x) = c$. Without loss of generality we may assume $f(s) < 0$, $c = 0$, and $f(t) > 0$, for otherwise we consider $F(x) = \text{sgn}(f(t) - f(s))[f(x) - c]$, since $F(x) = 0$ if and only if $f(x) = c$. Also we may change variables so that $s = 0$ and $t = 1$, $F(x) = F(s + x(t - s))$. We need only treat this normalized case: $f\colon [0, 1] \to \mathbf{R}$, $f(0) < 0, f(1) > 0$.

Assume f has no zeros. We form the telescoping product

$$\prod_{k=0}^{n-1} \frac{f((k+1)/n)}{f(k/n)} = \frac{f(1)}{f(0)} \qquad \text{for each} \quad n \geq 1.$$

Observe that since $f(1)/f(0)$ is negative one of the terms in the product is negative and therefore, letting $s_n = k/n$, $t_n = (k + 1)/n$, we know:

For each $n \in \mathbf{N}$ with $n \geq 1$ there exist points s_n and t_n with $0 \leq s_n < t_n \leq 1$ with $t_n - s_n = 1/n$ and $f(t_n)/f(s_n) < 0$.

We apply the $*$-transform of this statement to an infinite $\omega \in *\mathbf{N}$. Since s_ω and t_ω are in $*[0, 1]$, $\text{st}(s_\omega)$ and $\text{st}(t_\omega)$ both exist. Also, since $t_\omega - s_\omega = 1/\omega$ we know $\text{st}(s_\omega) = \text{st}(t_\omega) = x$, say. Now, by continuity at x, $f(s_\omega) \approx f(x) \approx$

$f(t_\omega)$ and by our hypothesis that $f(x)$ has no zeros,

$$0 \geq \frac{f(s_\omega)}{f(x)} \cdot \frac{f(x)}{f(t_\omega)} \approx 1,$$

a contradiction. Therefore f must have zeros and this proves the theorem.

Telescoping expressions like the product in this proof are a powerful elementary tool. We shall use them to prove the mean value theorem and the fundamental theorem of calculus.

EXERCISE: Prove the following limit reformulations. Let f be a real-valued function defined in a neighborhood of $a \in \mathbf{R}$.

1. The following are equivalent:

(SL in $\mathscr{R}$) For every $\varepsilon \in \mathbf{R}^+$, there exists a $\delta \in \mathbf{R}^+$ so that if $|x - a| < \delta$, then $|f(x) - b| < \varepsilon$, that is, $\lim_{x \to a} f(x) = b$.

(IL in $^*\mathscr{R}$) If $x \approx a$, then $f(x) \approx b$.

2. The following are also equivalent:

(SLS in $\mathscr{R}$) $\limsup_{x \to a} f(x) = b$ in the sense of epsilon–delta.

(ILS in $^*\mathscr{R}$) $b = \sup[\mathrm{st}(f(x)) : x \approx a]$.

3. State and prove the reformulation of "lim inf."

5.2 UNIFORM CONTINUITY

Now we look at the infinitesimal formulation of uniform continuity. Two examples to keep in mind here are $f(x) = x^2$ and the arctangent. We saw in Chapter 2 that at an infinite number x^2 fails to be S-continuous whereas $\mathrm{atn}(x)$ is S-continuous. We also want to discuss internal functions like $f(x) = (1 - (x^2/\omega))^\omega$, so we shall formulate our standard condition in the nonstandard model as follows:

(USC on A) A function f is uniformly standardly continuous on $A \subseteq {}^*\mathbf{R}$ if and only if for every $\varepsilon \in {}^\sigma\mathbf{R}^+$, there is a $\delta \in {}^\sigma\mathbf{R}^+$, so that whenever $|x - y| < \delta$ for x and y in A, then $|f(x) - f(y)| < \varepsilon$.

Our next result is deceptively simple (cf. Robinson [1966, especially historical remarks]).

(5.2.1) THEOREM *Let f be an internal function and A an internal set. Then f is uniformly S-continuous on A if and only if $f(x) \approx f(y)$ whenever $x \approx y$ for x and y in A, that is, if and only if f is S-continuous at each point of A.*

PROOF: The following sets are internal by the internal definition principle:

$$D(\varepsilon) = \{\delta \in {}^*\mathbf{R}^+ : \text{for every } x, y \in A, \text{ if } |x - y| < \delta, \\ \text{then } |f(x) - f(y)| < \varepsilon\}.$$

The infinitesimal condition says $D(\varepsilon)$ contains the infinitesimals when ε is standard, therefore it contains a standard δ and (USC on A) holds for that (arbitrary) ε.

When (USC on A) holds then surely f is S-continuous at each point of A, and we apply (5.1.1) to obtain the infinitesimal condition.

(5.2.2) COROLLARY *A standard function f is uniformly continuous on $B \subseteq \mathbf{R}$ if and only if ${}^*f$ is S-continuous everywhere on ${}^*B$.*

PROOF: For a fixed standard ε, the existence of δ is a standard statement subject to transfer back and forth. Hence ${}^*f$ is uniformly S-continuous if and only if f is uniformly continuous. Now apply (5.2.1).

EXERCISE: Describe the set $D(\varepsilon)$ from the proof of (5.2.1) in the case $f = {}^*f$ and $A = {}^*B$. Do this both in $\mathscr{R}$ and ${}^*\mathscr{R}$.

(5.2.3) COROLLARY *A standard continuous function is uniformly continuous on a compact set.*

PROOF: A compact set consists entirely of near-standard points by (5.1.8). If f is continuous, then it is S-continuous at each standard point by (5.1.3). Now apply (5.2.2).

5.3 BASIC DEFINITIONS OF CALCULUS

(5.3.1) DEFINITION Let $f: [a, b] \to \mathbf{R}$ be a real-valued function. We give a definition of $\int_a^b f(x)\,dx$ as follows: First, partition $[a, b]$ infinitesimally, this means, give an internal $*$-finite sequence $a = x_0 < x_1 < \cdots < x_\omega = b$ so that $x_i \approx x_{i-1}$ for $i = 1, \ldots, \omega$. For example, when $a = 0$, $b = 1$, take $x_i = i/\omega$ for $i = 1, \ldots, \omega$. Next, select an internal sequence of points y_i subordinate to the partition $y_i \in [x_i, x_{i-1}]$. Let $\Delta x_i = x_i - x_{i-1}$ and form the $*$-finite sum (see the last section of the previous chapter):

$$\sum_{i=1}^{\omega} f(y_i) \cdot \Delta x_i .$$

If this answer is infinitesimally nearly independent of the choices of infinitesimal partition and y_i's and if the sum is finite, *then the common standard part is the integral*, that is,

$$R\int_a^b f(x)\,dx = \mathrm{st}\left(\sum_{i=1}^{\omega} f(y_i)\,\Delta x_i\right),$$

provided the standard part exists and is independent of the particular infinitesimal partition and subordinate y_i's.

For example, if

$$f(x) = \begin{cases} 1, & x \in \mathbf{Q} \\ 0, & x \notin \mathbf{Q}, \end{cases}$$

then $\sum_{i=1}^{\omega} f(y_i)(1/\omega) = 0$ when y_i is irrational in $[k/\omega, (k+1)/\omega]$, whereas

$$1 = \sum_{i=1}^{\omega} \frac{1}{\omega} = \sum_{i=1}^{\omega} f\left(\frac{i}{\omega}\right)\cdot\frac{1}{\omega}.$$

Therefore $\mathrm{R}\int_a^b f(x)\,dx$ does not exist.

EXERCISE: $\mathrm{R}\int_a^b$ stands for the Riemann integral. State the standard definition as a single bounded sentence in $\mathscr{R}$. State the shortened limit definition in $\mathscr{R}$. Prove that the infinitesimal definition implies the standard one.

In the remainder of this section, $x_0, \ldots, x_\omega$ is a particular infinitesimal subdivision of $[a, b]$ and $f\colon [a, b] \to \mathbf{R}$ is *standard* and *continuous*. Let y_i and z_i be two internal sequences of evaluation points $y_i, z_i \in [x_i, x_{i-1}]$. We know by continuity of f that $f(y_i) \approx f(z_i)$, so the $*$-finite sequence $F_i = |f(y_i) - f(z_i)|$ is infinitesimal for each i. Now by $*$-transfer of the triangle inequality we know that

$$\left|\sum_{i=1}^{\omega} f(y_i)\,\Delta x_i - \sum_{i=1}^{\omega} f(z_i)\,\Delta x_i\right| \le \sum_{i=1}^{\omega} F_i\,\Delta x_i.$$

Next, the maximum of the F_i's is one of the F_i's by transfer of the fact that the maximum of finitely many numbers is attained. This is an infinitesimal m and

$$\left|\sum_{i=1}^{\omega} f(y_i)\,\Delta x_i - \sum_{i=1}^{\omega} f(z_i)\,\Delta x_i\right| \le m\cdot\sum_{i=1}^{\omega}\Delta x_i.$$

The term $\sum_{i=1}^{\omega} \Delta x_i$ telescopes to $b - a$ and $m\cdot(b - a)$ is infinitesimal. This shows that the infinite Riemann sum is independent of the (subordinate) evaluation points when f is continuous.

To compare infinite Riemann sums of different partitions simply subdivide. Given $a = x_0{}^1 < x_1{}^1 < \cdots < x_\omega{}^1 = b$ and $a = x_0{}^2 < x_1{}^2 < \cdots < x_\Omega{}^2 = b$ take the internal sequence $x_0{}^3 = a$, $x_1{}^3 = \min(x_1{}^1, x_1{}^2)$,

$$x_{\lambda+1}^3 = \min[\{x_0{}^1, x_1{}^1, \ldots, x_\omega{}^1, x_0{}^2, x_1{}^2, \ldots, x_\Omega{}^2\} \backslash \{x_0{}^3, \ldots, x_\lambda{}^3\}].$$

This is just the $*$-transform of the inductive definition of the common subdivision of two finite partitions.

Since we have already seen that the choice of $y_i{}^j$ barely matters so long as $y_i{}^j \approx z_i{}^j$, we assume $y_i{}^j = x_i{}^j$, $j = 1, 2, 3$, that is, we take the right evaluation in both sums and in the common subdivision. Consider a typical case as shown in Fig. 5.3.1 (infinite magnification), where m_i is the difference between $f(x_i{}^3)$ and $f(x_{i+1}^3)$ which is bounded by the maximum of the $*$-finite set $\{|f(x_i{}^j) - f(x_{i-1}^j)| : j = 1, 2, 3, \text{ all } i\text{'s}\}$. Call that maximum m and observe that it equals one of the $*$-finitely many numbers and thus is infinitesimal since $f(x_i{}^j) \approx f(x_{i-1}^j)$ by continuity. The difference between the first two sums is no more than if the maximum error is made at each subinterval of the common subdivision and that error is $m \cdot (b - a)$, an infinitesimal. Thus we have shown:

(5.3.2) THEOREM A (standard) *continuous real-valued function is Riemann integrable over a bounded interval* $[a, b]$.

The point of this proof is that up to an infinitesimal, the infinite sum of $f(y_i)\,\Delta x_i$ over an infinitesimal partition is independent of the particular infinite process. For discontinuous functions the sum may depend on the choice of infinitesimal partition or evaluation point, such dependent functions are simply not Riemann integrable! Stronger integrals than the Riemann integral are more conveniently discussed in the context of saturated models. Selecting a preferred partition and evaluation sequence and then declaring

$$\int_a^b f(x)\,dx = \operatorname{st}\left(\sum_{i=1}^{n} f(x_i)\,\Delta x_i\right)$$

enlarges the collection of integrable functions beyond Riemann integrable.

Now that we see that the integral of a continuous function is an infinite sum we consider a question of infinitesimal slicing: *the method of shells*. Let $f\colon [0, 1] \to \mathbf{R}^+$ be a continuous positive function and rotate the region bounded by the graph, the x-axis, the y-axis, and the line $x = 1$ about the y-axis. One calculates the volume by slicing infinitesimally along the x-axis

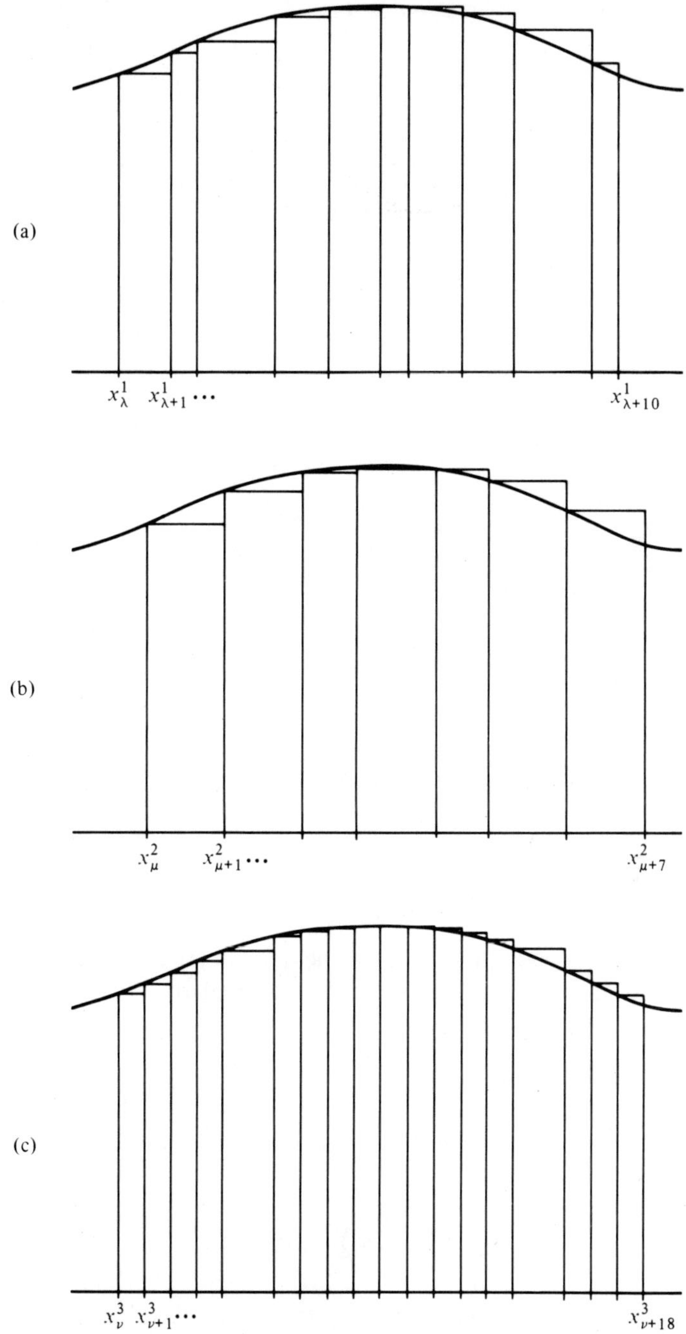

Fig. 5.3.1 Common refinement of two infinitesimal partitions. (a) First partition; (b) second partition; (c) combined partitions.

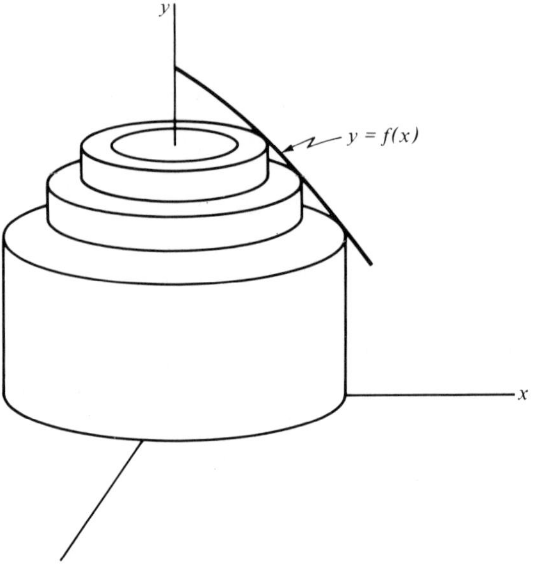

Fig. 5.3.2 The volume of the ith shell is height $\times$ π[(outer radius)^2 $-$ (inner radius)^2]; $h \times \pi[R^2 - r^2] = \Delta V_i$.

(Fig. 5.3.2). Now suppose we evaluate the height at $y_i \in [x_i, x_{i-1}]$, $h = f(y_i)$,

$$\begin{aligned}\Delta V_i &= f(y_i)\pi[(x_i + \Delta x_i)^2 - x_i{}^2] \\ &= 2\pi f(y_i)x_i\,\Delta x_i + \pi f(y_i)\,(\Delta x_i)^2.\end{aligned}$$

Then we may replace one Δx_i by the maximum m of the set $\{|\Delta x_i| : i = 1, \ldots, \omega\}$ in the second term in order to see that

$$\pi \sum_{i=1}^{\omega} f(y_i)\,(\Delta x_i)^2 \approx 0,$$

since $\int_0^1 f(x)\,dx$ exists and m times the integral is infinitesimal. Now we know that

$$\sum_{i=1}^{\omega} \Delta V_i \approx 2\pi \sum_{i=1}^{\omega} x_i\, f(y_i)\,\Delta x_i.$$

The expression $x_i\, f(y_i)$ is nearly $z_i\, f(z_i)$ for any $z_i \in [x_i, x_{i-1}]$, so by taking the $*$-finite maximum of the differences we see that $\sum_{i=1}^{\omega} \Delta V_i \approx 2\pi \sum_{i=1}^{\omega} z_i\, f(z_i)\,\Delta x_i$, which is an infinitesimal Riemann sum of the integral $2\pi \int_0^1 xf(x)\,dx$. Finally, first by taking z_i the points so that $\min[f(x) : x \in [x_i, x_{i-1}]] = f(z_i)$ and then so that $f(z_i)$ is the maximum, we see that the volume is nearly the sum $\sum_{i=1}^{\omega} \Delta V_i$. This proves

$$V = 2\pi \int_0^1 xf(x)\,dx,$$

since both sides are standard and infinitesimally close. When one fails to distinguish between "equals" and "is infinitesimally close to," the loss of the term $\pi f(x_i)\,(\Delta x_i)^2$ is somewhat mysterious—here we see that it simply sums to an infinitesimal.

(5.3.3) DEFINITION *Let f be a real-valued function defined in a neighborhood of $a \in \mathbf{R}$. If whenever $x \approx y \approx a$, we have*

$$\frac{f(x) - f(a)}{x - a} \approx \frac{f(y) - f(a)}{y - a}$$

and all such numbers are also finite, then

$$\frac{df}{dx}(a) = \operatorname{st}\left(\frac{f(x) - f(a)}{x - a}\right).$$

The point of the definition is that df/dx is nearly the ratio of an infinitesimal change in f over the corresponding infinitesimal change in x provided the ratio is finite and unique up to an infinitesimal. We saw positive examples in Chapter 2. The function

$$f(x) = \begin{cases} 0, & x = 0 \\ x \sin(\pi/x), & x \neq 0 \end{cases}$$

fails to have a derivative a $x = 0$ because when $1/x$ is an infinite integer, ω say, $\Delta f/\Delta x$ is zero, while when $1/x$ is $2\omega + \frac{1}{2}$, $\Delta f/\Delta x$ is one.

(5.3.4) ANOTHER FORMULATION OF DIFFERENTIABILITY *$f'(a)$ is the standard constant b provided that for each $x \approx a$ there is an infinitesimal ε so that $f(x) - f(a) = b \cdot (x - a) + \varepsilon \cdot (x - a)$.*

The proof of the equivalence simply amounts to writing

$$\frac{f(x) - f(a)}{x - a} = b + \varepsilon \text{ instead of } \approx b.$$

Note that ε depends on x, but it is always infinitesimal.

Another way to say this is

$$f(x) - f(a) = d \cdot (x - a)$$

where d is a finite constant nearly independent of x provided $x \approx a$; also see Section 5.6.

(5.3.5) DIFFERENTIABLE IMPLIES CONTINUOUS *If f has a derivative at $a \in \mathbf{R}$, then $f(x) - f(a) = d \cdot (x - a)$ for a finite d, therefore $f(x) \approx f(a)$ and f is continuous.*

(5.3.6) THE PRODUCT FORMULA *Let f and g be differentiable real-valued functions at $a \in \mathbf{R}$ and let $h(x) = f(x) \cdot g(x)$. Then*

$$\frac{dh}{dx}(a) = \frac{df}{dx}(a) \cdot g(a) + f(a) \cdot \frac{dg}{dx}(a).$$

Let $x \approx a$, $x \neq a$. Write the infinitesimal quotient:

$$\begin{aligned}
\frac{\Delta h}{\Delta x} &= \frac{f(x)g(x) - f(a)g(a)}{x - a} \\
&= \frac{f(x) \cdot g(x) - f(a) \cdot g(x) + f(a) \cdot g(x) - f(a) \cdot g(a)}{x - a} \\
&= \frac{(f(x) - f(a))}{x - a} \cdot g(x) + f(a) \frac{(g(x) - g(a))}{x - a} \\
&\approx \left(\frac{f(x) - f(a)}{x - a}\right) g(a) + f(a) \left(\frac{g(s) - g(a)}{x - a}\right) \approx f'(a) \cdot g(a) + f(a) \cdot g'(a).
\end{aligned}$$

So $\mathrm{st}(\Delta h/\Delta x)$ *is well defined as the last line.*

(5.3.7) CHAIN RULE *Let f and g be differentiable real-valued functions with domain and range so that $h(x) = f(g(x))$ can be formed (at least for x in a neighborhood of a). Then*

$$\frac{dh}{dx}(a) = \frac{df}{dg}(b) \cdot \frac{dg}{dx}(a)$$

where $b = g(a)$.

Let $x \approx a$ with $x \neq a$, let $y = g(x)$, and form the difference quotient

$$\frac{\Delta h}{\Delta x} = \frac{f(y) - f(b)}{x - a} = \frac{f(y) - f(b)}{y - b} \cdot \frac{y - b}{x - a} = \frac{f(y) - f(b)}{y - b} \cdot \frac{g(x) - g(a)}{x - a},$$

unless $y = b$ in which case $\Delta h = 0$. Since g is continuous, $y \approx b$ and $\Delta h/\Delta x \approx f'(b) \cdot g'(a)$.

5.4 THE MEAN VALUE THEOREM

We give a proof of the mean value theorem. The proof given here carries over to a very general setting (a complete convex metric space), whereas the usual proof does not.

(5.4.1) MEAN VALUE THEOREM *Let $f\colon [a, b] \to \mathbf{R}$ be continuous and differentiable on (a, b). There exists $c \in (a, b)$ so that*

$$\frac{f(b) - f(a)}{b - a} = f'(c).$$

PROOF: Let $n \geq 3$ be a natural number and define $g\colon [1/n, 1] \to \mathbf{R}$ by

$$g(t) = f(a + t[b - a]) - f(a + [t - (1/n)][b - a]).$$

The telescoping sum

$$f(b) - f(a) = \sum_{k=1}^{n} g(k/n)$$

gives us that

$$\frac{f(b) - f(a)}{b - a} = \frac{1}{n} \sum_{k=1}^{n} \frac{g(k/n)}{(b - a)/n}.$$

The right-hand side is an average of n numbers so that either they are all equal or one term is below the average and another is above the average. In the latter case apply Bolzano's intermediate value theorem to the term $g(t)/((b - a)/n)$ over the interval in question; if $g(x)/((b - a)/n)$ equals the intermediate value $(f(b) - f(a))/(b - a)$, then let $a_n = a + (x - (1/n))(b - a)$ and $b_n = a + x(b - a)$ so that there exist points $a_n < b_n$ in (a, b) with $b_n - a_n = (b - a)/n$ and

$$\frac{f(b_n) - f(a_n)}{b_n - a_n} = \frac{f(b) - f(a)}{b - a}.$$

The a_n and b_n here cannot be endpoints. Otherwise we can simply take

$$a_n = a + \frac{(k - 1)}{n}(b - a) \qquad \text{and} \qquad b_n = a + \frac{k}{n}(b - a)$$

for any k other than the end intervals—all the terms are equal in this case so any k between 2 and $n - 1$ works.

Apply this argument with $n = 3$ obtaining

$$a < a_3 < b_3 < b, \qquad b_3 - a_3 = (b - a)/3,$$

and

$$\frac{f(b) - f(a)}{b - a} = \frac{f(b_3) - f(a_3)}{b_3 - b_3}.$$

Proceed by induction applying this process on the interval $[a_n, b_n]$ to obtain sequences (a_n) and (b_n) so that "for every $n \in \mathbf{N}$ with $n > 3$,

(i) $a < a_n < a_{n+1} < b_{n+1} < b_n < b$,
(ii) $(b_{n+1} - a_{n+1}) = (b_n - a_n)/(n+1) \leq (b-a)/n$,

and

(iii) $$\frac{f(b)-f(a)}{b-a} = \frac{f(b_n)-f(a_n)}{b_n - a_n}\text{''}.$$

Now apply the $*$-transform of the last statement to the $*$-sequences $(a_n : n \in {}^*\mathbf{N})$ and $(b_n : n \in {}^*\mathbf{N})$ and to an infinite $\omega \in {}^*\mathbf{N}_\infty$. This says that $b_\omega \approx a_\omega$ and using the facts (proved below) that $a_\omega \leq \operatorname{st}(a_\omega) = c \leq b_\omega$, that c is finitely inside $[a, b]$, and that

$$f'(c) \approx \frac{f(b_\omega)-f(c)}{b_\omega - c} \approx \frac{f(c)-f(a_\omega)}{c - a_\omega}$$

(by the differentiability assumption) we see that

(A) $$\frac{f(b)-f(a)}{b-a} \approx f'(c),$$

because

$$\begin{aligned}\frac{f(b_\omega)-f(a_\omega)}{b_\omega - a_\omega} &= \frac{f(b_\omega)-f(c)+f(c)-f(a_\omega)}{b_\omega - a_\omega}\\ &= \frac{b_\omega - c}{b_\omega - a}\,\frac{f(b_\omega)-f(c)}{b_\omega - c} + \frac{c-a_\omega}{b_\omega - a_\omega}\,\frac{f(c)-f(a_\omega)}{c-a_\omega}\\ &= \frac{b_\omega - c}{b_\omega - a_\omega}[f'(c)+\varepsilon] + \frac{c-a_\omega}{b_\omega - a_\omega}[f'(c)+\delta]\\ &= f'(c) + \varepsilon\frac{b_\omega - c}{b_\omega - a_\omega} + \delta\frac{c-a_\omega}{b_\omega - a_\omega},\end{aligned}$$

where ε and δ are infinitesimal and both $(b_\omega - c)/(b_\omega - a_\omega)$ and $(c - a_\omega)/(b_\omega - a_\omega)$ lie between zero and one. Finally, since both sides of (A) are standard numbers if they are within an infinitesimal they must therefore be equal. This proves the theorem once we have the following:

(5.4.2) LEMMA *Let* $(a_n : n \in \mathbf{N})$ *be a standard increasing bounded sequence* (respectively, $b_n\!\downarrow$). *Then* $\lim_{n\to\infty} a_n = \operatorname{st}(a_\omega) \geq a_\omega$ (respectively, $\lim_{n\to\infty} b_n = \operatorname{st}(b_\omega) \leq b_\omega$) *for any infinite* $\omega \in {}^*\mathbf{N}_\infty$. (Here a_ω and b_ω denote the nonstandard extensions at ω.) *Moreover, strict inequality holds when the sequence is strictly increasing* (respectively, decreasing).

PROOF: Let $(a_n : n \in \mathbf{N})$ satisfy $a_n \le a_{n+1} \le B \in \mathbf{R}$, so it follows that when $m \le n$ in $\mathbf{N}$, $a_m \le a_n$. Then by the $*$-transform with ω infinite

$$a_m \le a_\omega \le B \qquad \text{for any finite} \quad m.$$

In particular, a_ω is finite and st $a_\omega = c$ exists.

Now $a_m \le c$ since $a_m \le a_\omega \approx c$ and both c and a_m are standard, ruling out $a_m \approx c$ unless $a_m = c$. Since m is an arbitrary finite index we actually have that "for every $m \in \mathbf{N}$, $a_m \le c$"; therefore by the $*$-transform $a_\lambda \le c$ for all $\lambda \in {}^*\mathbf{N}$.

Let $\varepsilon \in {}^\sigma R^+$. All the subscripts $\lambda \ge \omega$ satisfy $|c - a_\lambda| < \varepsilon$ so the statement "There exists an $n \in {}^*\mathbf{N}$ such that for every $k \ge n$, $|c - a_k| < \varepsilon$" is true in ${}^*\mathcal{R}$. The inverse $*$-transform shows that the limit is c since $\varepsilon \in {}^\sigma\mathbf{R}^+$ is arbitrary.

(5.4.3) INTERMEDIATE VALUE THEOREM FOR DERIVATIVES *Let f have a derivative f' everywhere on the interval $[a, b]$. Then f' takes on every value between $f'(a)$ and $f'(b)$ on the interval (a, b).*

PROOF: The two difference quotients

$$\frac{f(x) - f(a)}{x - a} \qquad \text{and} \qquad \frac{f(b) - f(x)}{b - x}$$

extend continuously to $[a, b]$. The first ranges between $[f(b) - f(a)]/(b - a)$ and $f'(a)$ while the second ranges between $[f(b) - f(a)]/(b - a)$ and $f'(b)$ by Bolzano's theorem. In any event, one of the two attains every value between $f'(a)$ and $f'(b)$ in the open interval. The mean value theorem in turn says $[f(t) - f(s)]/(t - s)$ is attained between t and s.

This proves Darboux's theorem, but let us consider the situation where $f'(x)$ is discontinuous at c. Discontinuity means that "there is an $\varepsilon \in {}^\sigma\mathbf{R}^+$ so that on any neighborhood of c, $f'(x)$ is at least ε from $f'(c)$." A jump discontinuity cannot occur by Darboux's theorem, but in fact we must have severe oscillation.

(5.4.4) COROLLARY *Let f be standard. Let $f'(x)$ be defined on a neighborhood of c, but be discontinuous at $c \in {}^\sigma\mathbf{R}^+$. On any infinitesimal neighborhood of c, $f'(x)$ has infinite variation, that is, there exists a $*$-finite sequence $x_1, \ldots, x_\omega$ all infinitesimally close to c satisfying $\sum_{i=1}^{\omega} |f(x_{i+1}) - f(x_i)| \in {}^*\mathbf{R}_\infty{}^+$.*

PROOF: Apply the $*$-transform of the statement above in quotes to the infinitesimal neighborhood. This says there is a point $x_1 \approx c$ so that $|f'(x_1) - f'(c)| \geq \varepsilon$ ($\in {}^{\sigma}\mathbf{R}^+$). By Darboux's theorem there exists x_2 between x_1 and c so that $|f'(x_1) - f'(x_2)| \geq \varepsilon/2$. Now, by the $*$-transform of the ε-discontinuity there is an x_3 between x_2 and c so that $f'(x_3)$ is greater than ε from $f'(c)$ and by Darboux's theorem an x_4 between x_3 and c so that $|f'(x_3) - f'(x_4)| \geq \varepsilon/2$. Continuing this by $*$-induction Ω times to the infinite number $\omega = 2\Omega$, we know then that

$$\sum_{k=1}^{\Omega} |f(x_{2k+1}) - f(x_{2k})| \geq \Omega\varepsilon/2,$$

which is infinite.

5.5 THE FUNDAMENTAL THEOREM OF CALCULUS

Suppose f: $[a, b] \to \mathbf{R}$ is continuous so that we can form its Riemann integral. Suppose also that we can find a function F: $[a, b] \to \mathbf{R}$ so that $dF/dx = f$. Then

$$\int_a^b f(x)\, dx = \int_a^b dF/dx\, dx = \int_a^b dF$$

and the right-hand term telescopes to $F(b) - F(a)$. Thus finding antiderivatives permits us to evaluate integrals. We now investigate the telescoping more carefully. First, the integral of f is nearly an infinite sum over an infinitesimal partition of terms $f(y_i)\,\Delta x_i$ for arbitrary $y_i \in [x_i, x_{i-1}]$. Now dF/dx is nearly $[F(x_i) - F(x_{i-1})]/\Delta x_i$, and the mean value theorem says we can select $y_i \in (x_i, x_{i-1})$ so that we have equality

$$\int_a^b f(x)\, dx \approx \sum_{i=1}^{\omega} f(y_i)\Delta x_i = \sum_{i=1}^{\omega} \Delta F_i/\Delta x_i\, \Delta x_i.$$

The latter sum actually telescopes to the standard number $F(b) - F(a)$ by cancellation of Δx_i's. (We could avoid the use of the mean value theorem here by investigating the infinitesimal difference quotients $\Delta F/\Delta x$ when dF/dx is continuous; see Theorems (5.7.4) and (5.7.9).)

The other half of the fundamental theorem considers the function $F(x) = \int_a^x f(t)\, dt$, when f is continuous at c and (Riemann integrable or simply) when f is continuous on $[a, b]$ containing $x \in {}^{\sigma}\mathbf{R}$.

By transfer of the facts that (1) $f \leq g$ implies $\int_a^b f(x)\, dx \leq \int_a^b g(x)\, dx$, and (2) $\int_a^b f(x)\, dx = \int_a^c f(x)\, dx + \int_c^b f(x)\, dx$ using $b = c + \Delta x$, where Δx is a nonzero infinitesimal, we see that $F(c + \Delta x) - F(c)$ lies between (Δx)

$\min[f(x) : x \in [c, c + \Delta x]]$ and $(\Delta x) \max[f(x) : x \in [c, c + \Delta x]$. The difference between that max and min (both are attained by transfer of (5.1.9) to the internal closed bounded interval) is infinitesimal by continuity of f at c; so $\Delta F/\Delta x \approx f(c)$, which proves that

$$\frac{d}{dx}\int_a^x f(t)\,dt\bigg|_{x=c} = f(c).$$

The telescoping of the integral of a derivative and the fact that the derivative of an integral is the integrand again form the fundamental connection between derivatives and integrals. Precisely, we have shown

(5.5.1) FUNDAMENTAL THEOREM OF CALCULUS *Let $f: [a, b] \to \mathbf{R}$ be continuous.*

(1) *If $F: [a, b] \to \mathbf{R}$ satisfies $dF/dx = f$, then*

$$\int_a^b f(x)\,dx = F(b) - F(a).$$

(2) *If $F(x) = \int_a^x f(t)\,dt$, and $c \in [a, b]$, then*

$$dF(c)/dx = f(c).$$

Part (1) of the theorem led Leibniz to the notation $\int f(x)\,dx$ for *any* function $F(x)$ whose derivative is $f(x)$. By the **mean value theorem** we know

$$F(x) = \int f(x)\,dx + C, \qquad \text{where} \quad C \text{ is constant,}$$

that is, on an interval, any two functions with the same derivative differ at most by a constant.

EXERCISE: Prove the standard properties (1) and (2) used in proving that the derivative of the integral is the integrand.

From the product formula (5.3.6) we know that $h'(x) = f(x)g'(x) + f'(x)g(x)$ when $h(x) = f(x)g(x)$ and both are differentiable. If we also assume f and g have continuous derivatives, then the indefinite integrals are related by $\int f(x)g'(x)\,dx = f(x)g(x) - \int f'(x)g(x)\,dx$ so we have:

(5.5.2) INTEGRATION BY PARTS *If f and g have continuous derivatives on an interval containing $[a, b]$, then*

$$\int_a^b f(x)g'(x)\,dx = f(b)g(b) - f(a)g(a) - \int_a^b f'(x)g(x)\,dx.$$

5.6 LANDAU'S "OH-CALCULUS"

This is an appropriate place to give the connection between our o and $\mathcal{O}$ and the "oh-calculus."

(5.6.1) THEOREM *Let f, g, and h be standard functions. Then*

(a) $f(x) = g(x) + o(h(x))$ *as $x \to y$ in the sense of Landau if and only if for each $\xi \approx y$ and $\xi \neq y$, $f(\xi) = g(\xi) + h(\xi) \cdot \varepsilon$ for some $\varepsilon \in o$, that is,*

$$f(\xi) = g(\xi) + o \cdot h(\xi) \qquad \text{for} \quad \xi \text{ near } y.$$

(b) $f(x) = g(x) + \mathcal{O}(h(x))$ *as $x \to y$ in the sense of Landau if and only if for each $\xi \approx y$ and $\xi \neq y$, $f(\xi) = g(\xi) + h(\xi) \cdot M$ for some $M \in \mathcal{O}$, that is,*

$$f(\xi) = g(\xi) + \mathcal{O} \cdot h(\xi) \qquad \text{for} \quad \xi \text{ near } y.$$

(In case the approximation is for $x \to \infty$, read "for each infinite ξ.")

PROOF: In Landau's notation, $f(x) = g(x) + o(h(x))$ means that $\lim_{x\to y} [f(x) - g(x)]/h(x) = 0$. This is equivalent to $[f(\xi) - g(\xi)]/h(\xi) = \varepsilon \approx 0$ by the nonstandard characterization of limit. ($\lim_{x\to y} F(x) = a$ iff for every ξ near y, $F(\xi) \approx a$; see Exercise 1 of Section 5.1.)

Similarly, Landau's $\mathcal{O}$-formula holds if and only if

$$\left|\frac{f(x) - g(x)}{h(x)}\right| \leq M \qquad \text{for some} \quad M \in \mathbf{R} \quad \text{and} \quad |x - y| < \varepsilon.$$

One way, the $*$-transform says the same M works since we have $|\xi - y| < \varepsilon$.

Assume the nonstandard formula. Consider the standard binary relation entity:

$$^*\Phi = \{(\varepsilon, M) : x \neq y \;\&\; |x - y| < \varepsilon \in {}^*\mathbf{R}^+ \text{ implies } |(f(x) - g(x))/h(x)| < M\}.$$

Then dom $^*\Phi \supseteq o$ and rng $^*\Phi \subseteq \mathcal{O}$ by the nonstandard assumption. Since dom $^*\Phi$ and rng $^*\Phi$ must also be standard sets, we conclude that dom Φ contains standard ε's and rng $^*\Phi$ is bounded. This proves the theorem.

Now that we have the "small oh" calculus we reiterate the alternate form of differentiability (5.3.4). For real-valued f defined in a neighborhood of $a \in \mathbf{R}$, *f is differentiable at a if and only if for every $x \approx a$ in $^*\mathbf{R}$, there is a standard constant $f'(a)$ so that*

$$f(x) - f(a) = f'(a) \cdot (x - a) + o \cdot (x - a)$$

or for all $dx \approx 0$,

$$f(a + dx) - f(a) = df(a)\, dx + o\, dx.$$

Consider the following natural generalization. Suppose that there is a standard constant $f'(a)$ so that whenever $x \approx y \approx a$ we know that

$$f(y) - f(x) = f'(a) \cdot (y - x) + (y + x) \cdot o.$$

For example, consider the functions $g(x) = x^2$ and $f(x) = x^2 \sin(\pi/x)$ extended to be zero at zero. In both cases, the derivative at zero is zero. Let ω be an infinite natural number and take x and y so that $(1/x) = 2\omega$, while $1/y = 2\omega - \frac{1}{2}$.

EXERCISE: Show that $g(y) - g(x) = (y - x) \cdot \varepsilon$ where $\varepsilon \approx 0$. Show that $f(y) - f(x) = (y - x) \cdot (k)$ where $k \approx \frac{1}{2}$.

This condition is equivalent to continuity of f' at a when $f'(x)$ exists in a neighborhood of a and we prove this for more general functions in the next section.

EXERCISE: Explain the formulas that follow in terms of infinitesimals:

(1) $1/(1 - x) = 1 + x + x^2 + o(x^2)$.
(2) $\log(1 + x) = x - (x^2/2) + (x^3/3) + o(x^3)$.
(3) $\sin(x) = x - (x^3/3!) + (x^5/5!) + o(x^6)$.

EXERCISE: Expand $x^2 + x + 1$ in powers of $(x - 1)$ by use of Taylor's theorem and the fact that quadratic polynomials differing by $o(x^2)$ are equal. Using polynomial uniqueness, give the Taylor approximation for $\sin^2 x$, which has a small error even compared with x^5 near zero.

EXERCISE: Show the connection between the algebra of o and $\mathcal{O}$ from the last chapter and the Landau calculus. To begin with, show that for standard f and g when x is near $a \in {}^\sigma\mathbf{R}$:

(1) $o(g(x)) \pm o(g(x)) = o(g(x))$;
(2) $o(cg(x)) = o(g(x))$ for $c \in \mathcal{O}$;
(3) $f(x) \cdot o(g(x)) = o(f(x)g(x))$;
(4) $o(o(g(x))) = o(g(x))$;
(5) $1/(1 + g(x)) = 1 - g(x) + o(g(x))$ if $g(x) \approx 0$ when $x \approx a$.

Devise similar properties for big oh.

EXERCISES

1. Apply the infinitesimal definition of integral to calculation of the arc length of a smooth curve, e.g., the graph $y = f(x)$ for continuously differentiable f. First, observe that even "in the small" we must account for Δx and Δy in the length $\Delta l \approx [(\Delta x)^2 + (\Delta y)^2]^{1/2}$. Then show that this leads to the formula $\int_a^b dl = \int_a^b [(dx)^2 + (dy)^2]^{1/2}$.
2. Derive the formula for surface area of a surface of revolution.

5.7 DIFFERENTIAL VECTOR CALCULUS

This section is a brief account of infinitesimal vector calculus. We will revisit the results given here for finite-dimensional spaces in Part 2 on infinitesimals in functional analysis. Many of the ideas carry over to the infinite-dimensional case intact, and we stress a general approach in our discussion of vector calculus results. We first give, however, the essential vector ingredients in the particular finite-dimensional case.

In the remainder of this section, **E** will denote the standard m-dimensional Euclidean space and **F** the standard n-dimensional space. The dimensions are fixed standard numbers. We need a description of $^*\mathbf{E}$ and $^*\mathbf{F}$, so we begin with a standard description. "Choose a set of m orthonormal vectors $e_1, \ldots, e_m$ and associate the m-tuple $(x^1, \ldots, x^m) \in \mathbf{R}^m$ with the vector $\sum_{j=1}^m x^j e_j$." Thus, by $*$-transform $^*\mathbf{E}$ is associated to $^*\mathbf{R}^m$—a point in $^*\mathbf{E}$ "is" an m-tuple of hyperreals. We now need to say which vectors are finite and infinitesimal. By extension, if $X = (x^1, \ldots, x^m)$ and $Y = (y^1, \ldots, y^m)$, we have

$$\langle X, Y \rangle = \sum_{j=1}^{m} x^j y^j \qquad \text{and} \qquad |X| = \left[\sum_{j=1}^{m} (x^j)^2\right]^{1/2}.$$

This gives us notions of distance and angle and we define

$$\mathcal{O}(\mathbf{E}) = \{X \in {}^*\mathbf{E} : |X| \text{ is a finite hyperreal}\},$$
$$o(\mathbf{E}) = \{X \in {}^*\mathbf{E} : |X| \approx 0\}.$$

EXERCISE: (a) Show that $\mathcal{O}(\mathbf{E}) = [\mathcal{O}(\mathbf{R})]^m$ with respect to the orthonormal basis $e_1, \ldots, e_m$, that is, the finite vectors are represented by m-tuples of finite scalars.

(b) Two vectors $X = (x^1, \ldots, x^m)$ and $Y = (y^1, \ldots, y^m)$ are infinitesimally close, $|X - Y| \approx 0$, if and only if $x^j \approx y^j$, $j = 1, \ldots, m$.

(c) Show that the factorization $\mathcal{O}^m/\approx$ is compatible with the linear structure. $\mathcal{O}^m$ is an $^\sigma\mathbf{R}$-module and a $\mathcal{O}$-module; you can do $\mathcal{O}$-linear algebra in $\mathcal{O}^m$ and ignore infinitesimal differences and obtain the same answer as factoring first and doing **R**-linear algebra in $\mathbf{R}^m$ (see Section 9.1).

"With respect to ordered bases $(e_1, \ldots, e_m)$ and $(f_1, \ldots, f_n)$ a linear transformation $\mathscr{A}: \mathbf{E} \to \mathbf{F}$ is represented by the matrix $A = [a_j{}^i]$, where

$$\mathscr{A}(e_i) = \sum_{j=1}^{n} a_i^j f_j;$$

moreover when X is represented by $(x^1, \ldots, x^m)$ the answer to $\mathscr{A}(X) = Y$

is computed by the matrix product

$$y^i = \sum_{j=1}^{m} a_j^{\ i} x^j."$$

The $*$-transpose simply says $*\mathscr{A}$ is represented by the matrix $[{}^*a_j^{\ i}]$. For example,

$$\begin{bmatrix} 1 & 2 & 3 \\ 2 & 4 & 1 \end{bmatrix} \begin{bmatrix} \omega \\ 1/\omega \\ \varepsilon \end{bmatrix} = \begin{bmatrix} \omega + 2/\omega + 3\varepsilon \\ 2\omega + 4/\omega + \varepsilon \end{bmatrix}.$$

Here $m = 3$ and $n = 2$, ω is infinite, and $1/\omega$ and ε infinitesimal.

Let Lin(**E**, **F**) denote the space of linear maps from **E** to **F**. This forms a vector space and a natural description of finite and infinitesimal maps is: $A \in {}^*\text{Lin}({}^*\mathbf{E}, {}^*\mathbf{F})$.

A is finite provided A takes finite vectors of $\mathcal{O}(\mathbf{E})$ to finite vectors of $\mathcal{O}(\mathbf{F})$.

A is infinitesimal provided A takes $\mathcal{O}(\mathbf{E})$ to $o(\mathbf{F})$.

The uniform operator norm of a linear map is defined by

$$\|A\| = \sup[|Ax| : |x| \leq 1 \text{ in } \mathbf{E}],$$

the radius of the ball containing the image of the unit ball of **E**.

EXERCISE: Let **E** and **F** be finite-dimensional vector spaces. Let $A \in {}^*\text{Lin}({}^*\mathbf{E}, {}^*\mathbf{F})$ and show that the following are equivalent:

(1) A is a finite linear transformation, that is, takes $\mathcal{O}(\mathbf{E})$ to $\mathcal{O}(\mathbf{F})$.
(2) $\|A\|$ is a finite scalar.
(3) A maps $o(\mathbf{E})$ to $o(\mathbf{F})$, that is, A is S-continuous at zero.
(4) A is S-continuous everywhere on $^*\mathbf{E}$.
(5) Let $(e_1, \ldots, e_m)$ and $(f_1, \ldots, f_n)$ be standard bases for **E** and **F**. Let $[a_j^{\ i}]$ be the matrix of A with respect to $^*e_1, \ldots, {}^*e_m; {}^*f_1, \ldots, {}^*f_n$. Each scalar $a_j^{\ i}$ is finite.

How would you define the standard part of a finite transformation?

EXERCISE: Give several equivalent forms of the statement that A is an infinitesimal transformation.

(5.7.1) THEOREM *Let $m \in {}^\sigma\mathbf{N}$ be a fixed finite dimension. If A and B are nonstandard finite $m \times m$ matrices which almost commute,* i.e., *$AB \approx BA$, then there exist commuting matrices $\hat{A}$ and $\hat{B}$ near A and B, $A \approx \hat{A}$, $B \approx \hat{B}$, and $\hat{A}\hat{B} = \hat{B}\hat{A}$. Almost commuting finite matrices are near standard matrices which commute.*

PROOF: Since A and B are finite, each component $a_j^{\ i}$ and $b_j^{\ i}$ is finite, so take $\hat{A} = [\mathrm{st}(a_j^{\ i})]$ and $\hat{B} = [\mathrm{st}(b_j^{\ i})]$. The commutator of $\hat{A}\hat{B}$ is $\hat{A}\hat{B} - \hat{B}\hat{A} = \hat{A}(\hat{B} - B) + (\hat{A} - A)B + (AB - BA) + B(A - \hat{A}) + (B - \hat{B})A$ and each term on the right-hand side is a finite times an infinitesimal. Hence $\hat{A}\hat{B} - \hat{B}\hat{A}$ is a standard infinitesimal, in other words, zero.

EXERCISE: Give a standard formulation of this theorem (see Rosenthal [1969], Luxemburg and Taylor [1970], and Lomonozov [1973]).

Now we give some nonlinear maps in coordinates:

(f): $(m = 1, n = 3)$ $\quad f(t) = (\cos t, \sin t, t/\pi)$,

(g): $(m = 2, n = 1)$ $\quad g(x, y) = (x^2 y)^{1/3}$,

(h): $(m = n = 2)$ $\quad h(r, \theta) = (r \cos \theta, r \sin \theta)$,

(β): $(m = 2, n = 1)$ $\quad \beta(x, y) = \begin{cases} 0 & \text{if } x = y = 0 \\ 2xy/(x^2 + y^2) & \text{if } (x, y) \neq 0, \end{cases}$

(γ): $(m = 2, n = 1)$ $\quad \gamma(x, y) = \begin{cases} 0 & \text{if } x = y = 0 \\ 2xy^2/(x^2 + y^4) & \text{if } (x, y) \neq (0, 0). \end{cases}$

The nonstandard extensions of these functions have the $*$-same descriptions and we can begin to ask questions like: What happens to the value of the function when the variable is moved infinitesimally? We deal with this question in the context of the examples given only in the following exercise, the reader should consult (5.1.1), (5.2.1), (6.10.1), (8.3.1), and (8.4.23) for generalities on continuity. We shall take up the question: What is the change in f like compared with an infinitesimal change in x?

EXERCISE: Show that f, g, and h are continuous at zero by showing $f(\varepsilon) \approx f(0)$, etc., for an arbitrary infinitesimal $\varepsilon \in {}^*\mathbf{E}$. Show that $\beta(\delta, \delta) \not\approx 0$ when $\delta \approx 0$ and $\neq 0$. Is γ continuous at zero?

(5.7.2) DEFINITION *Let $f: U \subseteq \mathbf{E} \to \mathbf{F}$ be a map between vector spaces. Then f is differentiable at $a \in U$ if there is a linear map $L_a: \mathbf{E} \to \mathbf{F}$ such that whenever $x \approx a$ in ${}^*\mathbf{E}$,*

$$(1) \quad f(x) = f(a) + L_a(x - a) + |x - a| \cdot \eta$$

for $\eta \approx 0$ in ${}^*\mathbf{F}$ or in other notation

$$(1') \quad f(x) = f(a) + L_a(x - a) + o \cdot |x - a|.$$

f is uniformly differentiable at $a \in \mathbf{E}$ if whenever $x \approx a$ and $y \approx a$ in ${}^*\mathbf{E}$,

$$(2) \quad f(x) - f(y) = L_a(x - y) + |x - y| \cdot \eta$$

for infinitesimal η in ${}^*\mathbf{F}$. The linear map L_a is denoted Df_a when it exists.

We can one-dimensionalize differentiability conditions by restricting the function to a line in **E**: If v is a unit vector, let $F(t) = f(a + tv)$. The derivative of F is called *the directional derivative of f in the direction of v.*

EXERCISE: Show by direct calculation that Dh from the above examples is represented by the matrix

$$Dh_{(r,\,\theta)} = \begin{bmatrix} \cos\theta & -r\sin\theta \\ \sin\theta & r\cos\theta \end{bmatrix}$$

so the approximate change $h(r + \rho, \theta + \varphi) - h(r, \theta)$ is

$$(\rho\cos\theta + r\varphi\sin\theta, \rho\sin\theta + r\varphi\cos\theta).$$

Recall what differentiability means for the coordinates in terms of infinitesimals and apply the addition formulas.

EXERCISE: Show that the directional derivatives of β and γ exist for $v = e_1$ and e_2. What happens when $v = (1/\sqrt{2})(e_1 + e_2)$? An arbitrary unit v? Show that neither of the functions is differentiable at zero and that the directional derivatives do not exist uniformly.

EXERCISE: If f is differentiable, show that the directional derivative is given by $(\Delta t) \cdot Df_a(v)$ as a linear function of Δt. If the directional derivatives are uniform, show that Df_a exists uniformly.

REMARK: Notice that Definition (5.7.2) is an infinitesimal but **standard** definition, that is, it refers to standard L_a, f, and a. We leave the ordinary standard definition in terms of epsilon and delta to the reader (see (5.6.1)). The operator $f \mapsto Df$ is an element of the superstructure $\mathscr{R}$ and thus has a nonstandard extension to the $*$-differentiable internal functions. Definition (5.7.2) is not the condition for $*$-differentiability of an internal function (see (5.7.5)).

Notice that our definition of "differentiable at a" includes the hypothesis that a is interior to the domain of f. (Simply transfer the statement "there exists r so that $\{x : |x - a| < r\}$ is contained in the domain of f" to the standard model.)

Another definition of Df_a goes as follows. Choose an infinitesimal scalar $\delta \neq 0$. Define the map δf_a on $\mathcal{O}(\mathbf{E})$ by taking $\delta f_a(x) = [f(a + \delta x) - f(a)]/\delta$. We want to factor the diagram $\mathcal{O}(\mathbf{E}) \xrightarrow{\delta f_a} \mathcal{O}(\mathbf{F})$ by $\approx$ and obtain a linear map from **E** to **F**; this means δf_a must be almost linear and into $\mathcal{O}(\mathbf{F})$. Last, after factoring $\mathcal{O}(\mathbf{E}) \xrightarrow{\delta f_a} \mathcal{O}(\mathbf{F})$ to $\mathbf{E} \xrightarrow{Df_a} \mathbf{F}$ our answer should be independent of the particular infinitesimal δ. (Note: δf_a is almost linear if whenever $\alpha, \beta \in \mathcal{O}$ and $x, y \in \mathcal{O}(\mathbf{E})$, $\delta f_a(\alpha x + \beta y) \approx \alpha(\delta f_a(x)) + \beta(\delta f_a(y))$ in $\mathcal{O}(\mathbf{F})$.) The reader may wish to apply this version to the calculation of Dh and in tangent spaces below.

Consider the function γ given above at $a = 0$. For $x = (\delta, 0)$ and $y = (0, 1)$ we have $\delta\gamma_0(x) = \delta\gamma_0(y) = 0$. When $z = x + y = (\delta, 1)$ we have $\delta\gamma_0(x + y) = 2\delta^2\delta^2/((\delta^2)^2 + (\delta)^4) = 1 \not\approx 0 + 0$, so $\delta\gamma_0$ is *not* almost linear and therefore not differentiable.

EXERCISE ON PRODUCT FORMULAS: Use infinitesimals to prove the following formulas (f, g: $\mathbf{E} \to \mathbf{F}$, α: $\mathbf{E} \to \mathbf{R}$, and $\langle\ ,\ \rangle$ bilinear).

(1) $D\langle f, g\rangle_a(h) = \langle Df_a(h), g(a)\rangle + \langle f(a), Dg_a(h)\rangle$.
(2) $D(\alpha f)_a(h) = D\alpha_a(h)f(a) + \alpha(a)Df_a(h)$.

(5.7.3) CHAIN RULE *Let g: $U \subseteq \mathbf{E} \to \mathbf{F}$ and f: $V \subseteq \mathbf{F} \to \mathbf{G}$ be maps so that the composite $h(x) = f(g(x))$ is defined. Let g be differentiable at a. Let f be differentiable at $b = g(a)$. Then $D(f \circ g)_a = Df_b \circ Dg_a$.*

DEMONSTRATION: Let $x \approx a$. By definition of Dg_a we know in particular $g(x) \approx b$ so

$$\begin{aligned} f(g(x)) - f(g(a)) &= Df_b(g(x) - g(a)) + |g(x) - g(a)| \cdot \eta \\ &= Df_b(Dg_a(x - a) + |x - a| \cdot \xi) \\ &\quad + |Dg_a(x - a) + |x - a| \cdot \xi| \cdot \eta \\ &= (Df_b \circ Dg_a)(x - a) + |x - a| \cdot \varepsilon, \\ &\quad \text{where } \varepsilon, \xi, \eta \approx 0 \quad \text{in} \quad {}^*\mathbf{G}. \end{aligned}$$

The term $|Dg_a(x - a) + (|x - a| \cdot \xi)| \le \|Dg_a\|\,|x - a| + |x - a| \cdot |\xi| \approx 0$, where $\|Dg_a\| = \sup[|Dg_a(z)| : |z| \le 1 \text{ in } {}^*\mathbf{E}]$, the operator norm. Now the finite scalar $\|Dg_a\| + |\xi|$ times η is an infinitesimal vector ε.

(5.7.4) THEOREM *Let f be an internally differentiable function* (i.e., lie in the extended set of differentiable functions). *Suppose f is defined for all $x \approx a$ in $^*\mathbf{E}$ and Df_a is a finite transformation. The mapping $x \mapsto Df_x$ is S-continuous at a if and only if whenever $z \approx a$ and $w \approx a$,*

$$f(z) - f(w) = Df_a(z - w) + |z - w| \cdot \eta$$

for some $\eta \approx 0$ in $^*\mathbf{F}$.

(5.7.5) DEFINITION *Let f: $U \to {}^*\mathbf{F}$ be an internal function defined for all $x \approx a$ in $^*\mathbf{E}$. We say f is S-uniformly differentiable at a if there is a finite linear map Df_a such that whenever $z \approx w \approx a$,*

$$f(z) - f(w) = Df_a(z - w) + |z - w| \cdot \eta$$

for some infinitesimal $\eta \in {}^*\mathbf{F}$ [in other words, when (5.7.4) holds at a alone].

We feel that the importance of this idea is not to be underestimated—it lies at the heart of the fundamental theorem, the change of variables theorem, Gauss's theorem, the inverse and implicit function theorems, and so on. Also see (5.7.9) with $k = 1$.

(5.7.6) COROLLARY *A standard differentiable f is continuously differentiable at $a \in \mathbf{E}$ if and only if f is uniformly differentiable at a.*

PROOF: First, the corollary simply specializes the theorem to the standard setting. S-continuity means $Df_x \approx Df_a$ as maps, or when d is a finite vector $Df_x(d) \approx Df_a(d)$.

Fix $w \approx a$ and consider values of z along a line in the direction of a unit vector d. By the $*$-Landau small oh differentiability condition we know that by taking z close enough to w we can make ξ infinitesimal where

$$f(z) - f(w) = Df_w(z - w) + |z - w| \cdot \xi.$$

Letting $Df_a(d) + \zeta = Df_w(d)$ we have

$$f(z) - f(w) = |z - w|[Df_w(d) + \xi] = |z - w|[Df_a(d) + \zeta + \xi],$$

and by the nonstandard condition $\eta = (\zeta + \xi)$ is infinitesimal forcing ζ to be infinitesimal. This proves S-continuity of Df_x.

Conversely, suppose Df_x is S-continuous at a so that the function

$$g(x) = f(x) - Df_a(x - a)$$

has an infinitesimal derivative near a

$$Dg_x = Df_x - Df_a.$$

By our next result then

$$|f(z) - f(w) - Df_a(z - w)| = |g(z) - g(w)| \leq |Dg_x(z - w)|,$$

and this shows that S-continuity implies the infinitesimal condition. The vector $\eta = Dg_x((z - w)/|z - w|)$ is infinitesimal and $|z - w| \cdot \eta$ is more than the difference between the change in f and its linear part.

(5.7.7) A NORM MEAN VALUE THEOREM *Suppose F is continuous on the closure* $\mathrm{cl}(U)$ *of a convex set $U \subseteq \mathbf{E}$ and differentiable on U. For any $a, b \in \mathrm{cl}(U)$, there exists $c \in U$ so that*

$$|F(b) - F(a)| \leq |DF_c(b - a)| \leq \|DF_c\| \cdot |b - a|.$$

PROOF: We shall apply the argument in the proof of (5.4.1) along the segment from a to b using the triangle inequality on the average of the n differences. Let $f(t) = F(a + t(b - a))$, $0 \leq t \leq 1$. The chain rule tells us that $DF_c(b - a)$ is the derivative of f.

Let $n \geq 3$ and define $g(t) = f(t) - f(t - (1/n))$. The telescoping sum

$$F(b) - F(a) = f(1) - f(0) = \sum_{k=1}^{n} g(k/n)$$

gives us the next inequality from the triangle inequality

$$\frac{|F(b) - F(a)|}{|b - a|} \leq \frac{1}{n} \sum_{k=1}^{n} \frac{|g(k/n)|}{|b - a|/n}.$$

The right-hand average of n numbers is either a sum of numbers all more than the left-hand side or one term is more and one less. In any event, by Bolzano's intermediate value theorem we may select an interior term $|g(t)|/(|b - a|/n)$ more than the left-hand side.

The argument proceeds as in (5.4.1)—we select a sequence $a_n \uparrow c$ and $b_n \downarrow c$ with $(|b_n - a_n|) \leq (|b - a|/n)$ and

$$\frac{|F(b_n) - F(a_n)|}{|b_n - a_n|} \geq \frac{|F(b) - F(a)|}{|b - a|}.$$

We conclude in a similar way except replacing $(b_\omega - a_\omega)$ with the scalar $|b_\omega - a_\omega|$, $(b_\omega - c)$ with $|b_\omega - c|$, etc. Notice that since a_ω, c, and b_ω lie along a line segment, $|a_\omega - c| + |c + b_\omega| = |a_\omega - b_\omega|$ (convexity of the norm).

REMARK: Notice that we really only require one-dimensional differentiability along the segment from a to b and then only absolute-value differentiability. Of course, nothing says that U cannot be a segment to begin with, but then differentiable in **E** requires something about all $x \approx c$, an open condition.

The hypotheses of the next result can also be one-dimensionalized in the domain space.

(5.7.8) AN INNER PRODUCT MEAN VALUE THEOREM *Suppose F is continuous on the closure* cl(U) *of a convex set $U \subseteq$* **E**, *and F is differentiable on U. Suppose there is a fixed vector $V \in$* **F** *so that $F(a)$ and $F(b)$ are orthogonal to V where $a, b \in$* cl(U). *Then there exists $c \in U$ so that $DF_c(b - a)$ is orthogonal to V in* **F**.

PROOF: Apply (5.4.1) to $\langle F(x), V\rangle$ along the line segment from a to b, that is, let

$$f(t) = \langle F(a + t(b - a)), V\rangle.$$

By the chain rule $f'(s) = \langle DF_c(b - a), V\rangle$, where $c = a + s(b - a)$.

Higher-order derivatives are obtained by considering the map $Df: U \subseteq \mathbf{E} \to \mathrm{Lin}(\mathbf{E}, \mathbf{F})$. Since this is also a map into a linear space, $D(Df)_a$ is a linear map from $\mathbf{E}$ into $\mathrm{Lin}(\mathbf{E}, \mathbf{F})$, an element of $\mathrm{Lin}(\mathbf{E}, \mathrm{Lin}(\mathbf{E}, \mathbf{F}))$, provided the condition of the definition above is met. We prefer to define $D^2 f_a$ as the bilinear function in $\mathrm{Lin}^2(\mathbf{E}; \mathbf{F})$ given by first letting $D(Df)_a(x) = B_x$ in $\mathrm{Lin}(\mathbf{E}, \mathbf{F})$ and then taking $D^2 f_a(x, y) = B_x(y)$. Thus $D^2 f: U \subseteq \mathbf{E} \to \mathrm{Lin}^2(\mathbf{E}, \mathbf{F})$ provided Df is differentiable. Next, $D^3 f$ is the trilinear function in $\mathrm{Lin}^3(\mathbf{E}; \mathbf{F})$ obtained by a similar process applied to $D^2 f$ and onward by induction $D^{(k+1)} f: U \to \mathrm{Lin}^{(k+1)}(\mathbf{E}, \mathbf{F})$ provided $D^k f$ is differentiable.

If $D^k f$ is continuous on U, we say $f \in \mathscr{C}^k(U, \mathbf{F})$, and if $D^k f$ exists for all $k \in \mathbf{N}$, we say f is *smooth* and $f \in \mathscr{C}^\infty(U, \mathbf{F})$. When $D^k f$ exists it is symmetric, that is, $D^k f_a(x_1, \ldots, x_k) = D^k f_a(x_{p(1)}, \ldots, x_{p(k)})$ for any permutation p of $1, \ldots, k$. We denote the symmetric k-linear functions by $\mathrm{SLin}^k(\mathbf{E}; \mathbf{F})$. When $A \in \mathrm{SLin}^k(\mathbf{E}; \mathbf{F})$ we write $A(x^{(k)})$ for $A(x, \ldots, x)$, k times. Also, let $D^0 f_a = f(a)$.

EXERCISE: In Example (g) show that $D^2 g_{(x, y)}(h, k)$ is the form

$$[h, k]\begin{bmatrix} -\frac{2}{9}x^{-4/3}y^{1/3} & \frac{2}{9}x^{-1/3}y^{-2/3} \\ \frac{2}{9}x^{-1/3}y^{-2/3} & -\frac{2}{9}x^{2/3}y^{-5/3} \end{bmatrix}\begin{bmatrix} h \\ k \end{bmatrix}$$

evaluated by matrix products.

EXERCISE ON LEIBNIZ' RULE: *Let $\langle\ ,\ \rangle: \mathbf{F}_1 \times \mathbf{F}_2 \to \mathbf{G}$ be a bilinear map and let $f: \mathbf{E} \to \mathbf{F}_1$ and $g: \mathbf{E} \to \mathbf{F}_2$ be smooth. Then*

$$D^k\langle f, g\rangle_x = \sum_{j=0}^{k} \binom{k}{j} \langle D^j f_x, D^{k-j} g_x\rangle.$$

How does one apply the right-hand side to $h_1, \ldots, h_k$? Note that $D\langle\ ,\ \rangle_{(f, g)}(h, k) = \langle f, k\rangle + \langle h, g\rangle$.

EXERCISE ON HIGHER-ORDER CHAIN RULE: *Let f and g be smooth and such that $f \circ g$ can be formed. Derive the formula*

$$D^k(f \circ g)_x = \sum_{j=1}^{k} \sum c_j D^j f_{g(x)} \circ (D^{i_1} g_x, \ldots, D^{i_j} g_x).$$

In particular, explain the inside sum and compute c_j.

Now we turn our attention to Taylor's local approximation for a k times continuously differentiable function. In this version we take advantage of another uniform limit. When $k = 1$ this further supplements (5.7.4); varying a gives us $\mathscr{C}^1$.

(5.7.9) TAYLOR'S UNIFORM SMALL OH FORMULA *Let U be open in* $\mathbf{E}$. *A standard function* $f: U \to \mathbf{F}$ *has continuous derivatives up to order k,* $f \in \mathscr{C}^k(U; \mathbf{F})$, *if and only if there exist unique standard maps*

$$L^h_{(\cdot)}: U \to \mathrm{SLin}^h(\mathbf{E}; \mathbf{F})$$

such that whenever $x \approx a$ *with a near standard in* $^*U$, *there is an infinitesimal* $\eta \in {}^*\mathbf{F}$ *satisfying*

$$f(x) = \sum_{h=0}^{k} (1/h!) L_a{}^h(x-a)^{(h)} + |x+a|^k \cdot \eta.$$

The unique maps $L^h = D^h f$.

PROOF: First, assume that $f \in \mathscr{C}^k(U, \mathbf{F})$. We must show that the derivatives are symmetric and uniquely satisfy Taylor's formula. We begin with the symmetry of the higher derivatives.

(5.7.10) LEMMA *If* $D^2 f_a$ *exists, it is symmetric,* i.e., $D^2 f_a(x, y) = D^2 f_a(y, x)$.

PROOF: To prove the lemma we obtain an infinitesimal estimate for the symmetric second difference $f(a + \delta + \varepsilon) - f(a + \delta) - f(a + \varepsilon) + f(a)$, first in terms of $D^2 f_a(\delta, \varepsilon)$ and then in terms of $D^2 f_a(\varepsilon, \delta)$. The estimate comes from the norm mean value theorem (5.7.7).

Let y be infinitesimal in $^*\mathbf{E}$ and define a function $g(x) = f(a + y + x) - f(a + x) - D^2 f_a(y, x)$ so that

$$\begin{aligned} Dg_x(\cdot) &= Df_{(a+y+x)}(\cdot) - Df_{(a+x)}(\cdot) - D^2 f_a(y, \cdot) \\ &= \{[Df_{(a+y+x)}(\cdot) - Df_a(\cdot) - D^2 f_a(y + x, \cdot)] \\ &\quad - [Df_{(a+x)}(\cdot) - Df_a(\cdot) - D^2 f_a(x, \cdot)]\} \end{aligned}$$

by linearity of $D^2 f_a$ in the first variable. The differentiability of $x \mapsto Df_x$ at a means that when x and y are infinitesimal,

$$Df_{(a+y+x)}(\cdot) - Df_a(\cdot) = D^2 f_a(y + x, \cdot) + |y + x| \cdot \xi(\cdot)$$

and

$$Df_{(a+x)}(\cdot) - Df_a(\cdot) = D^2 f_a(x, \cdot) + |x| \cdot \zeta(\cdot),$$

where $\xi(\cdot)$ and $\zeta(\cdot)$ are infinitesimal linear maps, so $\xi(z) = |z|\xi(z/|z|) = |z|\eta$ and $\zeta(z) = |z|\zeta(z/|z|) = |z|\varphi$ with η and φ infinitesimal.

The mean value theorem applied to g says that $|\Delta g| \leq |Dg_x(\Delta x)|$ with x in the Δx interval. We apply this in two cases $y = \varepsilon$ and $\Delta x = \delta$. Then

$y = \delta$ and $\Delta x = \varepsilon$ where ε and δ are arbitrary nonzero infinitesimals in $^*\mathbf{E}$, and

$$\Delta g_1 = f(a + \varepsilon + \delta) - f(a + \delta) - D^2 f_a(\varepsilon, \delta) - f(a + \varepsilon) \\ + f(a) + D^2 f_a(\varepsilon, 0)$$

and

$$|\Delta g_1| \le \big| |\varepsilon| \cdot (|\delta + \varepsilon|/|\varepsilon|) \cdot |\delta| \eta_1 + |\varepsilon| \cdot |\delta| \cdot \varphi_1 \big| = |\varepsilon| \cdot |\delta| \eta_0,$$

$$\Delta g_2 = f(a + \varepsilon + \delta) - f(a + \varepsilon) - D^2 f_a(\delta, \varepsilon) - f(a + \delta) \\ + f(a) + D^2 f_a(\delta, 0)$$

and

$$|\Delta g_2| \le \big| |\delta| \cdot (|\delta + \varepsilon|/|\delta|) \cdot |\varepsilon| \cdot \eta_2 + |\delta| \cdot |\varepsilon| \cdot \varphi_2 \big| = |\varepsilon| \cdot |\delta| \varphi_0,$$

where η_j and φ_j $(j = 0, 1, 2)$ are infinitesimal by the remarks above on differentiability of Df. We have now

$$f(a + \delta + \varepsilon) - f(a + \delta) - f(a + \varepsilon) - f(a) = D^2 f_a(\varepsilon, \delta) + |\varepsilon| \cdot |\delta| \cdot \eta_0 \\ = D^2 f_a(\delta, \varepsilon) + |\varepsilon| \cdot |\delta| \cdot \varphi_0.$$

Taking differences and dividing by $|\varepsilon| \cdot |\delta|$ we obtain

$$D^2 f_a(\varepsilon/|\varepsilon|, \delta/|\delta|) \approx D^2 f_a(\delta/|\delta|, \varepsilon/|\varepsilon|),$$

and since ε and δ are arbitrary infinitesimals, $\varepsilon/|\varepsilon|$ and $\delta/|\delta|$ run through all the unit vectors. At standard values of a, $D^2 f_a$ is standard and the $\approx$ is actually $=$. This proves the lemma.

Symmetry of $D^k f_a$ for $k > 2$ follows by induction since we can apply the lemma to $x \to D^{k-2} f_x(y_1, \ldots, y_{k-2})$. The inductive hypothesis says $D^{k-1} f_a(y_1, \ldots, y_{k-1})$ is symmetric and the lemma says $D^k f_a(y_1, \ldots, y_{k-2}, y_{k-1}, y_k) = D^k f_a(y_1, \ldots, y_{k-2}, y_k, y_{k-1})$. The permutation-symmetric group on $\{1, \ldots, k-1\}$ times transposition of $\{k-1, k\}$ generates the full symmetric group of permutations of $\{1, \ldots, k\}$, so these two things yield full symmetry of $D^2 f_a$.

Perhaps the simplest proof of Taylor's formula is to derive the integral form of the remainder. This will require an internal vector integral which we have not yet discussed; however, the finite-dimensional case could simply be done componentwise (and, moreover, the proofs of Riemann integrability of a continuous function in Section 5.3 of the fundamental theorem and of integration by parts from Section 5.5 can be readily carried over to maps from **R** into **F**). Vector integrals are discussed in more detail below.

So long as the line segment from a to x is in U, the remainder formula is:

$$f(x) = \sum_{h=0}^{k} \frac{D^h f_a}{h!} (x-a)^{(h)} + R_k(a, x),$$

where

$$R_k(a, x) = \int_0^1 \frac{(1-t)^{k-1}}{(k-1)!} g(t)\, dt$$

with

$$g(t) = [D^k f_{(a+t(x-a))}(x-a)^{(k)} - D^k f_a (x-a)^{(k)}].$$

The proof of this by induction uses integration by parts and the fundamental theorem.

For $k = 1$:

$$\begin{aligned} R_1(a, x) &= f(x) - f(a) - Df_a(x-a) \\ &= \int_0^1 Df_{(a+t(x-a))}(x-a)\, dt - \int_0^1 Df_a(x-a)\, dt \\ &= \int_0^1 (Df_{(a+t(x-a))} - Df_a)(x-a)\, dt \end{aligned}$$

by the fundamental theorem of calculus.

For $k \Rightarrow k+1$: Assume the formula holds for k, and by definition

$$\begin{aligned} R_{k+1}(a, x) &= R_k(a, x) - \frac{1}{(k+1)!} D^{(k+1)} f_a (x-a)^{(k+1)} \\ &= \int_0^1 \frac{(1-t)^{k-1}}{(k-1)!} [(D^k f_{(a+t(x-a))} - D^k f_a)(x-a)^k]\, dt \\ &\quad - \frac{D^{k+1}}{k!} f_a (x-a)^{(k+1)} \int_0^1 (1-t)^k\, dt \\ &= \int_0^1 \frac{(1-t)^k}{k!} D^{k+1} f_{(a+t(x-a))}(x-a)^{(k+1)}\, dt \\ &\quad - \int_0^1 \frac{(1-t)^k}{k!} D^{(k+1)} f_a (x-a)^{(k+1)}\, dt, \end{aligned}$$

using integration by parts and evaluation of the integral of $(1-t)^k$.

Notice that when $D^{k+1}f$ can be integrated, $R_k(a, x)$ takes the simpler form

$$\int_0^1 ((1-t)^k/k!)D^{k+1}f_{(a+t(x-a))}(x-a)^{(k+1)}\,dt$$

from integration by parts on $R_k(a, x)$ in the form we have given.

Now we take $x \approx a$ and observe that the maximum of $[D^k f_{(a+t(x-a))} - D^k f_a]$ in the integrand of the remainder is infinitesimal by the continuity. Therefore the remainder is smaller than

$$|x-a|^k m\left(\frac{x-a}{|x-a|}\right)^{(k)} \int_0^1 \frac{(1-t)^{k-1}}{(k-1)!}\,dt,$$

where $m(\cdot)$ is the max. This proves the formula.

Uniqueness is proved by calculating the derivatives of $\sum_{h=0}^{k} L_a{}^h(x-a)^{(h)} + o\cdot|x-a|^k$ at a, carrying along the remainder terms; in other words, it can be viewed as part of the converse.

The uniqueness in Taylor's formula makes it a flexible tool; one can use polynomial substitution, integration, or other tricks to obtain a formula instead of brute force calculation—see the exercises of Section 5.6. Uniqueness also shows that a Taylor polynomial of degree $k \le n$ which is $o(|x|^n)$ must in fact be the zero polynomial.

We now turn to the converse, which is sometimes useful in showing that a natural infinite dimensional map is $\mathscr{C}^k$, for example, the evaluation map of a $\mathscr{C}^k$ function. We will discuss this again when we return to calculus in Part 2; no doubt the reader has already noticed our avoidance of coordinates in this section. The infinitesimal condition at nonstandard a's is also useful (for example, in infinitesimal geometry).

The proof of the converse is by induction on k as follows.

For $k = 1$: Let x_0 and x_1 be unit vectors, δ a positive infinitesimal, and a standard. Then

$$f(a + \delta x_0 + \delta x_1) - f(a + \delta x_0) = L_{a+\delta x_0}(\delta x_1) + \delta\cdot\eta$$

and

$$\begin{aligned} f(a + \delta x_0 + \delta x_1) - f(a) + f(a) - f(a + \delta x_0) &= L_a(\delta x_0 + \delta x_1) \\ &\quad - L_a(\delta x_0) + \delta\cdot\zeta \\ &= L_a(\delta x_1) + \delta\cdot\zeta, \end{aligned}$$

hence, $L_{a+\delta x_0} \approx L_a = Df_a$, by the difference formula for derivatives at a standard point. By transform of the differentiability condition, when ε is sufficiently small, letting $x = a + \delta x_0$,

$$(1/\varepsilon)[f(x + \varepsilon z) - f(x)] = Df_x(z) + \eta' = L_x(z) + \zeta',$$

so $Df_x \approx Df_a$ and $f \in \mathscr{C}^1(U; \mathbf{F})$.

For simplicity, $k = 2$: Suppose x_0, x_1, x_2 are unit vectors, δ a positive infinitesimal, and a a standard point. Let $x = a + \delta x_0$. Then

$$\begin{aligned}\{f(x + \delta x_1 + \delta x_2) - f(x) + f(x) \\ - f(x + \delta x_1) - f(x + \delta x_2) + f(x)\} = L_x{}^2(\delta x_1, \delta x_2) + \delta^2 \cdot \eta\end{aligned}$$

and

$$\begin{aligned}&\{f(a + \delta x_0 + \delta x_1 + \delta x_2) - f(a) + f(a) - f(a + \delta x_0 + \delta x_1) \\ &\quad - f(a + \delta x_0 + \delta x_2) + f(a) - f(a) + f(a + \delta x_0)\} \\ &\quad = L_a{}^2(\delta x_1, \delta x_2) + \delta^2 \cdot \zeta,\end{aligned}$$

hence,

$$L_x{}^2 \approx L_a{}^2.$$

We also have

$$\begin{aligned}Df_{a+\delta x_0}(\delta x_1) - Df_a(\delta x_1) &= \{f(a + \delta x_0 + \delta x_1) - f(a + \delta x_0) \\ &\quad - \tfrac{1}{2}L^2_{a+\delta x_0}(\delta x_1)^{(2)} + \delta^2 \cdot \eta \\ &\quad - f(a + \delta x_1) + f(a) + \tfrac{1}{2}L_a{}^2(\delta x_1)^{(2)}\} \\ &= f(a + \delta x_0 + \delta x_1) - f(a + \delta x_0) \\ &\quad - f(a + \delta x_1) + f(a) + \delta^2 \cdot \zeta \\ &= L_a{}^2(\delta x_0, \delta x_1) + \delta^2 \cdot \theta\end{aligned}$$

which shows that $D^2 f_a = L_a{}^2$ when a is standard.

For $k \Rightarrow k + 1$: Suppose a is standard in $^*U$, δ a positive infinitesimal, and $x_0, x_1, \ldots, x_{k+1}$ unit vectors. The $(k + 1)$-difference

$$\begin{aligned}&\{f(a + \delta x_0 + \delta x_1 + \cdots + \delta x_{k+1}) \\ &\quad - \sum_{j=1}^{k+1} f(a + \delta x_0 + \cdots + \widehat{\delta x_j} + \cdots + \delta x_{k+1}) \\ &\quad + \sum_{1 \le i < j \le k+1} f(a + \delta x_0 + \cdots + \widehat{\delta x_i} + \cdots + \widehat{\delta x_j} + \cdots + \delta x_{k+1}) + \cdots \\ &\quad + (-1)^k \sum_{j=1}^{k+1} f(a + \delta x_0 + \delta x_j) + (-1)^{k+1} f(a + \delta x_0)\} \\ &\quad = L^{k+1}_{a+\delta x_0}(\delta x_1, \ldots, \delta x_{k+1}) + \delta^{k+1} \cdot \eta \qquad \text{expanding at } a + \delta x_0 \\ &\quad = L_a^{k+1}(\delta x_1, \ldots, \delta x_{k+1}) + \delta^{k+1} \cdot \zeta, \qquad \text{expanding at } a.\end{aligned}$$

Hence we see that

$$L_{a+\delta x_0}^{k+1} \approx L_a^{k+1}.$$

We assume for the induction that it is known that $L_x{}^h = D^h f_x$ when $h \le k$ and that the function can be expanded as above near a. Let ε and δ be comparable nonzero infinitesimals in $^*\mathbf{E}$, that is, $|\varepsilon|/|\delta|$ and $|\delta|/|\varepsilon|$ are finite. We expand $f(a+\delta+\varepsilon)$ two ways:

$$\begin{aligned} f(a+\delta+\varepsilon) &= \sum_{h=0}^{k} \frac{1}{h!} D^h f_{(a+\delta)}(\varepsilon)^{(h)} + \frac{1}{(k+1)!} L_{(a+\delta)}^{(k+1)}(\varepsilon)^{(k+1)} + |\varepsilon|^{(k+1)}\eta \\ &= \sum_{h=0}^{k} \frac{1}{h!} D^h f_a(\delta+\varepsilon)^{(h)} \\ &\quad + \frac{1}{(k+1)!} L_a^{(k+1)}(\delta+\varepsilon)^{(k+1)} + |\delta+\varepsilon|^{(k+1)}\varphi, \end{aligned}$$

where η and φ are infinitesimals in $^*\mathbf{F}$. We rewrite the second remainder as $|\varepsilon|^{(k+1)}\varphi'$ and form the difference. Letting $\eta - \varphi' = \zeta$ and dropping $(k+1)$ on the last L, we have

$$\begin{aligned} |\varepsilon|^{(k+1)}\zeta &= \sum_{h=0}^{k} \frac{1}{h!} [D^h f_{(a+\delta)}(\varepsilon)^{(h)} - D^h f_a(\delta+\varepsilon)^{(h)}] \\ &\quad + \frac{1}{(k+1)!} [L_{(a+\delta)}(\varepsilon)^{(k+1)} - L_a(\varepsilon+\delta)^{(k+1)}]. \end{aligned}$$

Now we wish to rewrite this as a polynomial in ε and we are especially interested in the terms $b_{(k+1)}(\delta)\varepsilon^{(k+1)}$ and $b_k(\delta)\varepsilon^{(k)}$, where the difference is $\sum_{h=0}^{k+1} b_h(\delta)\varepsilon^{(h)}$. By multilinearity we can expand $D^k f_a(\delta+\varepsilon)^{(k)}$ and $L_a(\delta+\varepsilon)^{(k+1)}$, e.g.,

$$\begin{aligned} \frac{1}{(k+1)!} L_a(\delta+\varepsilon)^{(k+1)} &= \sum_{j=0}^{k-1} \frac{1}{j!\,(k+1-j)!} L_a(\delta^{(j)}, \varepsilon^{(k+1-j)}) \\ &\quad + \frac{1}{k!} L_a(\delta, \varepsilon^{(k)}) + \frac{1}{(k+1)!} L_a(\varepsilon^{(k)}), \end{aligned}$$

so that

$$b_{k+1}(\delta)\varepsilon^{(k+1)} = (L_{a+\delta} - L_a)(\varepsilon^{(k+1)})/(k+1)!$$

and

$$b_k(\delta)\varepsilon^{(k)} = (1/k!)[D^k f_{a+\delta}(\cdot) - D^k f_a(\cdot) - L_a(\delta, \cdot)](\varepsilon)^{(k)}.$$

We know that $|\varepsilon|^{k+1} \cdot \zeta = \sum_{k=0}^{k+1} b_h(\delta)\varepsilon^{(h)}$ by the induction hypothesis. Since L_x is continuous $(L_{a+\delta} - L_a)(\varepsilon^{(k+1)}) = |\varepsilon|^{k+1} \cdot \zeta'$, for infinitesimal $\zeta' = (L_{a+\delta} - L_a)([\varepsilon/|\varepsilon|]^{(k+1)})$. We subtract across and obtain the following where $\xi = \zeta - \zeta'$:

$$|\varepsilon|^{k+1} \cdot \xi = \frac{1}{k!}[D^k f_{(a+\delta)}(\cdot) - D^k f_a(\cdot) - L_a(\delta, \cdot)](\varepsilon^{(k)}) + \sum_{h=0}^{k-1} b_h(\delta)\varepsilon^{(h)}.$$

Showing that the term including $\varepsilon^{(k)}$ is actually $o \cdot |\varepsilon|^{k+1}$ will complete the proof since then the continuous $L_a^{(k+1)} = D(D^k f_a)$. (Recall that "small compared with δ" is the same as "small compared with ε" by our assumption above.)

Let $\lambda_0, \ldots, \lambda_k$ be distinct standard scalars near 1. We use the homogeneity of our polynomial to observe that

$$\begin{pmatrix} 1 & \lambda_0 & \cdots & \lambda_0{}^k \\ 1 & \lambda_1 & \cdots & \lambda_1{}^k \\ \vdots & \vdots & \ddots & \vdots \\ 1 & \lambda_k & \cdots & \lambda_k{}^k \end{pmatrix} \begin{pmatrix} b_0(\delta) \\ b_1(\delta)\varepsilon \\ \vdots \\ b_k(\delta)\varepsilon^{(k)} \end{pmatrix} = \begin{pmatrix} |\lambda_0 \varepsilon|^{(k+1)}\xi \\ |\lambda_1 \varepsilon|^{(k+1)}\xi \\ \vdots \\ |\lambda_k \varepsilon|^{(k+1)}\xi \end{pmatrix} = |\varepsilon|^{k+1} \begin{pmatrix} \xi_0 \\ \xi_1 \\ \vdots \\ \xi_k \end{pmatrix}$$

where all the ξ_k are the infinitesimals $\lambda_k^{k+1} \cdot \xi$. The Vandermonde matrix of λ's is invertible (the determinant equals $\prod_{0 \le i < j \le k}(\lambda_j - \lambda_i)$). Applying the inverse matrix Λ to both sides gives us that

$$\begin{pmatrix} b_0(\delta) \\ b_1(\delta)\varepsilon \\ \vdots \\ b_k(\delta)\varepsilon^{(k)} \end{pmatrix} = |\varepsilon|^{k+1} \begin{pmatrix} \Lambda(\xi_0) \\ \Lambda(\xi_1) \\ \vdots \\ \Lambda(\xi_k) \end{pmatrix}$$

and, in particular, $b_k(\delta) = o \cdot |\delta|$ by the remarks above. This completes the proof.

It can do no harm to reiterate the warning that even in the smooth case Taylor's small oh formula, a useful tool in differentiability, is vastly weaker than the existence of a Taylor series expansion (which is called analyticity). When $f \in \mathscr{C}^\infty(U; \mathbf{F})$ it may well be the case that $f(x) \neq \sum_{h=0}^{\infty} (1/h!)D^h f_a(x-a)^{(h)}$, even if the series happens to converge.

EXERCISES ON FLAT FUNCTIONS

1. Show that the function

$$f(x) = \begin{cases} \exp(-1/x^2), & x \neq 0 \\ 0, & x = 0 \end{cases}$$

is smooth and *flat* at $x = 0$, that is, $D^h f_0 = 0$ for all h.

2. Show that

$$g(x) = \begin{cases} \exp(-1/x), & x > 0 \\ 0, & x \le 0 \end{cases}$$

is smooth and flat for $x \le 0$. Smooth functions thus are flabby—you can iron them out and they need not change where you do not iron. (Analytic functions are springy—bend them one place and they change nearly everywhere.)

3. Flatness is not compatible with the order structure on $\mathbf{R} = \mathbf{F}$. Show that $h(x) = \exp(-1/x^2)\sin(\exp(1/x^2))$ satisfies $-f(x) \le h(x) \le f(x)$, where f is flat and yet h is not flat. Is h a smooth function? What can you add if h is smooth and sandwiched between flat functions?

EXERCISE ON SMOOTH STANDARD PARTS: Let $U \subseteq \mathbf{E}$ be open. Suppose that f is an internal function (otherwise arbitrary) and suppose there are internal maps

$$L^h_{(\cdot)}: {}^*U \to {}^*\mathrm{SLin}({}^*\mathbf{E}; {}^*\mathbf{F}), \qquad 0 \le h \le k$$

so that $L_a{}^h$ is a finite transformation for each near-standard $a \in {}^*U$ and whenever $x \approx a$ in ${}^*\mathbf{E}$, there is an infinitesimal $\eta \in {}^*\mathbf{F}$ with

$$f(x) = \sum_{h=0}^{k} (1/h!) L_a{}^h (x-a)^{(h)} + |x-a|^k \cdot \eta.$$

Show that the function $\hat{f}(x) = \mathrm{st}(f(x))$, $x \in {}^\sigma U$, is a standard $\mathscr{C}^k$-function with $D^h f_a = \mathrm{st}(L_a{}^h)$.

How would you define a "near-standard infinitely smooth function"? You may assume $f \in {}^*\mathscr{C}^\infty$ without real loss of generality, but no restrictions on $D^\omega f$ are necessary when ω is infinite.

(5.7.11) THE INVERSE FUNCTION THEOREM *Let f be an internal function defined on a subset of* ${}^*\mathbf{E}$. *Suppose that Df_a is finite, has a finite inverse, and that whenever $x \approx y \approx a$ we have*

(S-UD) $$f(x) - f(y) = Df_a(x-y) + |x-y| \cdot \eta$$

for infinitesimal $\eta \in {}^*\mathbf{F}$. Then f has a single-valued inverse on a finite neighborhood of a, $D(f^{-1})_{f(a)} = (Df_a)^{-1}$, and f^{-1} satisfies S-uniform differentiability (S-UD) *at $f(a)$. When a and $f(a)$ are near standard, standard part* st *commutes with D and $(\ )^{-1}$.*

PROOF (cf. Behrens [1973c]): Since Df_a is finitely invertible, $|Df_a((x-y)/|x-y|)| \ge 1/\|(Df_a)^{-1}\| \not\approx 0$. Therefore, when $x \ne y$ in the infinitesimal neighborhood of a, $(f(x) - f(y))/|x - y| \not\approx 0$ and f is one-to-one or injective on the infinitesimals around a.

Consider the internal set of radii described by the property that f is injective on the ball with that radius about a. The set contains all the infinitesimal radii and thus contains a finite positive radius. This shows that f is invertible on a finite neighborhood of a.

We show that f maps *onto* the infinitesimals around $b = f(a)$ by constructing a $*$-Cauchy sequence infinitesimally close to a which tends to an arbitrary point near b. Let $c \approx b$ be fixed but arbitrary. Define an internal sequence by $*$-induction as follows: $x_0 = a$ and $x_{n+1} = x_n + (Df_a)^{-1}(c - f(x_n))$. Applying ($S$-UD) on the sequence we obtain

$$\begin{aligned} f(x_1) &= f(x_0) + Df_a(Df_a)^{-1}(c - f(x_0)) + |c - f(x_0)| \cdot \varepsilon_1 \\ &= c + |c - f(x_0)| \cdot \varepsilon_1, \\ f(x_1) - c &= |c - f(x_0)| \cdot \varepsilon_1, \\ f(x_2) - c &= |c - f(x_1)| \cdot \varepsilon_2 = |c - f(x_0)| \cdot |\varepsilon_1| \cdot \varepsilon_2, \\ |f(x_n) - c| &= |c - f(x_0)| \prod_{k=1}^{n} |\varepsilon_k|, \end{aligned}$$

and

$$|x_{n+1} - x_n| \le \|(Df_a)^{-1}\| \cdot |c - f(x_n)|$$

so that,

$$|c - f(x_n)| \le \frac{|c - f(a)|}{\|(Df_a)^{-1}\|} \cdot (\tfrac{1}{2})^{n+1}$$

and

$$|x_n - x_{n-1}| \le |c - f(a)|(\tfrac{1}{2})^n$$

for finite $n \ge 1$. We suppose these conditions hold for $n \le \omega$ and show them for $(\omega + 1)$. Thus $|c - f(x_{\omega+1})| \le |c - f(x_\omega)| \cdot \varepsilon$ with $\varepsilon \approx 0$ because $|x_\omega - a| \sum \Delta x_n \le 2|c - f(a)| \approx 0$ and clearly then $|\varepsilon| < 1/[2 \cdot \|(Df_a)^{-1}\|]$. This shows the first condition for $n = \omega + 1$. The difference $|x_{\omega+1} - x_\omega| \le |(Df_a)^{-1}(c - f(x_\omega))| \le \|(Df_a)^{-1}\| \cdot |c - f(x_\omega)| \le |c - f(a)|(\tfrac{1}{2})^{\omega+1}$, and this completes the internal induction. Since ***E** is $*$-complete $x_n \to y$ with $|a - y| \le 2|c - f(a)| \approx 0$, and by the $*$-continuity imposed by (S-UD) $[|f(x) - f(y)| \le 2 \cdot \|Df_a\| \cdot |x - y|]$, $f(y) = c$, which shows f is onto the infinitesimals around $f(a) = b$.

We saw above that f is injective on a finite neighborhood of a. It is easy to see that the image of such a neighborhood contains a finite neighborhood of $f(a)$ since it contains every infinitesimal ball around $f(a)$ and the set of radii which it contains is internal.

The S-uniform differentiability of f^{-1} at $f(a) = b$ requires that if we write the difference $(Df_a)^{-1}(z - w) - f^{-1}(z) + f^{-1}(w)$ as $|z - w| \cdot \beta$ for $\beta \in {}^*\mathbf{F}$, then we show that $\beta \approx 0$ when $z \approx b \approx w$ and $x \approx a \approx y$ with $f(x) = z$ and $f(y) = w$. (S-UD) for f says that $z - w - Df_a(x - y) = |x - y| \cdot \eta$ with infinitesimal η. Apply $(Df_a)^{-1}$ to this equation to obtain $(Df_a)^{-1}(z - w) - f^{-1}(z) + f^{-1}(w) = |x - y|(Df_a)^{-1}(\eta)$. Now, $(Df_a)^{-1}$ is a finite transformation so that

$$0 \approx (Df_a)^{-1}(\eta) = \frac{|z - w|}{|x - y|} \cdot \beta = \frac{|f(x) - f(y)|}{|x - y|} \cdot \beta$$
$$= \left| Df_a\left(\frac{x - y}{|x - y|}\right) + \eta \right| \cdot \beta$$

and β must be infinitesimal because $Df_a((x - y)/|x - y|)$ cannot be infinitesimal (otherwise $(Df_a)^{-1}$ is not finite).

Finally, suppose a and $b = f(a)$ are near standard. Cauchy's principle allows us to push the conditions we need out to a finite δ. First, recall from above that f is injective on a finite neighborhood of a and the image of that neighborhood contains a finite neighborhood of b. We restrict the discussion to that neighborhood. Take ε in ${}^\sigma\mathbf{R}^+$ and observe that

$$|g(x) - g(y) - Dg_c(x - y)| \leq |x - y| \cdot \varepsilon$$

and

$$|g(x) - g(y)| \leq 2 \cdot \|Dg_c\| \cdot |x - y|$$

are internal and hold on every infinitesimal ball around c, thus out to a standard δ for g either equal to f or f^{-1} and c equal to a and b, respectively. Standard parts do not disturb these epsilons and deltas, so that we know $\mathrm{st}(f)$ has derivative $\mathrm{st}(Df_a)$, $\mathrm{st}(f^{-1})$ has derivative $\mathrm{st}(Df_b^{-1}) = \mathrm{st}((Df_a)^{-1})$, and both are Lipschitz continuous on finite neighborhoods. We now look at the equation $f^{-1}(f(x)) = x$ and take standard parts: $\mathrm{st}(f^{-1}(\mathrm{st}\, f(\mathrm{st}(x)))) = \mathrm{st}(f^{-1}(\mathrm{st}\, f(x))) = \mathrm{st}(f^{-1}(f(x))) = \mathrm{st}(x)$ so $\mathrm{st}(f^{-1})$ is the inverse of $\mathrm{st}(f)$ on a finite neighborhood. Also $[\mathrm{st}(Df_a)]^{-1} = \mathrm{st}[(Df_a)^{-1}] = \mathrm{st}[D(f^{-1})_b]$ by the same kind of equations.

(5.7.12) THE STANDARD INVERSE FUNCTION THEOREM *Let $f \in \mathscr{C}^k(U, \mathbf{F})$ for $k \geq 1$ and let U be open in $\mathbf{E}$. Let Df_a be invertible for some $a \in U$. Then f is locally $\mathscr{C}^k$-invertible at a with $(Df_a)^{-1} = D(f^{-1})_{f(a)}$.*

PROOF: When $k = 1$, this follows from the last theorem applied to ${}^*f$ by using (5.7.4) to conclude that $f^{-1} \in \mathscr{C}^1$. We know now that $(Df^{-1})_x = (Df_{f^{-1}(x)})^{-1}$ and we can apply the chain rule since $(\ )^{-1}$ applied to a linear transformation is smooth where it is defined. "Locally invertible" in the standard theorem refers to the finite neighborhood we found in the proof of (5.7.11). Simple examples, like the complex exponential function, show that a global single-valued inverse need not exist, of course, but when we first fix a there is a standard neighborhood over which f is injective.

(5.7.13) THE IMPLICIT FUNCTION THEOREM *Let $U \subseteq \mathbf{E}$ and $V \subseteq \mathbf{F}$ be open. Suppose f: $U \times V \to \mathbf{F}$ is $\mathscr{C}^k$ for $k \geq 1$ and if $g(z) = f(a, z)$ that Dg_b is invertible where $(a, b) \in U \times V$. Let $c = f(a, b)$. The equation*

$$f(x, y) = c$$

locally defines y as a function of x near a, that is, there is a neighborhood of a, $N_a \subseteq U$, and a $\mathscr{C}^k$ function h: $N_a \to V$ so that $f(x, h(x)) = c$ on N_a.

REMARK: When $\mathbf{E}$ has dimension m and $\mathbf{F}$ has dimension n, $f(x, y) = c$ is n equations in $m + n$ unknowns. The linear problem of this size at best solves for n terms y in terms of the m terms x. In practice one usually has a function of many variables and has to decide which to solve for in terms of which others. Our condition on Dg_b helps decide.

PROOF: Normalize f by forming $(Dg_b)^{-1} \circ f$ so that if $g(z) = f(a, z)$, then Dg_b is the identity on $\mathbf{F}$. Define $F(x, y) = (x, f(x, y))$ and notice that in matrix blocks $DF_{(a,b)}$ is

$$\begin{bmatrix} I_{\mathbf{E}} & 0 \\ * & I_{\mathbf{F}} \end{bmatrix}$$

and hence invertible. Therefore we may apply the inverse function theorem to F, obtaining an inverse H locally. Now $(x, c) = F \circ H(x, c) = (x, f(x, h(x)))$ by the definitions of F and H.

EXERCISE ON S-$\mathscr{C}^k$ INVERSES: Show that f^{-1} has finite S-continuous derivatives up to order k when f does in (5.7.11).

5.8 INTEGRAL VECTOR CALCULUS

In this section we give an infinitesimal account of Riemann integrals. These integrals are not well suited to integration of discontinuous functions, in fact, Riemann showed that continuity almost everywhere (in the sense of Lebesgue measure) is required for ordinary Riemann integrability. In other words, to understand fully the weakest hypotheses for Riemann integrability one has to develop the Lebesgue theory (and this is not illuminating infinitesimally without stronger models). Hence we shall assume continuity or even smoothness where needed. Integration of smooth functions on smooth regions is still quite an interesting subject and Riemann integrals are more closely linked with the associated geometry than the Lebesgue integrals are. Once one wishes to relax continuity hypotheses, integrate over wild sets, or especially, integrate nonuniform limits of functions, then one should go all the way to the beautiful theory of Lebesgue, if for no other reason than that it is simpler when dealing with such problems.

Now we turn our attention to the n-dimensional Riemann integral in Euclidean n-space $\mathbf{E}^n$. We want our definition to be geometrical—not dependent on the choice of origin or particular orthonormal frame; rather only on orientation, the notion of angle and parallel, and the scale. The reflections through planes generate the rigid motions of Euclidean space, but a single reflection reverses orientation while the composition of an even number of reflections preserves orientation. The orientation-preserving rigid motions are the translations and rotations. Algebraically, the translations are very simple $x \mapsto x + t$. The other two indispensible algebraic facts that we shall need are: (1) *A linear map preserves orientation if it has positive determinant and reverses orientation if it has negative determinant.* This is an especially nice characterization since determinants are coordinate free and at the same time readily computable in a particular choice of coordinates. This tells us in particular that the rotations about a choice of origin are those (linear) orthogonal transformations with determinant plus one. (2) *The image of the unit cube under a linear transformation is a parallelepiped whose n-dimensional volume is the absolute value of the determinant of the transformation.* We shall consider this fact a theorem, which is left to the reader, meaning that it needs to be shown that the sum of infinitesimal cubes (discussed below) covering such an n-parallelepiped adds up nearly to the determinant. We do not want this swept under the rug, even though we do not provide the proof.

We said that we do not want our definition to depend on a choice of coordinates, but the way we shall operate is to show invariance under translation and rotation for one choice and thus justify the definition.

To each choice of origin and ordered orthonormal basis of $^*\mathbf{E}^n$ we can pick an infinitesimal scale δ and define *an oriented δ-paving of* $^*\mathbf{E}^n$ by cutting each axis by perpendicular $(n - 1)$-planes at intervals of length δ. The paving is actually a little more invariant than our description, since any vertex could have been the origin, with the directions and order left the same. Each finite parallel translation could be infinitesimally approximated by moving a $*$-finite number of δ's in each coordinate. A translation along a standard vector with coordinates $(t_1, \ldots, t_n)$ is approximated by ω_j steps in the jth coordinate where $\omega_j \delta \leq t_j < (\omega_j + 1)\delta$. The ordering of the directed edges of any one of the little frames determines the orientation (see Fig. 5.8.1).

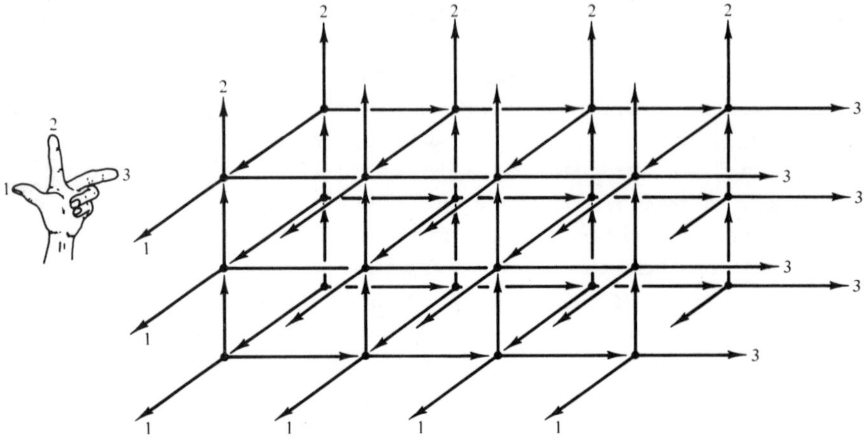

Fig. 5.8.1 Left-handed δ-paving of $^*\mathbf{E}^3$ through infinite microscope.

Now we select some notation to go with a particular paving. Say $e_1, \ldots, e_n$ is our ordered positively oriented basis, $x = (x^1, \ldots, x^n)$ are the coordinates of the vector $\sum_{j=1}^{n} x^j e_j$, and $x^j_{k_j}$ are the coordinates of $k_j \cdot \delta \cdot e_j$, $j = 1, \ldots, n$ for integer k_j. Every point of $^*\mathbf{E}^n$ lies in one little cube, so to each x, there is a multi-index $k = (k_1, \ldots, k_n)$ such that $x \in \Delta x_k$, where

$$\Delta x_k = \{x : x^j_{k_j} \le x^j < x^j_{(k_j+1)}\}.$$

The volume of each little cube is

$$\delta^n = v_n(\Delta x_k) = (x^1_{(k_1+1)} - x^1_{k_1}) \cdot (x^2_{(k_2+1)} - x^2_{k_2}) \cdots (x^n_{(k_n+1)} - x^n_{k_n}).$$

Notice that *reversing the orientation*, say, making our basis $f_1, \ldots, f_n$ with $f_1 = -e_1$ and $f_j = e_j$, $j > 1$, *makes the oriented product of the edges minus* $v_n(\Delta x_k)$ if you keep *the same arrangement of points* with the values of the new coordinates. In one dimension this gives us $\int_a^b f(x)\,dx = -\int_b^a f(x)\,dx$ and the reader may wish to verify this with a change of variables. This is clear if you reverse the order of the Riemann sum. We denote this ordered or oriented volume by $(\delta x^1 \wedge \delta x^2 \wedge \cdots \wedge \delta x^n)_k$.

Since the algebra of orientation is a little messy, let us verify that $\delta x^1 \wedge \delta x^2 = -\delta x^2 \wedge \delta x^1$ in $\mathbf{E}^2$. We have one basis (e_1, e_2) with coordinates (x^1, x^2) and a second basis (f_1, f_2) with $f_1 = e_2$, $f_2 = e_1$, and coordinates (y^1, y^2), as shown in Fig. 5.8.2.

The basic formula about orientation is

$$(\delta x^1 \wedge \delta x^2 \wedge \cdots \wedge \delta x^n) = -(\delta x^2 \wedge \delta x^1 \wedge \cdots \wedge \delta x^n).$$

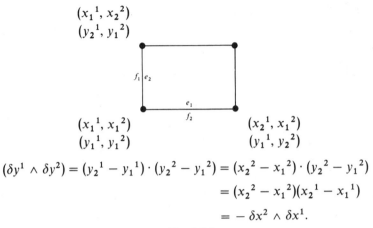

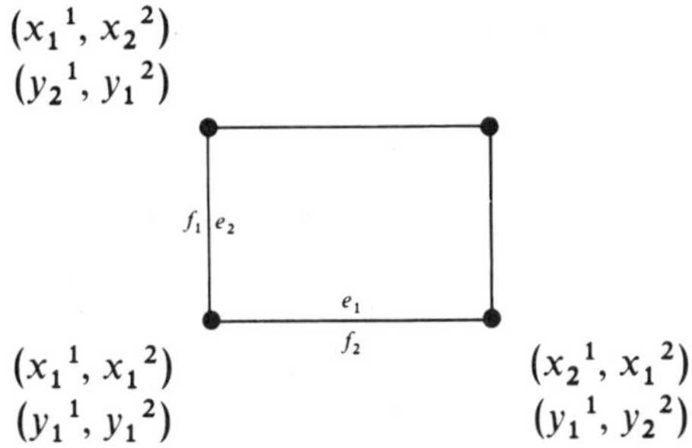

$$(\delta y^1 \wedge \delta y^2) = (y_2{}^1 - y_1{}^1)\cdot(y_2{}^2 - y_1{}^2) = (x_2{}^2 - x_1{}^2)\cdot(y_2{}^2 - y_1{}^2)$$
$$= (x_2{}^2 - x_1{}^2)(x_2{}^1 - x_1{}^1)$$
$$= -\,\delta x^2 \wedge \delta x^1.$$

Fig. 5.8.2

This generalizes to the statement that even permutations preserve and odd permutations reverse orientation.

When oriented volumes are involved we shall always sum over increasing k_j, summing k_1, then $k_2, \ldots$ and last k_n.

When A is an internal subset of $^*\mathbf{E}^n$, we shall denote the characteristic function of A by

$$\chi_A(x) = \begin{cases} 0, & x \notin A \\ 1, & x \in A. \end{cases}$$

(5.8.1) DEFINITION (*n*-dimensional oriented and absolute Riemann integrals in $\mathbf{E}^n$) *Fix an orientation for* $\mathbf{E}^n$. *Let A be a finitely bounded internal subset of* $^*\mathbf{E}^n$ *and let $f\colon A \to \mathbf{R}$ be an internal function. Then the Riemann integral of f is given by*

$$\iint_A \cdots \int f(x)\, dx^1 \wedge dx^2 \wedge \cdots \wedge dx^n$$
$$= \mathrm{st}\left(\sum_{k_n=-\infty}^{\infty} \cdots \sum_{k_1=-\infty}^{\infty} \chi_A(y_k) f(y_k)(\delta x^1 \wedge \delta x^2 \wedge \cdots \wedge \delta x^n)_k\right)$$

for a particular choice of coordinates provided all the oriented sums over points of infinitesimal δ-pavings give the same standard part with $y_k \in \Delta x_k$. The absolute integral of f is given by

$$\int_A f(x)\, dv_n(x) = \mathrm{st}\left(\sum_k \chi_A(y_k) f(y_k) v_n(\Delta x_k)\right)$$

with the same infinitesimal provisos, but without regard to orientation. The sums are $*$-finite since A is bounded. When A and f are standard, apply the definition to $*A$ and $*f$.

(5.8.2) The first problem we encounter is: What kind of sets A can we integrate over? We are willing to assume f is S-continuous, like the constant function $f(x) \equiv 1$. Examining the definition, we see that *A can be assigned "Jordan content" $\int_A dv_n(x)$ if and only if for each infinitesimal paving the sum $\sum v_n(\Delta x_k) \approx 0$ over those Δx_k which meet both A and its complement.* Thus we have the choice of picking $y_k \in A \cap \Delta x_k$ or $y_k \in \Delta x_k \backslash A$, so that in one case we add the contribution of the sum above, while in the other we do not. Piecewise smooth blobs can be assigned Jordan content and we outline this in the following exercises. The points of A with complementary points nearby are the topological boundary (see Chapter 8).

EXERCISE: Show that the volume of a finite internal rectangle with sides of lengths $s_1, \ldots, s_n$ obtained by the sum over a particular δ-paving is the product $s_1 \cdot s_2 \cdots s_n$. If L is a finite linear transformation, show that the sum over a δ-paving of the image of the unit cube $L(\{x : 0 \le x^j \le 1\})$ is nearly $\det(L)$.

EXERCISE: If A is a bounded internal subset of an m-dimensional affine subspace, or m-plane, of $*\mathbf{E}^n$, and $m < n$, show that $\int_A f(x)\, dv_n(x) = 0$ for any finite S-continuous f. Start with $f \equiv 1$.

EXERCISE: If g is internal and S-uniformly differentiable on a set containing A with A contained in an m-plane, $m < n$, show that $g(A)$ has zero content $\int_{g(A)} dv_n(x) = 0$. (*Hint*: See the proof of the change of variables formula.)

EXERCISE: Show that an n-dimensional ball can be assigned content by covering its boundary with the image of a finite number of flat $(n-1)$-disks under the action of smooth functions. Our point above is that any smoothly bordered blob can be assigned content; we had in mind this method of covering the boundary.

(5.8.3) THEOREM *If A* (a finitely bounded internal set) *can be assigned Jordan content* (5.8.2) *and if an internal f is finite and S-continuous for every x in $*\mathbf{E}^n$ with $x \approx a \in A$, then $\int_A f(x)\, dv_n(x)$ exists.*

The reader should be aware that we are essentially assuming that f is continuous on an open neighborhood of a compact set (see Chapter 8 for standard reformulations).

The proof of this theorem in a particular coordinate system is exactly like (5.3.2), you use refinements and the internal maximum of a $*$-finite family of infinitesimals. The coordinate-free requirements of the definition are then met by the proof of the next result taking G to be one of the rigid motions.

EXERCISE: Prove Riemann's theorem on necessity of almost everywhere continuity. *Hint*: Suppose f has the opposite property that there is an infinitesimal partition and a set of indices I such that if i is in I, then there are y_i and z_i in Δx_i with $|f(y_i) - f(z_i)| > \varepsilon$ and $\sum_I v_m(\Delta x_i) > \delta$; δ, ε standard and positive.

An internal mapping g restricted to an internal set U onto V will be called $\mathscr{C}^1$-*S-regular* provided g is S-uniformly differentiable at each point of U (in particular, U is properly contained in the S-open set of S-uniform differentiability) and g^{-1} is defined and S-uniformly differentiable on V. We will just say S-regular if g is S-smooth. We will say an internal set A is compactly contained in U if $a \in A$ implies $a \approx u \in {}^{\sigma}U$ (see Chapter 8 for the standard analog). A standard $\mathscr{C}^1$-regular map is S-regular on a compactly contained set. We make these remarks to indicate to the reader how our hypotheses in the change of variables formula are fulfilled in the special case of a standard mapping.

The S-regular transformations treat infinitesimal volume elements Δx_k very nicely, in fact, a whole paving uniformly transforms into a nonlinear paving which nearly is a parallelepiped paving (see Fig. 5.8.3). We need to make some remarks on what such maps do to orientation before we justify the "uniformly nearly parallelepiped" remark.

Now the orientation of $g(\Delta x)$ is the same as the orientation of Dg_x of the unit cube, since the δ scaled down error is infinitesimal, even compared with

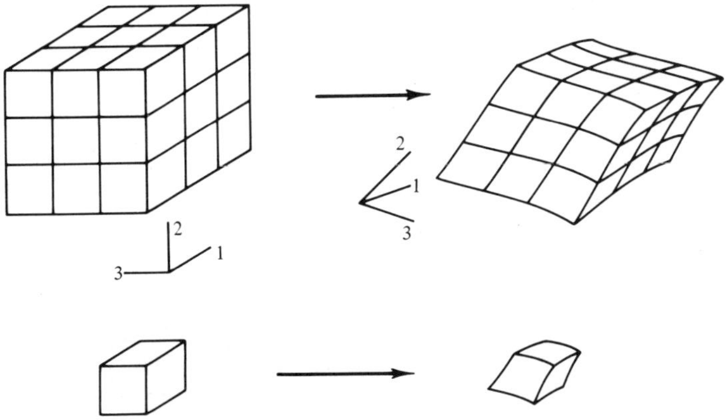

Fig. 5.8.3

δ by S-uniform differentiability. Hence when $\det(Dg_x)$ is negative we need to reverse orientation and otherwise leave it alone. We temporarily denote the change $(\delta x^1 \wedge \cdots \wedge \delta x^n)^g$.

(5.8.4) THE CHANGE OF VARIABLES FORMULA *Let $g: U \to V$ be a $\mathscr{C}^1$-S-regular transformation and let A be finitely bounded and contained in U. If f is internal, finite, and S-continuous on A and A can be assigned Jordan content, then*

$$\int \cdots \int_{g(A)} f(g)\, dg^1 \wedge \cdots \wedge dg^n = \int_A \cdots \int f(g(x)) \det(Dg_x)\, dx^1 \wedge \cdots \wedge dx^n,$$

where the left-hand integral carries the orientation induced by g.

PROOF: Pick an origin and properly oriented δ-paving. Consider one of the n-cubes Δx_k which meets A. Let

$$\eta_k = \sup[|g(x) - g(x_k) - Dg_{x_k}(x - x_k)|/|x - x_k| : x \in \Delta_k],$$

the maximum infinitesimal difference between a point of $g(\Delta x_k)$ and the point in the n-parallelepiped of $Dg_{x_k}(\Delta x_k)$. Now take the $*$-finite maximum of η_k's for $\Delta x_k \cap A \neq \varnothing$, say $\eta \approx 0$. Next, we take an oriented $\eta \cdot \delta$ paving in the same coordinate system we started in, but reversing orientation over the cells where g reverses it. Apply internally what we learned in the above exercises about the oriented volumes of parallelepiped images of linear maps, namely, that the sum of the interior $\eta \cdot \delta$ cubes is infinitesimally close to $\delta^n \cdot \det(Dg_{x_k})$, even compared with δ^n and including orientation $(\delta x^1 \wedge \cdots \wedge \delta x^n)^g$. The determinant is finite by S-regularity. (Scale up, apply the exercise, and scale down.) Now delete the $\eta \cdot \delta$-cubes that touch the boundaries of $g(\Delta_k)$, in fact, only keep those which are inside both the linear and nonlinear parallelograms. This adds up to no more than an infinitesimal times δ^n, since it is no more than an extra boundary layer on the paving of the parallelepiped (see Fig. 5.8.4). This proves

$$\sum \chi_{g(A)}(w_l) f(w_l) v_n(\Delta z_l) \approx \sum \chi_A(y_k) f(g(y_k)) \det(Dg_{y_k}) v_n(\Delta x_k)$$

when f is finite and S-continuous by (5.7.4) and the familiar estimate in terms of the $*$-finite max of infinitesimals of the form $f(w_l) - f(u_l)$; $w_l, u_l \in \Delta z_l$.

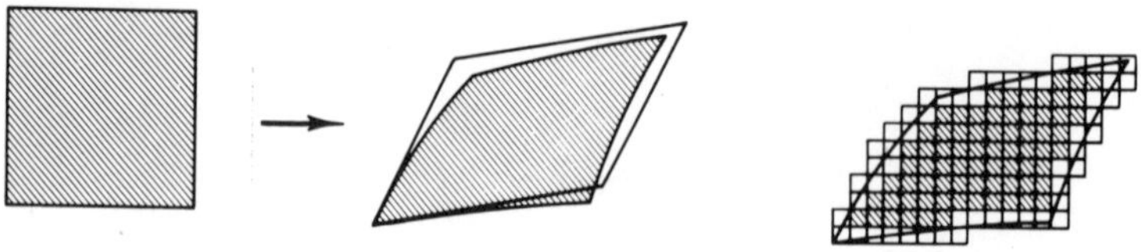

Fig. 5.8.4

We have arranged the orientations so the signs of the oriented sums agree. We will discuss the notation $dg^1 \wedge \cdots \wedge dg^n$ below with more general differential forms.

The moral of the proof of the change of variables formula was simply that parallelograms and even S-regular images of δ-pavings can be used to compute volume, the latter being uniformly infinitesimally approximated by their linear part. We shall now incorporate the linear part of this in our study of m-dimensional volume in n-space, that is, *δ-pavings can now be enlarged to include regular linear images of cubical δ-pavings. A regular linear map is finite and has a finite inverse.* Without further ado we turn to the algebraic study of these parallelepiped δ-paving elements in lower dimensions.

Let δ be a fixed positive infinitesimal. We would like to say an m-dimensional δ-volume element in $\mathbf{E}^n$ is a regular linear image of the canonical m-δ-cube:

$$\Delta^m x = \{x : 0 \le x^j \le \delta \text{ for } 1 \le j \le m \text{ and } x^j = 0 \text{ for } j > m\};$$

in other words, a finite parallelogram on the scale of δ. This allows the linear map to interfere with the variables that are out of our picture $x^{m+1}, \ldots, x^n$ and it does not keep adequate track of orientation. The two things are related since a reflection in x^{m+1} would reverse orientations overall while leaving our canonical m-dimensional element alone.

We can begin to understand the problem by first saying what we mean in the canonical m-space spanned by $e_1, \ldots, e_m$. A regular linear image of $\Delta^m x$ in this subspace makes sense now even including the orientation. The m-dimensional oriented volume is the associated determinant of the restricted linear map. We want to be able to treat volume elements like vectors, so we start with a coordinization of a typical one giving its corners $a_1 = (a_1{}^1, \ldots, a_m{}^1, 0, \ldots, 0)$ through $a_m = (a_1{}^m, \ldots, a_m{}^m, 0, \ldots, 0)$. We denote the new element by the *multivector*

$$a_1 \wedge a_2 \wedge \cdots \wedge a_m, \qquad \text{vertex at zero.}$$

Note that the oriented volume is $\det(a_j{}^i)$ with $i, j = 1, \ldots, m$ (the $m \times m$ determinant, not the $n \times n$ one). The crucial fact about orientation, which we already observed in n dimensions, is that reflections reverse it, so we set

$$a_2 \wedge a_1 \wedge \cdots \wedge a_m = -a_1 \wedge a_2 \wedge \cdots \wedge a_m,$$

and likewise for odd permutations while even ones are the same. This agrees with the assignment of volume. Now we say that $\wedge$ is an associative linear operation subject to alternation $[a \wedge b = -b \wedge a]$ and we can write

$$a_1 \wedge a_2 \wedge \cdots \wedge a_m = \left(\sum_{j=1}^m a_1{}^j e_j\right) \wedge \left(\sum_{j=1}^m a_2{}^j e_j\right) \wedge \cdots \wedge \left(\sum_{j=1}^m a_m{}^j e_j\right)$$

and reorganize this whole sum in terms of the element $e_1 \wedge e_2 \wedge \cdots \wedge e_m$. This is just formal algebra so far, so let us see what it could mean in terms of volume elements (Fig. 5.8.5). The calculation for part (a) of Fig. 5.8.5 is

$$(\delta e_1) \wedge (\delta e_1 + \delta e_2) = (\delta e_1) \wedge (\delta e_1) + (\delta e_1) \wedge (\delta e_2) = (\delta e_1) \wedge (\delta e_2)$$

and the "parallelogram" with both sides δe_1 is the zero multivector, since it has zero area and our minus sign rule requires that it equal its negative. The calculation for part (b) is $a \wedge b + b \wedge a = 0$.

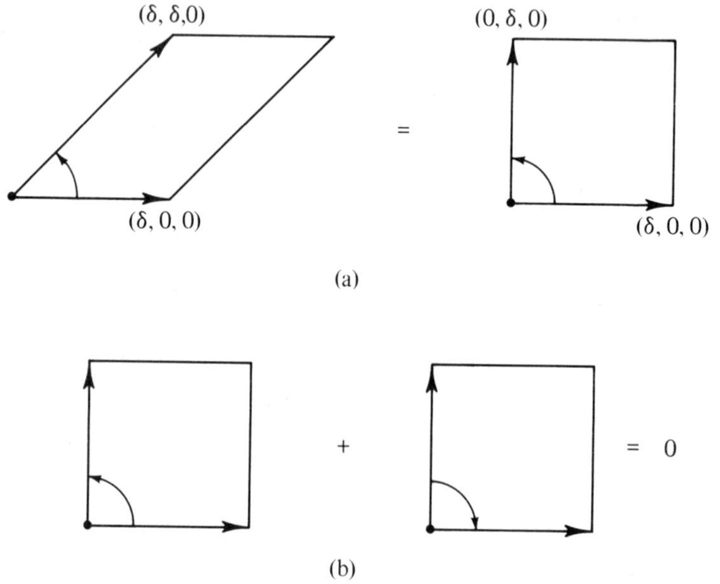

Fig. 5.8.5 (b) Annihilating volumes.

The point of the calculation is that these algebraic rules replace any m-volume element in the canonical subspace with the appropriate multiple of $e_1 \wedge \cdots \wedge e_m$, that is, with an m-cube of the same m-volume and orientation.

EXERCISE: Show that $a_1 \wedge \cdots \wedge a_m = \det(a_j^{\,i}) e_1 \wedge \cdots \wedge e_m$ in this case. (Note the alternating property of interchanging rows or columns for determinants.)

(5.8.5) DEFINITION *The Grassman algebra or exterior algebra* $\Lambda(\mathbf{E}^n)$ *is the linear algebra generated by* $\wedge$*-products of vectors in* $\mathbf{E}^n$ *subject to the rules that* $\wedge$ *is alternating* $a \wedge b = -b \wedge a$, *linear* $(\alpha a + b) \wedge c = \alpha(a \wedge c) + (b \wedge c)$, *and associative* $(a \wedge b) \wedge c = a \wedge (b \wedge c)$. *The algebra is graded into* *subspaces* $\Lambda^m(\mathbf{E}^n)$ *generated by m-fold products and for* $A \in \Lambda^r(\mathbf{E}^n)$, $B \in \Lambda^s(\mathbf{E}^n)$, $A \wedge B = (-1)^{rs} B \wedge A$. *The vectors* $e_k = e_{k_1} \wedge e_{k_2} \wedge \cdots \wedge e_{k_m}$ *with k an increasing sequence form a canonical linear basis for* $\Lambda^m(\mathbf{E}^n)$ *to which we associate an inner product*

$$\langle A, B \rangle = \sum a^k b^k, \qquad |A|^2 = \langle A, A \rangle,$$

where

$$A = \sum a^k e_k \qquad \text{and} \qquad B = \sum b^k e_k.$$

This defines algebra-of-volume elements in a simple way as a universal object so we can do linear calculations easily. We pay for the algebraic convenience by losing some geometry in the form of *indecomposable multivectors*, that is, ones which cannot be written as $a_1 \wedge \cdots \wedge a_m$ for vectors a_i. In a sense these are not mysterious. Take $A = e_{12} + e_{34}$ in $\mathbf{E}^4$. If this is supposed to be a single two-dimensional volume element, then the four-dimensional element $A \wedge A$ should have zero norm or 4-volume. It does not.

It is convenient to let $\Lambda^0(\mathbf{E}^n) = \mathbf{R}$, the scalars with $\wedge$ as scalar multiplication. The reader should observe that $\Lambda^n(\mathbf{E}^n) = \mathbf{R}(e_1 \wedge \cdots \wedge e_n)$ and that $\Lambda^{n+k}(\mathbf{E}^n) = \{0\}$ because $(a \wedge a) \wedge B = 0 \wedge B = 0$, and rewriting an $(n+k)$-fold product in terms of increasing e_j's leads to repetition.

(5.8.6) DEFINITION *An m-dimensional oriented volume element of scale* $\varepsilon > 0$, *or an m-ε-volume element is a decomposable m-multivector* $A = a_1 \wedge \cdots \wedge a_m$ *in* $\Lambda^m(^*\mathbf{E}^n)$ *with* $|a_i|/\varepsilon$, $\varepsilon/|a_i|$, $|A|/\varepsilon^m$, *and* $\varepsilon^m/|A|$ *all finite scalars.* The most important volume elements are on an infinitesimal scale, but with $\varepsilon = 1$ we are talking about a finite m-parallelepiped with finite volume and vertex zero.

PROPOSITION Two m-dimensional volume elements $A = a_1 \wedge \cdots \wedge a_m$ and $B = b_1 \wedge \cdots \wedge b_m$ are equal in $\Lambda^m(^*\mathbf{E}^n)$ if and only if $(a_1, \ldots, a_m)$ and $(b_1, \ldots, b_m)$ are bases of the same m-dimensional subspace of $^*\mathbf{E}^n$ with the same orientation (det of the linear map generated by $a_j \to b_j$ is positive) and span parallelepipeds with vertex at 0 of the same m-dimensional volume equal to $|A| = |B|$.

PROOF: **Not given.** If you do not believe it, forget about the Grassman algebra and concentrate on volume elements in a fixed m-dimensional subspace. The proof does not have anything to do with infinitesimals, so we do not feel that it is appropriate here. The interested reader is referred to a text on multilinear algebra or differential forms.

Multivectors are a little complicated because they essentially are needed only in dimension four and above. In three dimensions we could use ordinary vectors tangent to curves for 1-volume and normal to surfaces for 2-volume. The cross product is the oriented volume of a 2-parallelogram represented on a normal vector. In four dimensions we could use tangents and normals for 1- and 3-volumes, but 2-volumes would be left over.

The *adjoint* of a 0-vector $a \in \mathbf{R}$ is the n-vector $ae_{1 \cdots n}$ and the adjoint of $ae_{1 \cdots n}$ is a. The adjoint of an m-vector, $A = \sum a^k e_k$, for k an increasing sequence, is the vector ${}^{\alpha}A$ whose components are ${}^{\alpha}a^h = a^k \cdot \operatorname{sgn}(h, k)$, where h is the increasing sequence of indices not in k, so $(h, k) = (h_1, \ldots, h_{n-m}, k_1, \ldots, k_m)$ is a permutation of $(1, \ldots, m)$ and sgn refers to the oddness (-1) or evenness $(+1)$ of the permutation. When A is a decomposable m-vector, ${}^{\alpha}A$ is the positively oriented unit normal frame, that is, ${}^{\alpha}A \wedge A$ is a positively oriented n-frame with $|{}^{\alpha}A \wedge A| = |A|$. The general connection with the inner product is

$$(A \wedge {}^{\alpha}B) = (-1)^{m(n-m)}\langle A, B\rangle$$

since

$$^{\alpha\alpha}B = (-1)^{m(n-m)}B.$$

A differential form is a rule for assigning weights times oriented volumes so it is convenient to have the linear dual space of the Grassman algebra and since decomposable multivectors generate the full algebra we define a *multicovector of rank m* to be an *alternating r-linear map*,

$$\varphi\colon \mathbf{E}^n \times \cdots \times \mathbf{E}^n = (\mathbf{E}^n)^m \to \mathbf{R},$$

$$\varphi(x_1, x_2, \ldots, x_m) = -\varphi(x_2, x_1, \ldots, x_m),$$

$$\varphi(ax_1 + by_1, x_2, \ldots, x_m) = a\varphi(x_1, \ldots, x_m) + b\varphi(y_1, x_2, \ldots, x_m).$$

This means $\varphi(x_1 \wedge x_2 \wedge \cdots \wedge x_m)$ is defined compatibly with the algebra of $\wedge$; specifically, it is well defined.

We let e^k be the dual basis so that $e^k(e_l) = \delta_l{}^k = \delta_{l_1 \cdots l_m}^{k_1 \cdots k_m}$ and let $e^k \wedge e^l = e^{k,l}$, extending linearly to all multicovectors. Let $\langle\ ,\ \rangle$ denote the induced inner product $\sum a_k b_k$. We denote these spaces by $\Lambda_m(\mathbf{E}^n)$ for the dual of $\Lambda^m(\mathbf{E}^n)$.

The *adjoint* of an m-covector can be defined in the dual to the way that adjoint of m-vector was defined: $\omega = \sum \omega_k e^k$, increasing k,

$$({}^{\alpha}\omega)_h = \omega_k \operatorname{sgn}(h, k),$$

${}^{\alpha}\omega = \sum {}^{\alpha}\omega_h e^h$, increasing h, where (h, k) is a permutation of $(1, \ldots, n)$. Again, ${}^{\alpha}(\omega \wedge {}^{\alpha}\Omega) = (-1)^{m(n-m)}\langle\omega, \Omega\rangle$ and ${}^{\alpha\alpha}\omega = (-1)^{m(n-m)}\omega$.

When a linear map $L\colon \mathbf{E}^n \to \mathbf{E}^n$ is given, there are *induced* maps L:

$\Lambda^m(\mathbf{E}^n) \to \Lambda^m(\mathbf{E}^n)$ given on decomposable vectors by $L(a_1 \wedge \cdots \wedge a_m) = (La_1) \wedge \cdots \wedge (La_m)$ and dual maps $L'\colon \Lambda_m(\mathbf{E}^n) \to \Lambda_m(\mathbf{E}^n)$ given by

$$[L'(z), a] = [z, L(a)]$$

for multicovectors z applied to multivectors a.

Generally speaking, when R is an orthogonal transformation from one inner product space with ordered basis (orientation) onto a second, the following diagrams can be completed preserving all the above structure:

$$\begin{array}{ccc} V^+ \xrightarrow{R} W^+ & \qquad & V_+' \xleftarrow{R'} W_+', \\ \Lambda^m(V^+) \overset{R}{\dashrightarrow} \Lambda^m(W^+), & \qquad & \Lambda_m(V_+') \overset{R'}{\dashleftarrow} \Lambda_m(W_+'). \end{array}$$

(5.8.7) DEFINITION (Standard) *Let U be a subset of $\mathbf{E}^n$. A differential form of rank m or m-form on U is a smooth map from U into $\Lambda_m(\mathbf{E}^n)$.* In more descriptive terms, it is a smooth assignment of weights to oriented linear volume elements with vertex at a point of U. (Smooth means $\mathscr{C}^\infty$, so U must be open.)

The form $f(x)\,dx^1 \wedge \cdots \wedge dx^n$ under the sign of the Riemann integral assigns the ordinary oriented volume times $f(x)$ for each parallelepiped at x, $dx^1 \wedge \cdots \wedge dx^n$ is the coordinate expression for $e^1 \wedge \cdots \wedge e^n$ above. The terms in the Riemann sums $f(x_k)\,dx^1 \wedge \cdots \wedge dx^n\,(x^1_{(k_1+1)} - x^1_{k_1}, \ldots, x^n_{(k_n+1)} - x^n_{k_n})$ equal $f(x_k)(\delta x^1 \wedge \cdots \wedge \delta x^n)_k$ in the notation of the definition of the n-dimensional integral.

An internal differential m-form is an internal map as above; this means that if we write

$$\omega_x = \sum \omega_{i_1 \cdots i_m}(x) e^{i_1 \cdots i_m} = \sum \omega_{i_1 \cdots i_m}(x)\, dx^{i_1} \wedge \cdots \wedge dx^{i_m},$$

the coordinate functions $\omega_{i_1 \cdots i_m}(x)$ are smooth internal functions. We could similarly talk about an S-uniformly differentiable m-form, if the coordinate functions satisfy that property, an S-continuous m-form, and so on.

(5.8.8) DEFINITION The *exterior derivative* d is defined by: A 0-form is a smooth scalar function $f(x)$, and so we define

$$df_x = \frac{df}{\partial x_1}(x)\,dx^1 + \cdots + \frac{\partial f}{\partial x_n}(x)\,dx^n$$

in this case, while

$$d\omega = \sum d\omega_{i_1 \cdots i_m} \wedge dx^{i_1} \wedge \cdots \wedge dx^{i_m}$$

when

$$\omega = \sum \omega_{i_1 \cdots i_m}\, dx^{i_1} \wedge \cdots \wedge dx^{i_m}.$$

The *codifferential* of an m-form ω is given by

$$\delta\omega = (-1)^{n(m+1)+1}[^{\alpha}(d^{\alpha}\omega)].$$

The *Laplace–Beltrami* operator is $\Delta = \delta d + d\delta$, and on 0-forms it reduces to $-\sum_{i=1}^{n} \partial^2 f/(\partial x^i)^2$ as the reader should check by calculation.

Since our forms are smooth, the second derivatives are symmetric and $\partial^2\omega/\partial x^i\,\partial x^j \cdots dx^i \wedge dx^j \cdots$ cancels $\partial^2\omega/\partial x^j\,\partial x^i \cdots dx^j \wedge dx^i \cdots$, making $d(d\omega) = 0$ for every form ω.

EXERCISE: Compute this for $f(x, y)$, a 0-form on $\mathbf{E}^2$.

Now suppose ω is a finite $\mathcal{S}$-uniformly differentiable m-form and $a_1 \wedge \cdots \wedge a_{m+1}$ is an $(m+1)$-ε-volume element with a vertex at zero. Then $d\omega_0(a_1 \wedge \cdots \wedge a_{m+1})$ is defined and we wish to connect its value to the value of $\omega(\partial[a_1 \wedge \cdots \wedge a_{m+1}])$ where $\partial[\ ]$ denotes the operation of taking the boundary with its induced orientation: $\partial[a_1 \wedge \cdots \wedge a_{m+1}]$ consists of

$-a_2 \wedge \cdots \wedge a_{m+1}$	with vertex at 0,
$a_2 \wedge \cdots \wedge a_{m+1}$	with vertex at a_1,
$a_1 \wedge a_3 \wedge \cdots \wedge a_{m+1}$	with vertex at 0,
$-a_1 \wedge a_3 \wedge \cdots \wedge a_{m+1}$	with vertex at a_2,
$\vdots$	$\vdots$
$(-1)^m a_1 \wedge \cdots \wedge a_m$	with vertex at 0,
$(-1)^{m+1} a_1 \wedge \cdots \wedge a_m$	with vertex at a_{m+1}.

The evaluation of ω on the boundary takes the vertex into account. When a_1 is a 1-vector, we let $\partial[a_1]$ consist of

$-[0]$,	the point at zero
$[a_1]$,	the point at a_1.

Then $\omega(\partial[a_1]) = \omega_{a_1} - \omega_0$ when ω is a 0-form. This sign convention is the orientation induced on the boundary by saying the boundary component plus the outward normal (in the m-space) should have the same orientation as $a_1 \wedge \cdots \wedge a_{m+1}$. In general we have,

$$\begin{aligned}\omega(\partial[a_1 \wedge \cdots \wedge a_{m+1}]) &= \sum_{i=1}^{m+1} (-1)^i\{\omega_{a_i}(a_1 \wedge \cdots \wedge [a_i] \wedge \cdots \wedge a_{m+1}) \\ &\quad - \omega_0(a_1 \wedge \cdots \wedge [a_i] \wedge \cdots \wedge a_{m+1})\} \\ &= \sum_{i=1}^{m+1} (-1)^i\{D\omega_0(a_i)(a_1 \wedge \cdots \wedge [a_i] \wedge \cdots \wedge a_{m+1}) \\ &\quad + |a_i| \cdot \eta(a_1 \wedge \cdots \wedge [a_i] \wedge \cdots \wedge a_{m+1}]\},\end{aligned}$$

where $\eta \approx 0$ in $\Lambda_m(^*\mathbf{E}^n)$, by the uniform differentiability. This proves

(5.8.9) THEOREM *If ω is a finite S-uniformly differentiable internal m-form, $a_1 \wedge \cdots \wedge a_{m+1}$ an $(m+1)$-ε-volume element with vertex at 0, and ε infinitesimal, then*

$$\omega(\partial[a_1 \wedge \cdots \wedge a_{m+1}]) = d\omega_0(a_1 \wedge \cdots \wedge a_{m+1}) + \varepsilon^{(m+1)} \cdot \zeta_0$$

where ζ_0 is an infinitesimal scalar.

We now wish to discuss Riemann-type integrals of m-forms. We saw above that the Riemann integral of a finite S-continuous n-form exists over sets which can be assigned Jordan n-content, the analogous details for what follows are left to the reader.

Let an m-1-volume element $a_1 \wedge \cdots \wedge a_m$ be given. This spans an m-dimensional subspace V of $^*\mathbf{E}^n$. An oriented ε-paving of V is the family $(\varepsilon a_1) \wedge \cdots \wedge (\varepsilon a_m)$ with vertex zero and all its translates to successive vertices. A subset A of V can be assigned m-content provided it is finitely bounded and the sum of ε^m over the number of elements of an ε-paving which meet both A and $V \backslash A$ is infinitesimal.

(5.8.10) DEFINITION Let $a_1 \wedge \cdots \wedge a_m$, V, and A be as above. *Let ω be an internal finite S-continuous m-form defined in a neighborhood of A. The oriented integral of ω*

$$\int_{A+} \omega = \mathrm{st}\left(\sum_{\varepsilon\text{-paving}} \chi_A(y_k)\omega_{y_k}((\varepsilon a_1) \wedge \cdots \wedge (\varepsilon a_m))\right),$$

where y_k lies anywhere inside the kth element of the paving. The answer does not depend on ε or $a_1 \wedge \cdots \wedge a_m$ except for orientation, that is, if $b_1 \wedge \cdots \wedge b_m$ spans V and has the same orientation, the sum is nearly the same.

(5.8.11) FUNDAMENTAL THEOREM OF CALCULUS FOR PARALLELEPIPEDS *Let P be a parallelepiped with vertex at a_0 spanned by the m-1-volume element $a_1 \wedge \cdots \wedge a_m$. Let ω be an internal finite SU-differentiable $(m-1)$-form. Then*

$$\int_{P^+} d\omega = \int_{\partial P} \omega.$$

PROOF: Take an ε-partition of P, write the Riemann sum for $\int_{P^+} d\omega$, and apply (5.8.9). All the terms in the resulting sum cancel except for the Riemann sums of $\int_{\partial P} \omega$ and an error term no bigger than $\varepsilon \cdot \mathrm{vol}(P)$.

The argument in the change of variables formula could be easily extended to include $\mathscr{C}^1$-S-regular maps from one m-dimensional affine subspace into another. When the image is not an affine subspace, but a curved m-dimensional manifold embedded in $\mathbf{E}^n$, the m-dimensional integral is not yet defined and with good reason. Even for a cylinder (with only one direction of curvature) in $\mathbf{E}^3$ one cannot say

$$\iint_C f(x, y)\, dS(x, y) = \lim_{|\Delta S| \to 0} \sum f(x, y)\, \Delta S$$

(as is frequently seen in "applied" calculus texts) without provisos on how the "mesh" ΔS tends to zero. The reader should prove that the infinitesimal mesh we are about to describe (**Schwarz' accordion**) is not locally obtainable by an S-regular map applied to a paving.

Let $C = \{(x, y, z) \mid x^2 + y^2 = 1, 0 \leq z \leq 1\}$. Let $\Omega \in {}^*\mathbf{N}_\infty$. Slice along the z-axis Ω^Ω times, equally far apart. Cut around the top of the first slice Ω times and around the bottom half way and continue down, staggering each time. In polar coordinates we have points on C at

$$(r, \theta_h, z_k) = \begin{cases} \left(1, \dfrac{h}{\Omega} 2\pi, \left(1 - \dfrac{k}{\Omega^\Omega}\right)\right); & k \text{ even} \\[2ex] \left(1, \dfrac{h + \frac{1}{2}}{\Omega} 2\pi, \left(1 - \dfrac{k}{\Omega^\Omega}\right)\right); & k \text{ odd,} \end{cases}$$

$1 \leq h \leq \Omega, 0 \leq k \leq \Omega^\Omega$. Now connect these points by affine triangles in $\mathbf{E}^3$.

EXERCISE: Draw the picture of this from front and top. Show that each triangle nearly lies in a horizontal plane. Compute the unit normal and show that it is nearly $(0, 0, 1)$.

The area of one of these triangles is

$$\begin{aligned} \frac{1}{2} b \cdot h &= \left(\left[1 - \cos\left(\frac{2\pi}{\Omega}\right)\right]^2 + \left[\frac{1}{\Omega^\Omega}\right]\right)^{1/2} \left(\sin\left(\frac{2\pi}{\Omega}\right)\right) \\ &= \left[\frac{1}{2}\left(\frac{\pi}{\Omega}\right)^2 + \left(\frac{1}{\Omega}\right)^4 [a + \varepsilon]\right]^{1/2} \left[\frac{2\pi}{\Omega} + \left(\frac{1}{\Omega}\right)^2 \cdot \delta\right] \\ &= \sqrt{2}\left(\frac{\pi}{\Omega}\right)^2 + \left(\frac{1}{\Omega}\right)^3 \cdot b = \sqrt{2}\left(\frac{\pi}{\Omega}\right)^2 + \mathcal{O}\left(\frac{1}{\Omega}\right)^3. \end{aligned}$$

Since there are more than Ω^Ω triangles to total area of the inscribed triangles exceeds $\sqrt{2}\,\pi^2\Omega^{\Omega-2}$, not a very good approximation to π to say the least. The moral of this example is that cutting C into infinitesimal triangles does not generally lead to an approximation by linear triangles, instead we must require that the cutting be done so that the triangles are nearly tangent. Obtaining area correctly in this way leads to some fascinating problems (see Whitney [1957] and Almgren [1966]), but the change of variables formula provides an easy way around the difficulties and is essentially forced on us if we begin with an abstract manifold. This easy approach does present some difficulty in showing that the area of an embedded manifold is invariant under rigid motions of the big space. Perhaps the reader might be interested in developing infinitesimal tangential approximation for smooth surfaces directly. Notice that an S-regular image of the vertices of an infinitesimal ε-paving produces nearly tangent infinitesimal rectangles whose m-volume is ε^m-nearly given by the change of variables formula.

The n-dimensional invariance of m-dimensional integrals obtained through change of variables requires a generalized Pythagorean theorem:

(*Geometric version*): The volume of an m-simplex σ in $\mathbf{E}^n$ is equal to $[\sum \mathrm{vol}_m(\sigma_k)^2]^{1/2}$, where σ_k is the projection of σ onto the plane spanned by $(e_{k_1}, \ldots, e_{k_m})$ and the sum is over all increasing m-sequences.

(*Algebraic version*): Let A be an $n \times m$ matrix and $B = RA$ where R is orthogonal. The sum of the squares of the $m \times m$ subdeterminants (there are $\binom{n}{m}$ of them) of A is the same as sum of the squares of the subdeterminants of B. This is quite interesting, but off the subject of infinitesimals, so we omit the details.

We take the easy route in defining integrals in the next section.

(5.8.12) CHANGE OF VARIABLES FOR DIFFERENTIAL FORMS: PULL-BACKS Let $g\colon U \to V \subseteq \mathbf{E}^n$ be a C^1 map defined on an open set. Let ω be an m-form on V, and ω^g the differential form on U defined by replacing dx^i in the expression for ω by $dg^i(x)$. More precisely, $\omega^g(x) = (Dg_x)'[\omega(g(x))]$, where $(Dg_x)'$ is the induced linear transformation described above. For 0-forms we take $\omega(g(x))$. In terms of the matrix of mixed partial derivatives,

$$(dx^{i_1} \wedge \cdots \wedge dx^{i_m})^g = \sum_j \det(\partial g^{i_h}/\partial x^{j_k})\, dx^{j_1} \wedge \cdots \wedge dx^{j_m}.$$

The change of variables formula now reads $\int_{g(A)} \omega = \int_{A^+} (\omega)^g$ with the induced orientation, when g is an S-regular transformation taking one m-dimensional affine space into another, when A can be assigned m-content, and when ω is S-continuous and finite. The notation is cryptic, but the ideas are simple enough—we can apply the proof of the n-dimensional change of variables formula in the subspaces.

EXERCISE: Show that the iterated integrals formula

$$\int_R \cdots \int f(x)\, dx^1 \wedge \cdots \wedge dx^n = \int_{a_n}^{b_n} \left(\cdots \left(\int_{b_n}^{b_1} f(x)\, dx^1 \right) \cdots \right) dx^n$$

holds for a positively oriented rectangle R and S-continuous finite f.

EXERCISE: *Leibniz' rule for differentiating under the integral* says that if $\partial f(x, y)/\partial y$ is continuous and standard and A is compact, then

$$D\left[\int_A f(x, y)\, dm(x)\right]_y (h) = \int_A \partial f(x, y)(h)/\partial y\, dm(x),$$

where dm represents any of our previous finite-dimensional integrals. The main idea is to estimate $\eta(x)$ given by $(1/|h|)[f(x, y+h) - f(x, y) - \partial f(x, y)(h)/\partial y]$ uniformly in x as $|h| \to 0$. Give a nonstandard proof of Leibniz' rule and generalize to internal f with the appropriate S-continuity assumptions.

5.9 CALCULUS ON MANIFOLDS

Let M be a paracompact connected $\mathscr{C}^k$-*manifold* modeled on $\mathbf{R}^m$ with $1 \le k \le \infty$. This means M is a connected metric space covered by an *atlas* of *charts* which are $\mathscr{C}^k$-related. A chart (x, U) is a homeomorphism $x: U \to \mathbf{R}^m$ from an open subset U of M. Two charts (x, U) and (y, V) are $\mathscr{C}^k$-related provided $y \circ x^{-1}: x(U \cap V) \to y(U \cap V)$ and $x \circ y^{-1}: y(U \cap V) \to x(U \cap V)$ are $\mathscr{C}^k$ maps in $\mathbf{R}^m$. The reader who is unfamiliar with this abstract definition may wish to think of a surface $(m = 2)$ embedded in $\mathbf{E}^3$ with the charts being projections onto a plane which we move around (see Fig. 5.9.1). To keep the projections smooth and invertible we restrict the domain to a part of the surface and this gives us a chart.

We have said M is a metric space from the start in order to avoid some unnecessary abstraction; to begin with, what we really require is a notion of "infinitesimally near a standard point," and we may as well measure this in standard coordinates. When $p \in {}^\sigma M$ there is a standard chart such that $p \in {}^\sigma U$ and $q \approx p$ in $^*M$ if and only if $x(p) \approx x(q)$ in $^*\mathbf{R}^m$. (We refer the reader to Chapter 8 for general topology via monads in their full-blown generality.) Throughout this section we will focus primarily on the near-standard part of $^*M$ (where the notion of "infinitesimally close" is actually topological in Hausdorff spaces). The near-standard points of $^*M$ are denoted $\mathrm{ns}(^*M)$.

The way one tests a map $f: M \to N$ between abstract manifolds for differentiability properties is to "pull down" to coordinates on both sides,

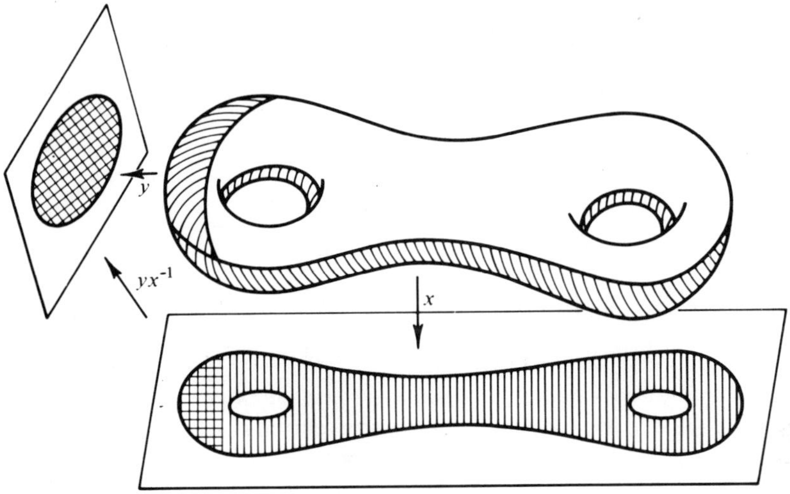

Fig. 5.9.1

that is, whenever (x, U) is a chart on M and (y, V) is a chart on N we can test $y \circ f \circ x^{-1}$ as a map from $\mathbf{R}^m$ to $\mathbf{R}^n$. To test S-differentiability we must require that the charts be standard (unless there is a preferred metric). A diffeomorphism on M has a $\mathscr{C}^k$-inverse.

(5.9.1) DEFINITION *Let δ be a fixed positive infinitesimal. A first-order δ-infinitesimal transformation on $^*M$ is an internal diffeomorphism $X\colon {}^*M \to {}^*M$ such that given standard charts $(x, {}^*U)$ and $(y, {}^*V)$, the map*

$$\bar{X}(u) = (1/\delta)\{y \circ X \circ x^{-1}(u) - y \circ x^{-1}(u)\}$$

is a k times SU-differentiable function on the near standard $u \approx a \in {}^\sigma U$ (that is, the small oh Taylor formulas have finite S-continuous terms). Two infinitesimal transformations X and Y are first-order δ-equivalent at u if $yXx^{-1}(u) = yYx^{-1}(u) + \delta \cdot \eta$ with η infinitesimal and x, y, u as above. We denote the set of infinitesimal transformations by $\delta\mathscr{X}M$.

EXERCISE: Show that first-order equivalence is actually an equivalence relation.

If $G\colon M \to N$ is S-regular, a δ-infinitesimal transformation on M can be transferred to N by $X \to G \circ X \circ G^{-1}$. The scale δ is unchanged, because of finite growth in G.

The idea here is simple enough; we pull down in standard coordinates and require $\overline{X}$ to have a $\mathscr{C}^k$-standard part term by term. In other words, X moves things $S\mathscr{C}^k$-smoothly on the scale of δ.

Standardly, X can be thought of as a vector field on M and in fact if we pick a standard chart $(x, {}^*U)$ around p in ${}^\sigma M$, then xXx^{-1} is defined on a finite neighborhood of $x(p)$, since X only moves points an infinitesimal amount. Since $\overline{X}$ is SU-differentiable, we can solve

$$du/dt = \mathrm{st}[\overline{X}(u)]$$

for the local flow

$$\Phi(t, u)$$

on a neighborhood of $x(p)$. Posed this way as a standard differential equation, the reader could apply the conventional existence–uniqueness theorem for $\mathscr{C}^1$ first-order equations. (Consult Hirsch and Smale [1974] and compare this with Peano's theorem below.)

Since $\overline{X}$ is an S-uniformly differentiable function we can apply that property to obtain immediately an infinitesimal approximation to $\Phi(t, u)$ as follows. Partition the time axis by multiples of $\delta, \ldots, -2\delta, -\delta, 0, \delta, 2\delta, \ldots$. To each $t \in {}^*\mathbf{R}$ assign a nearest $n \cdot \delta$, call the integer part $n = n_t$ (take the left side at the midpoint). The map $t \mapsto n_t$ is internal by the internal definition principle. Now define

$$\varphi(t, u) = \underline{X}^{n_t}(u) = \underline{X} \circ \cdots \circ \underline{X}(u) \qquad (n_t \text{ times}),$$

where $\underline{X} = xXx^{-1}$. This is defined for a finite time interval for each standard u by Cauchy's principle, since it is defined for all infinitesimal t.

Show that $\Phi(t, u) = \mathrm{st}\, \varphi(t, u)$. Show that Φ has the local group property and the correct derivative. Note that $[\underline{X}^{n_t}(u) - u] = [\underline{X}^{n_t}(u) - \underline{X}^{n_t - 1}(u)] + \cdots + [\underline{X}(u) - u]$ with n_t terms in the sum and apply (5.7.5). Prove the form of the existence–**uniqueness** theorem in the work of Hirsch and Smale using (5.7.5).

We lift the local flow onto M by taking $\Psi(t, q) = x^{-1}\Phi(t, x(q))$ where it is defined. We now show that $\Psi(\delta, q)$ is locally equivalent to $X(q)$:

$$d\Phi(o, u)/dt \approx (1/\delta)[\Phi(\delta, u) - u] \approx \overline{X}(u) = (1/\delta)[xXx^{-1}(u) - u]$$

and thus, $\Phi(\delta, u) = xXx^{-1}(u) + \delta \cdot \eta$.

One preliminary result is as follows, since $\overline{X}$ is SU-differentiable, whenever $v \approx u \approx a \in {}^\sigma x(U)$, then

$$(1/\delta)[xXx^{-1}(v) - xXx^{-1}(u) - (v - u)] = D\overline{X}_a(v - u) + |v \cdot u| \cdot \eta,$$

where $D\overline{X}_a$ is a finite linear transformation. Therefore, the left-hand side is infinitesimal.

Infinitesimal transformations naturally form a group under composition of maps. The composition $X \circ Y$ is equivalent to $Y \circ X$, and in fact, both correspond to $\overline{X} + \overline{Y}$ in coordinates. Let $(x, {}^*U)$ be a standard chart and write $\underline{X}$ for $x \circ X \circ x^{-1}$ and $\underline{Y}$ for $x \circ Y \circ x^{-1}$ on a finite neighborhood of $a \in {}^{\sigma}U$. We have then for $X \circ Y$,

$$\begin{aligned}(1/\delta)[x \circ X \circ Y \circ x^{-1}(u) - u] &= (1/\delta)[\underline{X} \circ \underline{Y}(u) - u] \\ &= (1/\delta)[\underline{X} \circ \underline{Y}(u) - \underline{Y}(u) + \underline{Y}(u) - u] \\ &= (1/\delta)[\underline{X}(v) - v] + (1/\delta)[\underline{Y}(u) - u]\end{aligned}$$

with $v = \underline{Y}(u)$, and therefore, by the above calculation,

$$\begin{aligned}&\approx (1/\delta)[\underline{X}(u) - u] + (1/\delta)[\underline{Y}(u) - u] \\ &= \overline{X}(u) + \overline{Y}(u).\end{aligned}$$

Similarly, $Y \circ X$ is also locally equivalent to $\overline{X}(u) + \overline{Y}(u)$.

We can include scalar multiplication by finite scalars in the structure of infinitesimal transformations by the local definition $aX = Y$, where $\overline{Y}(u) = a\overline{X}(u)$.

The last part in the verification of the above statement that δ-infinitesimal transformations form a group is a demonstration that $(1/\delta)[\underline{X}^{-1}(u) - u]$ has finite S-continuous terms in the Taylor formula of order k. (This is essentially the exercise at the end of Section 5.7.)

Since $D\underline{X}_u = I + \delta D\overline{X}_u$ with $D\overline{X}$ finite and S-continuous, we may apply the $*$-inverse function theorem and obtain the fact that $\underline{X}^{-1}$ is ${}^*\mathscr{C}^k$. Next, $D\underline{X}_b^{-1} \approx I$, since

$$I = D\underline{X}_b^{-1} \circ D\underline{X}_a = D\underline{X}_b^{-1} \circ (I + \delta A) = D\underline{X}_b^{-1} + \delta B,$$

with B finite by (5.7.11).

The Taylor polynomial for $\underline{X}^{-1} \circ \underline{X}(u) = a + I(u - a) +$ zero, whereas, the composite formula [from the exercise before (5.7.9)] gives

$$0 = D^2(X^{-1} \circ X)_a = c_1 D\underline{X}_b^{-1}(D^2\underline{X}_a(\ )) + c_2 D^2\underline{X}_b^{-1}(D\underline{X}_a(\ ), D\underline{X}_a(\ )).$$

Now let u and v be finite vectors and

$$D\underline{X}_a(u - \zeta) = u, \qquad D\underline{X}_a(v - \theta) = v$$

where ζ and θ are infinitesimal since $D\underline{X} \approx I$. Now,

$$(1/\delta)[D^2\underline{X}^{-1}(u, v)] = -(c_1/c_2)[D\underline{X}_b^{-1}((1/\delta)D^2\underline{X}_a(u - \zeta, u - \theta))]$$

with the right-hand side finite. Generally,

$$0 = D^k(\underline{X}^{-1} \circ \underline{X})_a = \sum_{j=1}^{k} \sum c_j D^j \underline{X}_b^{-1} \circ D^{i_1}\underline{X}_a \circ \cdots \circ D^{i_j}\underline{X}_a$$

so that we may proceed inductively showing $D^k\underline{X}_b^{-1}$ is δ-finite. This proves that X^{-1} is a first-order δ-infinitesimal transformation, and from the fact that it is the inverse in the full group of diffeomorphisms, it follows that it is the inverse in the infinitesimal trånsformations as well.

Let δ be a positive infinitesimal.

(5.9.2) DEFINITION *The δ-finite tangent bundle of* $^*M$ *is the set* $\delta TM = \{(p, r) : p$ *is near standard on* $^*M$ *and* $r = X(p)$ *with* $X \in \delta\mathscr{X}M$, *the set of first-order δ-infinitesimal transformations*$\}$. *In* δTM *two points are δ-infinitesimally close,* $(p, r) \approx (q, s)$, *provided* $p \approx q$ *in* $^*M$ *and* $r = X(p)$, $s = Y(q)$ *with* X *and* Y *δ-equivalent at* p *and* q. The projection $\pi\colon \delta TM \to {}^*M$ is defined by $\pi(p, r) = p$ and $\pi^{-1}(p)$ is denoted by $\delta T_p M$. The tangent bundle of M denoted TM is $\delta TM/\approx$ or $\widehat{\delta TM}$, the infinitesimal hull, with the inherited structure of a $\mathscr{C}^k$-$\mathbf{R}^m$-vector bundle over M.

Our foregoing remarks were to the effect that each fiber $\delta T_p M$ is nearly $\mathscr{O}(\mathbf{R}^m)$ with the vector space structure coming from composition, better perhaps, $\delta T_p M/\approx\ \simeq \mathbf{R}^m$.

The local product structure of TM comes from the map $(p, r) \mapsto (x(p), \overline{X}[(x, r)]) \xrightarrow{\text{st}} x(U) \times \mathbf{R}^m$, where infinitesimal perturbations produce no change in the standard part.

(5.9.2) CONTINUED FOR MAPS *When* $f\colon M \to N$ *is SU-differentiable, the map* $Tf\colon TM \to TN$ *is given by picking a representative* $(p, X(p)) \in \delta TM$ *and sending it to the infinitesimal hull of the point* $\widehat{(f(p), f(X(p)))}/\approx$ *in* $\delta TM/\approx$, *or in the hull notation,* $Tf(\hat{p}, \hat{r}) = (\widehat{f(p)}, \widehat{f(r)})$ [but note that the caret in the first coordinate is different from the caret in the second].

(5.9.3) DEFINITION (Lie derivatives) *Let f be an SU-differentiable internal map with values in a normed space* $\mathbf{F}$. *The action of a δ-infinitesimal transformation X on f is given by*

$$X_p f = Xf(p) = (1/\delta)[f(X(p)) - f(p)]$$

for near-standard p on $^*M$ *and*

$$\hat{X}f(p) = \text{st}(Xf(p))$$

for $p \in M$.

The standard interpretation here is that X acts as a weighted directional derivative in the direction of $\overline{X}$, i.e.,

$$\begin{aligned}(1/\delta)[f(X(p)) - f(p)] &= D(f \circ x)_{x(p)}(\overline{X}) + |xX(p) - x(p)| \cdot \eta \\ &\approx |\overline{X}| D(f \circ x)_{x(p)}(\overline{X}/|\overline{X}|).\end{aligned}$$

(5.9.3) CONTINUED FOR SCALAR FIELD MULTIPLICATION OF X *If α is an S-continuous function on M, we write $\alpha \cdot X$ to mean the action* $\alpha \cdot [Xf](p) = \alpha(p)(Xf)(p)$.

EXERCISE: Show that this is compatible with the scalar multiplication given locally by $a \cdot \overline{X}$ above in the case where α is S-k-differentiable, that is, $\alpha X = Y$ where $\overline{Y}(u) = \alpha(x(u))\overline{X}(u)$ in coordinates.

EXERCISE: Show that when f and g are real valued and SU-differentiable, a and b are scalars, and X and Y infinitesimal transformations, that

(1) $X(af + bg) = a(Xf) + b(Xg)$,
(2) $X(f \cdot g) \approx f \cdot (Xg) + g \cdot (Xf) \approx (fX)g + (gX)f$,
(3) $(X \circ Y)f \approx (Y \circ X)f \approx Xf + Yf$, hence it makes sense to write $X + Y$ for $(X \circ Y)$ considered as a derivation. Finally, you can apply carets any way you like and still end up with the same answer.

One principal application of the Lie derivative of a scalar function is in the first integration theorem on a manifold—the Gauss divergence theorem, which we shall discuss below.

Now we study the noncommutativity of the group of infinitesimal transformations. By ignoring "very infinitesimal" errors we managed to turn the group into a vector space, and we simply do not want to lose all that structure (a geometric reason is given in Section 6.6).

The commutator of two transformations is $Y^{-1} \circ X^{-1} \circ Y \circ X$; up to first-order equivalence this is $\overline{X} + \overline{Y} - \overline{X} - \overline{Y} \approx 0$ by a calculation above, so we examine the terms in detail. To do this we shall need more than $\mathscr{C}^1$-smoothness and for brevity we shall simply assume $\mathscr{C}^k = \mathscr{C}^\infty$ for the remainder of the chapter. Smoothness, S-differentiability, etc., now refer to all (finite) derivatives.

Our calculation is no more than a careful trip around the infinitesimal "parallelogram" of Fig. 5.9.2. The first step in the calculation shows that the parallelogram fails to be closed by δ^2 times a finite amount.

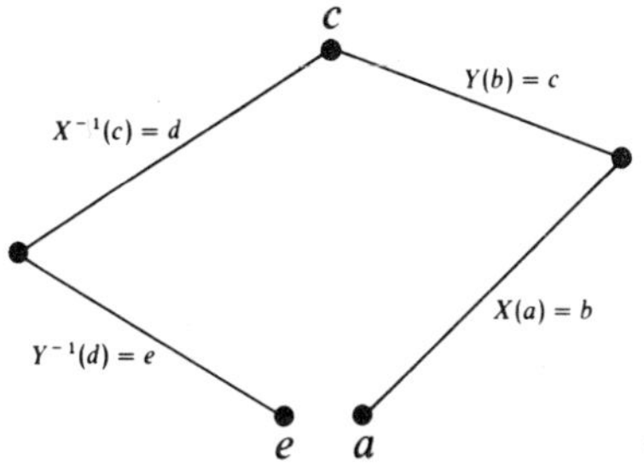

Fig. 5.9.2

If X and Y are δ-infinitesimal transfomations and a is near standard, then

(5.9.4) $$[\underline{Y}(\underline{X}(a)) - \underline{X}(a)] = [\underline{Y}(a) - a] + \delta^2 \cdot \mathcal{O}.$$

This follows from the norm mean value theorem applied to $\overline{Y}$, noting that $|\underline{X}(a) - a| = \delta \cdot \mathcal{O}$ and thus

$$(1/\delta)|\underline{Y}(\underline{X}(a)) - \underline{X}(a) - \underline{Y}(a) + a| \leq |D\overline{Y}_*||\underline{X}(a) - a|.$$

Using this we show that

(5.9.5) $$\underline{Y}^{-1}\underline{X}^{-1}\underline{Y}\underline{X}(a) = a + \delta^2 \cdot \mathcal{O}$$

by altering the sides of the parallelogram:

$$\begin{aligned}\underline{Y}^{-1}\underline{X}^{-1}\underline{Y}\underline{X}(a) - a &= (\underline{Y}^{-1}(d) - d) + (\underline{X}^{-1}(c) - c)\\ &\quad + (\underline{Y}(b) - b) + (\underline{X}(a) - a)\\ &= (\underline{Y}^{-1}(c) - c) + (\underline{X}^{-1}(b) - b)\\ &\quad + (\underline{Y}(b) - b) + (\underline{X}(a) - a) + \delta^2 \cdot \mathcal{O}\\ &= (b - \underline{Y}(b)) + (a - \underline{X}(a))\\ &\quad + (\underline{Y}(b) - b) + (\underline{X}(a) - a) + \delta^2 \cdot \mathcal{O}\\ &= \delta^2 \cdot \mathcal{O}.\end{aligned}$$

EXERCISE: Apply the mean value theorem to show that

(5.9.6) $$D^h\underline{X}_{Y(b)} = D^h\underline{X}_b + \delta^2 \cdot \mathcal{O}$$

when h is finite and X and Y are δ-infinitesimal transformations. Note that $D^{(h+1)}\overline{X} = (1/\delta)D^{(h+1)}\underline{X}$.

When X and Y are δ-infinitesimal transformations,

(5.9.7) $$D\underline{X}_a \circ D\underline{Y}_b = D\underline{Y}_b \circ D\underline{X}_a + \delta^2 \cdot \mathcal{O},$$

because

$$\begin{aligned}D\underline{X} \circ D\underline{Y} &= [I + \delta A] \circ [I + \delta B]\\ &= I + \delta A + \delta B + \delta^2 \cdot \mathcal{O}\\ &= I + \delta B + \delta A + \delta^2 \cdot \mathcal{O}\\ &= [I + \delta B] \circ [I + \delta A] + \delta^2 \cdot \mathcal{O} = D\underline{Y} \circ D\underline{X} + \delta^2 \cdot \mathcal{O}\end{aligned}$$

Now we combine (5.9.6) and (5.9.7) taking $\underline{X}(a) = b$, $\underline{Y}(b) = c$, and $\underline{X}^{-1}(c) = d$, to show

(5.9.8) $$D\underline{Y}_d^{-1} \circ D\underline{X}_c^{-1} \circ D\underline{Y}_b \circ D\underline{X}_a = I + \delta^2 \cdot \mathcal{O},$$

as follows:

$$\begin{aligned} D\underline{Y}_d^{-1} \circ D\underline{X}_c^{-1} \circ D\underline{Y}_b \circ D\underline{X}_a &= D\underline{Y}_c^{-1} \circ D\underline{X}_b^{-1} \circ D\underline{Y}_b \circ D\underline{X}_a + \delta^2 \cdot \mathcal{O} \\ &= D\underline{Y}_c^{-1} \circ D\underline{Y}_b \circ D\underline{X}_b^{-1} \circ D\underline{X}_a + \delta^2 \cdot \mathcal{O} \\ &= I + \delta^2 \cdot \mathcal{O}. \end{aligned}$$

(5.9.9) $$D^2(\underline{Y}^{-1}\underline{X}^{-1}\underline{Y}\underline{X})_a = \delta^2 \cdot \mathcal{O},$$

by a similar trick with a commutation estimate and the composite formula for $D^2(\underline{Y}^{-1}\underline{Y}\underline{X}^{-1}\underline{X})_a = D^2(I) \equiv 0$. The almost commutant is

(5.9.10) $$\begin{aligned} D\underline{Y} \circ D^2\underline{X} &= D^2\underline{X}(D\underline{Y}, D\underline{Y}) + \delta^2 \cdot \mathcal{O}, \\ D\underline{Y} \circ D^2\underline{X}(u, v) &= (I + \delta A) \circ D^2\underline{X}(u, v) \\ &= D^2\underline{X}(u, v) + \delta^2 \cdot \mathcal{O}, \\ D^2\underline{X}((I + \delta A)(u), (I + \delta A)(v)) &= D^2\underline{X}(u, v) + \delta^2(D^2\underline{X}/\delta)(*, *) \\ &= D^2\underline{X}(u, v) + \delta^2 \cdot \mathcal{O}. \end{aligned}$$

We take $\underline{X}(a) = b$, $\underline{Y}(b) = c$, $X^{-1}(c) = d$, and $Y(a) = e$. Then

$$\begin{aligned} 0 &= D^2(\underline{Y}^{-1}\underline{Y}\underline{X}^{-1}\underline{X})_a \\ &= \{c_1 D\underline{Y}_e^{-1}\{D\underline{Y}_a[D\underline{X}_b^{-1}(D^2\underline{X}_a) + D^2\underline{X}_b^{-1}(D\underline{X}_a)^2] + D^2\underline{Y}_a[D\underline{X}_b^{-1} \circ D\underline{X}_a]^2\} \\ &\quad + c_2 D^2\underline{Y}_e^{-1}[D\underline{Y}_a \circ D\underline{X}_b^{-1} \circ D\underline{X}_a]^2\} \\ &= \{c_1 D\underline{Y}_d^{-1}\{D\underline{X}_c^{-1}[D\underline{Y}_b(D^2\underline{X}_a) + D^2\underline{Y}_b(D\underline{X}_a)^2] + D^2\underline{X}_c^{-1}[D\underline{Y}_b \circ D\underline{X}_a]^2\} \\ &\quad + c_2 D^2\underline{Y}_d^{-1}[D\underline{X}_c^{-1} \circ D\underline{Y}_b \circ D\underline{X}_a]^2\} + \delta^2 \cdot \mathcal{O} \end{aligned}$$

by (5.9.10), (5.9.7), (5.9.6), and rearrangement. This proves (5.9.9), using the composite formula for $D^2(\underline{Y}^{-1}\underline{X}^{-1}\underline{Y}\underline{X})$. The calculation from (5.9.4) to here proves the following.

(5.9.11) THEOREM *The commutator $Y^{-1} \circ X^{-1} \circ Y \circ X$ of two δ-infinitesimal transformations (where X and Y are at least $\mathscr{C}^3$-finite) is a δ^2-infinitesimal transformation (with $\mathscr{C}^2$-finiteness).*

The proof that the commutator is δ^2-$\mathscr{C}^\infty$-finite when X and Y are δ-$\mathscr{C}^\infty$-finite is too horrible to contemplate, so naturally we leave it to the reader! (This is partly facetious because of the following standardization.)

The calculation of the zeroth, first, and second derivatives of the commutator above was somewhat technical, but by no means mysterious—it is simply a calculation aimed at placing the noncommutativity of the group of infinitesimal transformations on a finite scale. This motivation is even more compelling in the setting of Lie algebras because one wants to know finitely what the infinitesimal group is like [see Section 6.6]. This approach via direct infinitesimal commutators was *not* taken classically despite the terminology "infinitesimal transformation," which is the classical name for vector fields. The connection with that approach lies in the action of the commutator as a derivation on S-smooth scalar-valued functions as follows (we use at least $S\mathscr{C}^2$):

$$\begin{aligned}
\frac{1}{\delta^2}\{f[Y^{-1}X^{-1}YX(a)] - f(a)\} &= \frac{1}{\delta^2}\left\{f\Big|_d^e + f\Big|_b^c\right\} + \frac{1}{\delta^2}\left\{f\Big|_c^d + f\Big|_a^b\right\} \\
&= \frac{1}{\delta}\left\{\frac{1}{\delta}[f(Y^{-1}X^{-1}(c)) - f(X^{-1}(c))]\right. \\
&\qquad \left. - \frac{1}{\delta}[f(Y^{-1}(c)) - f(c)]\right\} \\
&\quad + \frac{1}{\delta}\left\{\frac{1}{\delta}[f(X^{-1}Y(b)) - f(Y(b))]\right. \\
&\qquad \left. - \frac{1}{\delta}[f(X^{-1}(b)) - f(b)]\right\} \\
&= X_c^{-1} \circ (Y_{\cdot}^{-1} \circ f) + Y_b \circ (X^{-1} \circ f) \\
&= X_c(Yf) - Y_b(Xf),
\end{aligned}$$

since

$$(Y^{-1} \circ f) = (1/\delta)[f(Y^{-1}(\cdot)) - f(\cdot)]$$

and

$$X_c^{-1}(Y^{-1}f) = \frac{1}{\delta}\left\{\frac{1}{\delta}[f(Y^{-1}X^{-1}(c)) - f(X^{-1}(c))] - \frac{1}{\delta}[f(Y^{-1}(c)) - f(c)]\right\};$$

similarly for $Y \circ (X^{-1} \circ f)$. Finally, since f is S-smooth, $[Y^{-1}X^{-1}YX]_a \circ f \approx X_a(Yf) - Y_a(Xf)$, that is, you can move b and c.

(5.9.12) THEOREM *The bracket operation of smooth vector fields on S-smooth functions is nearly the action of the infinitesimal commutator of the infinitesimal transformations, when X and Y are first-order smooth δ-infinitesimal transformations $[X, Y]_a f$ given by $X_a(Yf) - Y_a(Xf)$ is infinitesimally close to $[Y^{-1}X^{-1}YX]_a f$ applied as a δ^2-infinitesimal transformation.*

EXERCISE: Show that $[X, Y] \approx -[Y, X]$. Prove Jacobi's identity:

$$[X, [Y, Z]] + [Y, [Z, X]] + [Z, [X, Y]] \approx 0.$$

Derive a local expression for $[X, Y]$ as a derivation of Y with respect to X:

$$\lim_{t\to 0} (1/t)\{\varphi_{\hat{t}}[\bar{Y}] - [\bar{Y}]\},$$

where φ is the local flow associated with X and $\varphi_{\hat{t}}$ gives $\overline{(X^{-1}YX(a) - a)}$, so a difference quotient looks like

$$(1/\delta)\{[X^{-1}YX(a) - a] - [Y(a) - a]\}.$$

(*Hint*: Draw the parallelogram of infinitesimal displacements.)

The rest of this section merely sketches the role of infinitesimals in three of the main theorems of differentiation and integration on manifolds. Unfortunately, a complete account of these topics would require a separate book, but the reader familiar with these results will see how naturally infinitesimals enter the proofs. The "naturalness" of the proofs rather than the technical ease is what we hope to explicate. (Fine sheaves in de Rham's theorem may be easier, for example, but the geometry enters the proof less clearly.)

(5.9.13) DE RHAM'S THEOREM *The real-valued smooth rectangular cohomology of a compact (bordered) smooth manifold is isomorphic to the de Rham cohomology of closed modulo derived differential forms.*

PROOF: First, a smooth manifold can be embedded in a Euclidean space and "rectangulated" (see Whitney [1957]). By "rectangulated" we mean that there is a complex K composed of a finite number of m-dimensional rectangles, meeting in faces, edges, etc., if at all. The complex is mapped homeomorphically onto M with the map being diffeomorphically extendable to the infinitesimal neighborhood of each rectangle taken one at a time.

Subdivision does not change the rectangular cohomology, so we divide each rectangle of K into $*$-finitely many subrectangles so that each is an m-ε-volume element, that is, has sides finite on a scale of $\varepsilon \approx 0$. Specifically, add a new vertex at the center of each rectangle and fill in with faces. Do this ω times; then if ℓ_1 is the original length in one direction, $\ell_1/2^{\omega}$ is the new length of each offspring of that particular rectangle and $\varepsilon = 2^{-\omega}$ makes each rectangle an m-ε-volume element in the complex. Call the subdivision K_ε. **This ε-paving of K lifts onto** M via the rectangulation map, which since it is standard and diffeomorphic on individual rectangles, is S-regular. (Recall from the beginning of the section what S-regular means on an abstract manifold.)

The rectangular cohomology is obtained by dualizing the chain sequence of boundary maps

$$\cdots \leftarrow C_{h-1}(K, \mathbf{R}) \overset{\partial}{\leftarrow} C_h(K, \mathbf{R}) \overset{\partial}{\leftarrow} C_{h+1}(K, \mathbf{R}) \leftarrow \cdots$$

$$\cdots \rightarrow C^{h-1}(K, \mathbf{R}) \overset{\delta}{\rightarrow} C^h(K, \mathbf{R}) \overset{\delta}{\rightarrow} C^{h+1}(K, \mathbf{R}) \rightarrow \cdots$$

to the coboundary sequence between the spaces of linear functionals on C_h's which are denoted by C^h. When $*$'s are applied everywhere on these sequences we obtain more than we actually want, namely, infinite linear functionals on the various chain spaces. In this case we know that the cohomology of the hull sequence

$$\cdots \rightarrow \mathcal{O}(C^{h-1})/\approx \overset{\delta}{\rightarrow} \mathcal{O}(C^h)/\approx \overset{\delta}{\rightarrow} \mathcal{O}(C^{h+1})/\approx \rightarrow \cdots$$

is the same as that obtained from

$$\cdots \rightarrow {}^{\sigma}C^{h-1} \overset{\delta}{\rightarrow} {}^{\sigma}C^h \overset{\delta}{\rightarrow} {}^{\sigma}C^{h+1} \rightarrow \cdots$$

and hence equal to the standard cohomology. We also know that the full sequences

$$\cdots \rightarrow {}^*C^{h-1}(K, \mathbf{R}) \overset{\delta}{\rightarrow} {}^*C^h(K, \mathbf{R}) \overset{\delta}{\rightarrow} {}^*C^{h+1}(K, \mathbf{R}) \rightarrow \cdots$$

and

$$\cdots \rightarrow C^{h-1}(K_\varepsilon, {}^*\mathbf{R}) \overset{\delta}{\rightarrow} C^h(K_\varepsilon, {}^*\mathbf{R}) \overset{\delta}{\rightarrow} C^{h+1}(K_\varepsilon, {}^*\mathbf{R}) \rightarrow \cdots$$

produce the same cohomology by the $*$-transform of the subdivision lemma. We want to define the right notions of "finite" and "infinitesimal" for cochains over K_ε so that the infinitesimal hull of the finite sequence produces the standard cohomology. The natural choice is to weight the elements of the h-chains of K_ε by ε^h so that the value of the unit cochain is finite over the original h-rectangles after subdivision. With this choice, finite cochains over K become finite over K_ε by extending them to be constant on the subrectangles of the original rectangles of K. Our weights of ε^h make the "volume" of the original rectangles, written as a $*$-finite sum, come out to the same finiteness. *A $*$-h-cochain φ of K_ε is finite provided*

$$\sum_k \varphi_{a_0{}^k}(a_1{}^k \wedge \cdots \wedge a_h{}^k) \cdot \varepsilon^h$$

is finite for every unit chain $\sum_k {}_{a_0{}^k}\langle a_1{}^k \wedge \cdots \wedge a_h{}^k\rangle$ consisting of internal sums of rectangles ${}_{a_0}\langle a_1 \wedge \cdots \wedge a_h\rangle$ from K_ε. Since $\delta\varphi(c) = \varphi(\partial c)$, δ takes finite cochains to finite ones, hence is S-continuous. The infinitesimal hull sequence thus produces the ordinary rectangular cohomology, that is, we may work with finite elements and ignore infinitesimal differences:

$$\cdots \rightarrow C^{h-1}(K_\varepsilon, {}^*\mathbf{R}) \overset{\delta}{\rightarrow} C^h(K_\varepsilon, {}^*\mathbf{R}) \overset{\delta}{\rightarrow} C^{h+1}(K_\varepsilon, {}^*\mathbf{R}) \rightarrow \cdots.$$

(5.9.14) DEFINITION A smooth differential h-form on M is a smooth mapping from M into $\Lambda^h(TM)$ such that $\pi(\omega) = id$. We leave a complete nonstandard description in terms of the tangent bundle construction given above to the reader. What we need is that a standard h-form ω smoothly and S-continuously assigns weighted volume to h-dimensional linear volume elements.

We use the rectangulation map (plus the idea in the change of variables formula from the last section) to define the value of an h-form over an element of K_ε, taking the value $\omega_{a_0}(a_1 \wedge \cdots \wedge a_h)$ for the h-ε-volume element $a_1 \wedge \cdots \wedge a_h$ with vertex at a_0, that is, the image of the rectangle ${}_{a_0}\langle a_1 \wedge \cdots \wedge a_h \rangle$ in K_ε. Notice that the image of the infinitesimal vectors $\overline{a_0 a_1}, \overline{a_0 a_2}, \cdots, \overline{a_0 a_h}$ all lie in $\varepsilon T_{a_0} M$.

Again we encounter the question of what is finite and what is infinitesimal, which we resolve by assigning the h-volume of $a_1 \wedge \cdots \wedge a_h$ in K_ε to its image in M. That is, some finite multiple of ε^h, so if we consider the S-smooth differential h-forms,

$$\sum \omega_{a_0{}^k}(a_1{}^k \wedge \cdots \wedge a_h{}^k)$$

is finite whenever the sum extends over a $*$-chain (we obtain something resembling volume over the original rectangles embedded in M. Our next topic will be actual integration on manifolds.)

Theorem (5.8.9) says that δ and d differ by an infinitesimal (mod ε^h), thus the infinitesimal hulls of the two sequences

$$\cdots \to C^{h-1} \xrightarrow{\delta} C^h \xrightarrow{\delta} C^{h+1} \to \cdots$$

and

$$\cdots \to \mathscr{C}^\infty(M, \Lambda^{h-1}) \xrightarrow{d} \mathscr{C}^\infty(M, \Lambda^h) \xrightarrow{d} \mathscr{C}^\infty(M, \Lambda^{h+1}) \to \cdots$$

produce the same cohomology. *The isomorphism evaluates the differential form at vertices of the ε-paving, forgets that it is defined elsewhere, and thus considers it a cochain.* This proves the theorem.

A form ω is said to be *closed* if $d\omega = 0$ and *derived* if $d\zeta = \omega$ for some other form ζ. The h-cohomology $H^h(M)$ is the space of closed forms modulo the derived forms $H^h(M) = \ker d/\mathrm{im}\ d$.

(5.9.15) A (SMOOTH) VOLUME DENSITY ON M This has the following infinitesimal description: Let ε be a fixed positive infinitesimal and consider the ε-finite tangent bundle. An ε-density on M is an internal smooth assignment of volume to infinitesimal parallelepipeds. More specifically, an internal map so that, given x in ns($*M$) and m distinct ε-infinitesimal transformations $X_1, \ldots, X_m$, the real-valued map

$$\rho_x(X_1(x), \ldots, X_m(x))/\varepsilon^m$$

is S-smooth and, moreover, S-continuous for ρ_x as a map on $(\varepsilon T_x M)^m$; that is, if Y_j is ε-equivalent to X_j for $j = 1, \ldots, m$, then

$$\rho_x(X_1(x), \ldots, X_m(x))/\varepsilon^m \approx \rho_y(Y_1(y), \ldots, Y_m(y))/\varepsilon^m$$

when $x \approx y$ is in ns($^*M$). This much means that the hull of ρ_x is a map on $T_x M$. To say that ρ_x is a function of infinitesimal parallelepipeds we require that when A is an internal *S-continuous almost linear map defined on* $\varepsilon T_x M$ (that is, so that $\hat{A}$: $T_x M \to T_x M$ is defined and linear) *that*

$$|\det \hat{A}|\rho_x(t_1, \ldots, t_m) \approx \rho_x(At_1, \ldots, At_m) \pmod{\varepsilon^m}$$

for $t_1, \ldots, t_m$ *in* $\varepsilon T_x M$. This means ρ satisfies the change of variables formula.

As an example suppose M is a surface smoothly embedded in $\mathbf{E}^3$, so points of $\varepsilon T_x M$ are simply points t in $^*M$ with $|x - t|/\varepsilon$ finite. A volume element is given by an infinitesimal parallelogram in $T_x M$ spanned by x and any two points t_1, t_2 on $^*M$ with $|t_1 - t_2|$, $|x - t_1|$, $|x - t_2|$ all finite and noninfinitesimal compared with ε. The affine parallelogram with vertex x and sides $\overline{xt_1}$, $\overline{xt_2}$ is nearly tangent to $^*M$ and the linear area in $^*\mathbf{E}^3$ is an ε-density. Notice that SU-differentiable maps induce almost linear maps on the tangent spaces $\varepsilon T_x M$, $(x, t) \to (A(x), A(t))$, since in coordinates

$$\underline{A}(\underline{t}) = \underline{A}(\underline{x}) + |\underline{t} - \underline{x}| \cdot D\underline{A}_{\underline{x}}((\underline{t} - \underline{x})/|\underline{t} - \underline{x}|) + |\underline{t} - \underline{x}| \cdot \mathcal{O}.$$

An important generalization of the embedded case is when there is an internal assignment of infinitesimal scale of the kind which Riemann called the "simplest intrinsic case"—what is known now as a "Riemannian metric." The ε-infinitesimal description of this is a map that continuously assigns inner products to the tangent spaces, where $\varepsilon g_x\langle\ ,\ \rangle$ is an ε-finite internal function so that the nonstandard hull of

$$[g_x\langle\ ,\ \rangle]\hat{\ } = G_{\hat{x}}\langle\ ,\ \rangle$$

is a positive definite inner product on $T_x M$. By the ε-finiteness of $\varepsilon g_x\langle\ ,\ \rangle$ an $\approx$-orthonormal cube of length ε has volume ε^m in the $\approx$-inner product space, thus g induces a density. In coordinates, letting $\partial(x)/\partial x^i$ be the ε-length infinitesimal in the direction of x^i, the density is given by

$$\varepsilon^m |\det[g_x\langle\partial(x)/\partial x^i, \partial(x)/\partial x^j\rangle]|^{1/2}.$$

To define the integral of a density we take a standard rectangulation of M. Since M need not be compact, this is a possibly infinite complex K of rectangles meeting in faces edges, etc., together with a homeomorphism into M which has a diffeomorphic extension to a neighborhood of any single rectangle (see Whitney [1957]). In $^*K$ we subdivide each rectangle from the center ω times and call the image of the new complex K_ε, where $\varepsilon = 2^{-\omega}$. The infinitesimal subrectangles of the standard elements of $^*K$ are m-ε-volume elements—that is, have edges of finite length on a scale of ε and finite volume on a scale of ε^m—and their images lie completely within $\varepsilon T_x M$ when x is a point of the rectangle. We denote a typical such image subrectangle by ${}_{a_0}\langle a_1 \wedge \cdots \wedge a_m\rangle_k$ with vertex $a_0{}^k$ and sides $a_0{}^k a_1{}^k, \ldots, a_0{}^k a_m{}^k$.

(5.9.16) THE RIEMANN INTEGRAL Of an internal S-continuous function f over an internal set $A \subseteq \mathrm{ns}(^*M)$ with respect to ρ,

$$\int_A \cdots \int f(x)\rho = \mathrm{st}\left[\sum_{K_\varepsilon} \chi_A(y_k) f(y_k) \rho_{a_0}(a_1, \ldots, a_m)_k\right],$$

where $y_k \in {}_{a_0}\langle a_1 \wedge \cdots \wedge a_m\rangle_k$, provided

$$\left[\sum_{\partial A} \rho_{a_0}\langle a_1, \ldots, a_m\rangle_k\right] \approx 0,$$

where this sum extends over those infinitesimal rectangles that meet both A and its complement.

The sums are $*$-finite since A is internal and contained in the standard original rectangles. The integral is independent of coordinates by the change-of-variables theorem.

(5.9.17) ***REMARK:*** We have used the common ε-paving of a rectangulation of M because we feel that this is the most geometrical approach to integration. (Naturally, de Rham's theorem requires it.) Partitions of unity are technically a much simpler localization tool, but there is nothing infinitesimal about that approach beyond the work of the last section on rectangles. Partitions of unity are also the easiest technique to use in producing a Riemannian metric when a manifold is not embedded in a Euclidean space. When it is embedded, an infinitesimal description is quite natural—as we believe it must also be in other concretely given examples of Riemann manifolds.

(5.9.18) ***REMARK:*** The *divergence* of an ε-infinitesimal transformation X with respect to a (smooth) ε-density ρ on M is the ε-density given by

$$(\mathrm{div}_\rho X)_x(X_1(x), \ldots, X_m(x))$$
$$= (1/\varepsilon)\{\rho_{X(x)}[X_1 \circ X(x), \ldots, X_m \circ X(x)] - \rho_x[X_1(x). \ldots, X_m(x)]\}.$$

This produces a finite answer first of all, since on a standard chart,

$$(1/\varepsilon^m)\rho_x(\underline{X}_1(\underline{x}), \ldots, \underline{X}_m(\underline{x})) = r(\underline{x})$$

is S-smooth (write the small oh formula). When A is almost linear on $\varepsilon T_x M$, $Ax \approx x \bmod \varepsilon$, so $r(x) = r(A(x)) + \varepsilon \cdot o$ and $A(X_1 \circ X(x)) \approx A(X_1(x)) + A(X(x))$ in the tangent space addition. This gives the formula

$$(\operatorname{div}_\rho X)_x[At_1, \ldots, At_m] \approx |\det \hat{A}|(\operatorname{div}_\rho X)_x[t_1, \ldots, t_m],$$

mod ε^m, as the reader can easily check.

(5.9.19) THE DIVERGENCE THEOREM *Let N be a compact bordered submanifold of M, let X be a vector field, and ρ a density on M. Then*

$$\int \underset{N}{\cdots} \int \operatorname{div}_\rho X = \int_{\partial N} n_X(\rho).$$

PROOF: The term in the boundary integral is defined in the course of the proof. Let K be a rectangulation of M that has a subcomplex L forming a rectangulation of N. Let K_ε and L_ε be the images of infinitesimal central subdivisions of size ε. Let X and ρ be ε-representatives of the vector field and density.

A Riemann sum for the left-hand integral is

$$(1/\varepsilon)\{\sum_{L_\varepsilon} [\rho_{X(x)}(Xt_1, \ldots, Xt_m) - \rho_x(t_1, \ldots, t_m)]\},$$

where ${}_x\langle t_1, \ldots, t_m\rangle$ is a typical element of L_ε. Replace the terms $\rho_{X(x)}(Xt_1, \ldots, Xt_m)$ with the value of ρ on ${}_{x'}\langle t_1', \ldots, t_m'\rangle$ in K_ε, where $X(x)$ is in ${}_{x'}\langle t_1', \ldots, t_m'\rangle$. The error in this replacement is infinitesimal compared with ε^m by the S-continuity of ρ. The integral therefore nearly equals

$$(1/\varepsilon)\{\sum [\rho_{x'}(t_1', \ldots, t_m') - \rho_x(t_1, \ldots, t_m)]\}.$$

Whenever ${}_{x'}\langle t_1', \ldots, t_m'\rangle$ is in L_ε, there is a term canceling it so the expression reduces to

$$(1/\varepsilon)[\sum \rho_{x'}(t_1', \ldots, t_m')] + (1/\varepsilon)[\sum - \rho_x(t_1, \ldots, t_m)],$$

where the first sum runs over rectangles ${}_{x'}\langle t_1', \ldots, t_m'\rangle$ that X maps out of L_ε and the negative one runs over terms that X fails to cover in L_ε or ones that X^{-1} would remove from L_ε.

What we have said so far really only requires that N can be assigned Jordan content, but now to interpret our reduced sum as a Riemann sum we use smoothness. Then X moves points a finite amount compared with ε, so we look at the faces of L_ε on ∂N and define

$$(n_X \rho)_x\langle t_1, \ldots, t_{m-1}\rangle$$

to be $(1/\varepsilon)\rho_x(t_1, \ldots, t_{m-1}, X(x))$ when $X(x) \notin L_\varepsilon$, $-(1/\varepsilon)\rho_x(t_1, \ldots, t_{m-1}, X(x))$ when $X^{-1}(x) \notin L_\varepsilon$, and zero otherwise.

The replacement of t_m by $X(x)$ counts the number of cells near the face ${}_x\langle t_1, \ldots, t_{m-1}\rangle$ that move the same way across the boundary (with only infinitesimal ε^m error). The sum $\sum n_x \rho$ is therefore the integral of the right-hand side.

When ρ is a standard density the replacement is with the vector $\overline{X}$, $\hat{\rho}_{(\,)}[(\,), \ldots, (\,), \overline{X}] \approx (1/\varepsilon)\rho_{(\,)}[(\,), \ldots, (\,), X]$ and the plus, minus, or zero determined by whether $\overline{X}$ points out of N, into N, or tangent to ∂N. This makes sense since $\overline{X}$ is S-smooth.

When N of the last theorem is orientable, the rectangles of L can be consistently arranged so that meeting faces have the orientation induced by either side. The ε-paving L_ε inherits the orientation, and the hull of the multivector given by ${}_{a_0}\langle a_1 \wedge \cdots \wedge a_m\rangle$ for all x in the half-open rectangle is a nonzero m-multivector on N.

The proof of the fundamental theorem for rectangles (5.8.11) now carries over to L_ε, giving

(5.9.20) STOKES' THEOREM *If N is an orientable smooth bordered submanifold of M, and ω is an SU-differentiable m-form*

$$\int_{N^+} d\omega = \int_{\partial N} \omega.$$

EXERCISES ON $\Lambda(TM)$: The most interesting case to study differential forms on a manifold M is when M is an orientable Riemannian manifold, so suppose a function $g_{(\,)}\langle\ ,\ \rangle$ is given as above (so that the hull of $g_x = G_{\mathring{x}}$ is an inner product on $T_x M$.)

For a fixed base point x we can carry out the construction of $\Lambda^k(\varepsilon T_x M)$ as in Section 5.8 except on a scale of ε^k. The inner product structure is inherited from $g\langle\ ,\ \rangle$ so adjoints and codifferentials are defined and ${}^{\alpha}A \wedge A$ is an integrable quantity over bounded subsets of M. This view of $\Lambda(TM)$ in εTM still allows Riemann integrals to be viewed as sums. The change of variables formula makes integrals independent of coordinates. Orientation allows the possibility that all of the determinants are positive, so cancellation does not occur in the sums of form integrals.

The interested reader should work out some of the details of integration of forms used above or suggested in these remarks. Some of these details may turn out to be rather interesting.

CHAPTER 6 TOPICS IN INFINITESIMAL CALCULUS

We saw in the last chapter that the main basic results of infinitesimal calculus are what Leibniz or Euler might have said they were modulo the provisos that the infinitesimal operations should be nearly well defined (independent of the particular choice of infinitesimal) and that "infinitesimally close" simply is not "equality."

In this chapter we treat a variety of topics from "advanced calculus" or the "cours d'analyse." The chapter is not a systematic development of a subject, only an attempt to exhibit the use of infinitesimals in some appropriate contexts. Modern analysis is treated later with the machinery of saturated models.

6.1 PEANO'S EXISTENCE THEOREM REVISITED

This is a continuation of the last section of Chapter 2.

(6.1.1) THEOREM *If $f: I \times \mathbf{R} \to \mathbf{R}$ is continuous and bounded, $|f(t, x)| < M$, where I is a compact interval, and if $U_0 \in \mathbf{R}$ is given, then there exists $u: I \to \mathbf{R}$ with $u(0) = U_0$ and*

$$du(t)/dt = f(t, u(t)).$$

PROOF: We begin as in Chapter 2 by constructing an infinitesimally fine equal partition of I and a piecewise linear function defined on the subintervals so that the slope of the linear function almost equals the desired answer $(f(t, u(t)))$.

"For each $n \in \mathbf{N}$ we divide I into n equal parts $0 = t_0,\ t_1 = 1/n, \ldots, t_i = i/n, \ldots, t_n = 1$ and define a function $v_n(t)$ on I inductively on the t_i's so $v_n(t_0) = U_0$:

$$(1) \qquad (v_n(t_{i+1}) - v_n(t_i))/(t_{i+1} - t_i) = f(t_i, v_n(t_i)),$$

filling in linearly between the t_i's."

The $*$-transform of this statement says we may do this for each $n \in {}^*\mathbf{N}$, in particular, for infinite $\lambda \in {}^*\mathbf{N}_\infty = {}^*\mathbf{N} \setminus {}^\sigma\mathbf{N}$.

Now a formal property of all the v_n's is as follows: for each $n \in \mathbf{N}$, for $1 \leq \omega \leq n$,

$$v_n(t_\omega) = U_0 + \sum_{i=1}^{\omega-1} (v_n(t_{i+1}) - v_n(t_i))$$

a telescoping sum, so

$$v_n(t_\omega) = U_0 + \sum_{i=1}^{\omega-1} f(t_i, v_n(t_i))(t_{i+1} - t_i)$$

by use of (1). Again, the $*$-transform applied to $\lambda \in {}^*\mathbf{N}_\infty$ says

(2) $$v_\lambda(t_\omega) = U_0 + \sum_{i=1}^{\omega-1} f(t_i, v_\lambda(t_i))(t_{i+1} - t_i).$$

Next, by (1), $|v_\lambda(t) - v_\lambda(s)| \leq M|t - s|$ so that $t \approx s$ implies $v_\lambda(t) \approx v_\lambda(s)$. We form the standard function u by giving the following external definition:

$$u(t) = \mathrm{st}(v_\lambda({}^*t)) \qquad \text{for} \quad t \in I.$$

Now $u(t)$ is continuous since for standard t and s

$$|u(t) - u(s)| \approx |v_\lambda(t) - v_\lambda(s)| \leq M|t - s|.$$

Since I is compact, each $t \in {}^*I$ satisfies $\mathrm{st}(t) \in I$ and since continuity of a standard function u at $\mathrm{st}(t)$ is equivalent to

$${}^*u(t) \approx u(\mathrm{st}(t))$$

we have

$${}^*u(t) \approx u(\mathrm{st}(t)) \approx v_\lambda(\mathrm{st}(t)) \approx v_\lambda(t)$$

by the above remark that when $t \approx s$ then $v_\lambda(t) \approx v_\lambda(s)$.

Next consider the rectangle $I \times [U_0 - M, \; U_0 + M] = R_B$, which is compact. The functions v_λ and u stay inside R_B. Also, if $(t, x) \in {}^*R_B$, $(\mathrm{st}(t), \mathrm{st}(x)) \in R_B$, and since f is continuous and standard, when $t \approx s$ and $x \approx y$, all in ${}^*R_B$,

$$f(t, x) \approx f(\mathrm{st}(t), \mathrm{st}(x)) = f(\mathrm{st}(s), \mathrm{st}(y)) \approx f(s, y).$$

Putting these remarks together we see that

$$f(t, {}^*u(t)) \approx f(t, v_\lambda(t)) \qquad \text{for} \quad t \in {}^*I.$$

Consider the set

$$\{\delta : \delta \in {}^*\mathbf{R}^+ \;\&\; |f(t, u(t)) - f(t, v_\lambda(t))| < \delta \text{ for all } t \in {}^*I\}.$$

By the **internal definition principle** this set is internal and since it contains every standard positive δ it must contain an infinitesimal $\zeta \approx 0$. In that case, taking $t \in {}^*I$ and $t_\omega \approx t$,

$$u(t) \approx v_\lambda(t) \approx v_\lambda(t_\omega) = U_0 + \sum_{i=1}^{\omega-1} f(t_i, v_\lambda(t_i))(t_{i+1} - t_i)$$

$$\approx U_0 + \sum_{i=1}^{\omega-1} f(t_i, u(t_i))(t_{i+1} - t_i) + \zeta \cdot 1$$

$$\approx U_0 + \sum_{i=1}^{\omega-1} f(t_i, u(t_i))(t_{i+1} - t_i).$$

Now since $f(t, u(t))$ is continuous, the standard part of the Riemann sum on the right-hand side is the integral and

$$u(t) = U_0 + \int_a^t f(s, u(s))\, ds$$

since both sides are standard and within an infinitesimal. This proves Peano's theorem by use of the fundamental theorem of calculus.

The answer here is not unique. The reader should observe that we have choices in constructing v_λ. In particular we could use the slope $\max[f(t, x) : t \in [t_i, t_{i+1}]$ and $x \in [v_n(t_i) - (M/n), v_n(t_i) + (M/n)]$ or the min or numbers in between. Interpreting this proof in terms of standard proofs, part of the reason it works easily compared with the standard sequence arguments is that the Ascoli theorem has a particularly simple infinitesimal formulation (see (8.4.43)). A discussion on that level is best left until we look at topological notions infinitesimally. This is related to the property of compactness used above, namely that each point in $^*I$ or $^*R_B$ has a standard part in I or R_B. (In fact, a topological space is compact if and only if each point of the nonstandard extension is near standard.)

(6.1.2) EXERCISE: Generalize Peano's theorem to a vector setting. Let $\mathbf{E}$ be a vector space (with infinitesimal relation $\approx$ on $^*\mathbf{E}$) and $f: I \times \mathbf{E} \to \mathbf{E}$. What do you need to assume about f so that v_λ has a standard part? (See Part 2 for the infinite-dimensional conditions.)

(6.1.3) EXERCISE: Obtain the two solutions $u = 0$ and $u = t^3$ to $du/dt = 3u^{2/3}$, $u(0) = 0$, by use of infinitesimal partitions.

6.2 INTERCHANGING LIMITS

A question which frequently arises in analysis is "When can the order of taking limits be interchanged?" For example, if $a(k, h)$ is a double sequence (see Robinson [1966] for the basics of double sequences), we may form $\sum_{k=0}^{\infty} (\sum_{h=0}^{\infty} a(k, h))$, $\sum_{h=0}^{\infty} (\sum_{k=0}^{\infty} a(k, h))$, and $\sum_{k,h=0}^{\infty} a(k, h)$. Now since $*$-finite sums can be rearranged at will one might jump to the conclusion that by approximating with infinite partial sums these three limits are the same as long as one of them exists. This would contradict the well-known fact that this need not happen. For example, let $a(k, h) = \{(-1)^k/[(k+1)/2]\}(1/h)$, $k, h \geq 1$ (the square bracket denotes greatest integer), then $\sum_{k=1}^{\infty} (\sum_{h=1}^{\infty} a(k, h))$ does not converge since $\sum_{h=1}^{\infty} a(k, h)$ does not (see Chapter 2 for $\sum (1/h)$). On the other hand,

$$\sum_{k=1}^{\infty} a(k, h) = (-1/h) + (1/h) - \tfrac{1}{2}(1/h) + \tfrac{1}{2}(1/h) - \cdots$$

$$= (1/h)(-1 + 1 - \tfrac{1}{2} + \tfrac{1}{2} \cdots) = 0$$

and

$$\sum_{h=1}^{\infty} 0 = 0$$

so

$$\sum_{h=1}^{\infty} \left(\sum_{k=1}^{\infty} a(k, h) \right) = 0.$$

Let us examine what is wrong with the hasty conclusion drawn above. That $\sum_{k=1}^{\infty} (\sum_{n=1}^{\infty} a(k, h))$ exists means first that $\sum_{h=1}^{\infty} a(k, h) = A(k)$ is defined and second that $\sum_{k=1}^{\infty} A(k)$ exists. The latter condition happens if and only if each infinite partial sum $\sum_{k=1}^{\kappa} A(k)$ is nearly equal to the limit. Also, when λ is infinite, $A(k) \approx \sum_{h=1}^{\lambda} a(k, h)$; let us say $\varepsilon_k(\lambda) = A(k) - \sum_{h=1}^{\lambda} a(k, h)$. Now,

$$\sum_{k=1}^{\infty} A(k) \approx \sum_{k=1}^{\kappa} A(k) = \sum_{k=1}^{\kappa} \left(\sum_{h=1}^{\lambda} a(k, h) + \varepsilon_k(\lambda) \right)$$

$$= \sum_{k,h=1}^{\lambda} a(k, h) + \sum_{k=1}^{\kappa} \varepsilon_k(\lambda),$$

the first sum in any internal ordering. The thing to point out here (as in Cauchy's classic error above) is that an infinite sum of infinitesimals need not be infinitesimal or even finite, hence $\sum_{k=1}^{\kappa} \varepsilon_k(\lambda)$ may not be negligible.

For the rest of this section we shall look at a double sequence $s(m, n)$ which could be the partial sums $\sum_{k,h=1}^{m,n} a(k, h)$.

(6.2.1) THEOREM *Let $s(m, n)$ be a standard double sequence. In order that* $\lim_m (\lim_n s(m, n)) = \lim_{m,n} s(m, n) = \lim_n (\lim_m s(m, n))$ *it is necessary and sufficient that the following three conditions be satisfied for each* μ, μ', ν, $\nu' \in {}^*\mathbf{N}_\infty$, *infinite subscripts*:

(a) $s(m, \nu) \approx s(m, \nu')$ *for all* $m \in {}^*\mathbf{N}$,
(b) $s(\mu, n) \approx s(\mu', n)$ *for all* $n \in {}^*\mathbf{N}$,
(c) $s(\mu, \nu) \approx s(\mu', \nu')$.

In particular, (c) *implies that the double limit exists.*

PROOF: Assume first that the limits (exist and) are equal. Part (c) is the double sequence Cauchy condition (Robinson [1966]). Similarly, if m is fixed and finite, (a) is the Cauchy condition for $\lim_n s(m, n)$, but in the form given, (a) is stronger than this. This corresponds to a uniform (in m) standard convergence requirement. It holds nonetheless, because when m is infinite, part (c) gives the answer. Part (b) is symmetric to (a) in m and n.

Now suppose the three infinitesimal conditions are met. Existence of the double limit comes from (c), the corresponding Cauchy condition, say $s(m, n) \to S$.

Applying the Cauchy conditions for fixed m and n we define $S^m = \lim_n s(m, n)$ and $S_n = \lim_m s(m, n)$ in the standard model. Now consider ${}^*(S^m : m \in \mathbf{N})$ and ${}^*(S_n : n \in \mathbf{N})$. We know "$(\forall \varepsilon \in {}^*\mathbf{R}^+)(\exists n \in {}^*\mathbf{N})$ such that if $\nu > n$, then $|s(m, \nu) - S^m| < \varepsilon$" holds for each $m \in {}^*\mathbf{N}$, though a priori n may be infinite even for standard ε. This is by the $*$-transform of the standard definition of S^m, that is, S^m is the $*$-limit of $s(m, n)$ as $n \to \infty$. Next, take m infinite, then $S^m \approx s(m, \nu)$ for sufficiently large ν by taking $\varepsilon \approx 0$ and $S^m \approx S$, the double limit, by (c). The argument is symmetric in m and n for S_n. This proves the theorem.

Another result on interchanging limits is Dini's theorem:

(6.2.2) THEOREM *Let $s(m, n)$ be a standard double sequence such that* $\lim_m (\lim_n s(m, n)) = \lim_n (\lim_m s(m, n))$. *Suppose that with one variable fixed, say for each m, $s(m, n)$ is monotone in n, the other variable. In this case the double limit exists and equals the two repeated limits.*

PROOF: Let us assume that $s(m, n) \geq s(m, n + 1)$. The proof that there is no loss in generality to assume that $\lim s(m, n) = 0$ we leave to the reader. This means that for $\mu \in {}^*\mathbf{N}_\infty$ infinite, $s(\mu, n) \approx 0$ for fixed finite n and since also $0 \leq s(\mu, \nu) \leq s(\mu, n)$, we see that $s(\mu, \nu) \approx 0$ for μ and ν infinite.

There are many interesting interchanges of limit theorems. In particular, the powerful integration theorems (dominated convergence and Fubini's) remain to be explored infinitesimally.

We close the section with a theorem and some exercises.

(6.2.3) THEOREM *Let $s(m, n)$ be a standard double sequence. Suppose both of the repeated limits* $\lim_m (\lim_n s(m, n))$ *and* $\lim_n (\lim_m s(m, n))$ *exist. Show that*

(a) $\lim_{m,n} s(m, n) = \lim_m (\lim_n s(m, n))$ *if and only if for all* $\nu, \nu' \in {}^*\mathbf{N}_\infty$ *infinite,* $s(m, \nu) \approx s(m, \nu')$ *for all* $m \in {}^*\mathbf{N}$.

(b) $\lim_{m,n} s(m, n) = \lim_n (\lim_m s(m, n))$ *if and only if for all* $\mu, \mu' \in {}^*\mathbf{N}_\infty$ *infinite,* $s(\mu, n) \approx s(\mu', n)$ *for all* $n \in {}^*\mathbf{N}$.

PROOF: Left as an exercise.

(6.2.4) EXERCISE: Replace (a) and (b) in (6.2.3) by equivalent standard conditions. Also, try (6.2.3) out on $s(k, h) = (k + h)/(k^2 + h)$, as well as the partial sums of the example in the beginning of the section.

6.3 EULER'S PRODUCT FOR THE SINE FUNCTION

One of the many beautiful formulas that was discovered by Euler is the product formula for the sine function:

(6.3.1) $\sin z = z \prod_{k=1}^{\infty}(1 - (z^2/k^2\pi^2))$, z complex and $z \neq k\pi$, for $k = \pm 1, \pm 2, \ldots$.

It is interesting to examine how Euler proved his formula. As far as we know, Euler's original proof is contained in his book, "Introduction ad Analysin Infitorum," which appeared in 1748. It proceeds as follows. The mathematical expressions such as "infinitely large" and "infinitely close" that occur in it are Euler's. This formula was a key lemma in several of Euler's famous results.

For infinitely large values of n we have

(6.3.2) $$2 \sinh x = (1 + (x/n))^n - (1 - (x/n))^n.$$

We are now going to factor the polynomial occurring on the right-hand side of (6.3.2) by observing that $a^n - b^n = (a - b)(a - \varepsilon_1 b)\cdots(a - \varepsilon_{n-1} b)$, where $1, \varepsilon_1, \ldots, \varepsilon_{n-1}$ are the nth roots of unity. We combine the pairs of complex conjugate roots to obtain the real quadratic polynomials

$$\left(a - b\exp\left(\frac{2k\pi i}{n}\right)\right)\left(a - b\exp\left(-\frac{2k\pi i}{n}\right)\right) = a^2 + b^2 - 2ab\cos\frac{2k\pi}{n},$$

and so, since $a^2 + b^2 = 2 + (2x^2/n^2)$ and $2ab = 2 - (2x^2/n^2)$, we obtain

$$2\left(1 - \cos\frac{2k\pi}{n}\right) + 2\left(1 + \cos\frac{2k\pi}{n}\right)\frac{x^2}{n^2} = 4\sin^2\frac{k\pi}{n}\left(1 + \frac{x^2}{n^2\tan^2(k\pi/n)}\right).$$

It follows that the polynomial is divisible by x and for all values of $k = 1, 2, \ldots$ by $1 + (x^2/n^2 \tan^2 (k\pi/n))$. Since n is infinitely large, this factor is infinitely close to $1 + (x^2/k^2\pi^2)$. Furthermore, it is easy to see that the coefficient of x is equal to 2, and so we obtain

$$\sinh x = x \prod_{k=1}^{\infty} (1 + (x^2/k^2\pi^2)).$$

Finally, by applying it for $x = iz$, the required formula is obtained.

The reader will agree that Euler's proof is an example of the way infinitely large and infinitely small numbers were used with success in the early stages of the development of the calculus. It is, however, no wonder that the inability to give the theory of infinitely large and infinitely small numbers a firm foundation led to the unacceptability of such proofs. Of course now it is no problem to make Euler's proof precise.

It follows that for all finite $x \in \mathcal{O}$ and for infinitely large natural numbers $\omega \in {}^*\mathbf{N}_\infty$ we have

$$\text{(6.3.3)} \qquad 2 \sinh x \approx (1 + (x/\omega))^\omega - (1 - (x/\omega))^\omega;$$

see Exercise (6.4.2). Factoring the polynomial as before leads to the formula:

$$\left(1 + \frac{a}{m}\right)^m - \left(1 - \frac{a}{m}\right)^m$$
$$= \frac{4^{m/2}}{m} \left(\prod_{k=1}^{[(m-1)/2]} \sin^2 \frac{k\pi}{m}\right) a \prod_{k=1}^{[(m-1)/2]} \left(1 + \frac{a^2}{m^2 \tan^2(k\pi/m)}\right)$$

for all $a \in {}^*\mathbf{R}$ and for all $m \in {}^*\mathbf{N}$, and where $[(m - 1)/2]$ as before denotes the largest natural number less than or equal to $(m - 1)/2$. Dividing by $a \neq 0$ and letting $a \to 0$ shows that

$$\frac{4^{m/2}}{m} \prod_{k=1}^{[(m-1)/2]} \sin^2 \frac{k\pi}{m} = 2 \qquad \text{for all} \quad m \in {}^*\mathbf{N}.$$

Thus we obtain finally that

$$\text{(6.3.4)} \quad \left(1 + \frac{x}{\omega}\right)^\omega - \left(1 - \frac{x}{\omega}\right)^\omega = 2x \prod_{k=1}^{[(\omega-1)/2]} \left(1 + \frac{x^2}{\omega^2 \tan^2(k\pi/\omega)}\right)$$

for all $x \in \mathcal{O}$ and for all $\omega \in {}^*\mathbf{N}_\infty$.

We shall now prove the following lemma.

(6.3.5) LEMMA *If $x \in \mathbf{R}$ is standard, then for all infinitely large $\omega \in {}^*\mathbf{N}_\infty$ we have*

$$\operatorname{st}\left(\prod_{k=1}^{[(\omega-1)/2]}\left(1+\frac{x^2}{\omega^2\tan^2(k\pi/\omega)}\right)\right)=\prod_{k=1}^{\infty}\left(1+\frac{x^2}{k^2\pi^2}\right).$$

PROOF: Since for all $x \in \mathbf{R}$, the infinite product $\prod_{k=1}^{\infty}(1+(x^2/k^2\pi^2))$ is convergent, it follows that

$$\prod_{k=1}^{\infty}\left(1+\frac{x^2}{k^2\pi^2}\right)\approx\prod_{k=1}^{[(\omega-1)/2]}\left(1+\frac{x^2}{k^2\pi^2}\right)$$

for all $x \in {}^\sigma\mathbf{R}$ and for all $\omega \in {}^*\mathbf{N}_\infty$.

Since $\omega^2\tan^2(k\pi/\omega) \geq k^2\pi^2$ for all $1 \leq k \leq [(\omega-1)/2]$ we obtain that

$$\sum_{k=1}^{[(\omega-1)/2]}\left(\log\left(1+\frac{x^2}{k^2\pi^2}\right)-\log\left(1+\frac{x^2}{\omega^2\tan^2(k\pi/\omega)}\right)\right)\geq 0,$$

for all $x \in {}^\sigma\mathbf{R}$.

The following sequence is internal:

$$h_n=\sum_{k=1}^{n}\left(\log\left(1+\frac{x^2}{k^2\pi^2}\right)-\log\left(1+\frac{x^2}{\omega^2\tan^2(k\pi/\omega)}\right)\right)\geq 0,$$

for $n \in {}^*\mathbf{N}$ and $x \in {}^\sigma\mathbf{R}$. If n is finite, then by the continuity of the log function and $n/\omega \approx 0$,

$$\log\left(1+\frac{x^2}{\omega^2\tan^2(k\pi/\omega)}\right)\approx\log\left(1+\frac{x^2}{k^2\pi^2}\right),\qquad x \in {}^\sigma\mathbf{R},$$

and so $h_n \approx 0$ for all finite $n \in {}^\sigma\mathbf{N}$. There exists an infinitely large natural number $\nu \leq [(\omega-1)/2]$ such that $h_n \approx 0$ for all $n \leq \nu$ by the **internal definition principle**. This is because $\{n \in {}^*\mathbf{N} : (\forall k \in {}^*\mathbf{N})[k \leq n \Rightarrow k|h_k| \leq 1]\}$ is internal and contains ${}^\sigma\mathbf{N}$ (see the next section).

Observing that

$$\log\left(1+\frac{x^2}{\omega^2\tan^2(k\pi/\omega)}\right)\geq 0$$

for all $1 \leq k \leq [(\omega-1)/2]$, we obtain that

$$\begin{aligned}0 \leq h_{[(\omega-1)/2]} &\leq h_\nu+\sum_{k=\nu+1}^{[(\omega-1)/2]}\log\left(1+\frac{x^2}{k^2\pi^2}\right)\\&\approx\sum_{k=\nu+1}^{[(\omega-1)/2]}\log\left(1+\frac{x^2}{k^2\pi^2}\right).\end{aligned}$$

From Cauchy's criterion it follows, however, that

$$\sum_{k=\nu+1}^{[(\omega-1)/2]} \log\left(1 + \frac{x^2}{k^2\pi^2}\right) \approx 0 \qquad \text{for all } x \in {}^\sigma\mathbf{R},$$

and so we obtain that $h_{[(\omega-1)/2]} \approx 0$. Finally, the lemma follows from the continuity of the log function.

In order to complete the proof, observe that from (6.3.3) and (6.3.4) it follows that for all standard $x \in {}^\sigma\mathbf{R}$ we have

$$\sinh x \approx x \prod_{k=1}^{[(\omega-1)/2]} \left(1 + \frac{x^2}{\omega^2 \tan^2(k\pi/\omega)}\right), \qquad \omega \in {}^*\mathbf{N}_\infty,$$

and so by taking standard parts and using the lemma we obtain finally that

$$\sinh x = x \prod_{k=1}^{\infty} \left(1 + \frac{x^2}{k^2\pi^2}\right)$$

for all $x \in \mathbf{R}$.

6.4 ROBINSON'S LEMMA AND GENERALIZED LIMITS

In the course of proving Euler's product formula for the sine function we proved the following important lemma (which was named by Laugwitz [1974]):

(6.4.1) ROBINSON'S SEQUENTIAL LEMMA *Let $(a_n : n \in {}^*\mathbf{N})$ be an internal sequence. If $a_n \approx 0$ for standard $n \in {}^\sigma\mathbf{N}$. Then a_n is in fact infinitesimal out to some infinite subscript, that is, there exists $\Omega \in {}^*\mathbf{N}_\infty$ so that $a_k \approx 0$ for $0 \le k \le \Omega$.*

PROOF: The following set is internal:

$$\{n \in {}^*\mathbf{N} : \text{for all } k \le n \text{ in } {}^*\mathbf{N},\ k|a_k| \le 1\}.$$

(6.4.2) EXERCISE: Verify the steps in the following proof that

$$(1 + (z/\omega))^\omega \approx e^z, \qquad z \in \mathcal{O}$$

for finite $z \in {}^*\mathbf{C}$, where the number Ω is infinite:

$$\begin{aligned}\left[1 + \frac{z}{\omega}\right]^\omega &= \sum_{k=0}^{\omega} \binom{\omega}{k} \left[\frac{z}{\omega}\right]^k \\ &= \sum_{k=0}^{\omega} \left[\frac{\prod_{m=0}^{k-1}(\omega - m)}{\omega^k}\right]\left[\frac{z^k}{k!}\right] \\ &\approx \sum_{k=0}^{\Omega} \frac{z^k}{k!} + \sum_{k=\Omega}^{\omega} \binom{\omega}{k}\left[\frac{z}{\omega}\right]^k \approx \sum_{k=0}^{\Omega} \frac{z^k}{k!} \approx \sum_{k=0}^{\lambda} \frac{z^k}{k!}\end{aligned}$$

for any infinite $\lambda \in {}^*\mathbf{N}_\infty$. From Chapter 2, $e^z \approx \sum_{k=0}^{\lambda} (z^k/k!) \approx [1 + (z/\omega)]^\omega$.

Hints: (a) Ω comes from Robinson's lemma applied to part of the sum.

(b) $$0 \le \sum_{k=\Omega}^{\omega} \binom{\omega}{k} \left|\frac{z}{\omega}\right|^k \le \sum_{k=\Omega}^{\omega} \frac{|z|^k}{k!} \le \frac{|z|^\Omega}{\Omega!} \sum_{h=0}^{\omega-\Omega} \left|\frac{z}{\Omega}\right|^h = \frac{|z|^\Omega}{\Omega!}\left(\frac{1-|z/\Omega|^p}{1-|z/\Omega|}\right).$$

There are instances in analysis where one wants to assign values to divergent series. (This is especially important in Fourier analysis where Cesaro $(C, 1)$-methods play a fundamental role.) As a simple example, suppose we were considering the equation

$$1/(1-x) = 1 + x + x^2 + x^3 + \cdots = \sum_{k=0}^{\infty} x^k.$$

On the left-hand side if $x \to -1$, then $1/(1-x) \to \frac{1}{2}$, and we might conclude that we should have

$$1 - 1 + 1 - 1 + 1 - 1 \cdots = \tfrac{1}{2}.$$

On the other hand, considering the equation

$$(1 - x^2)(1/(1 - x^3)) = (1 - x^2)(1 + x^3 + x^6 + \cdots)$$

we obtain

$$(1 + x)/(1 + x + x^2) = 1 - x^2 + x^3 - x^5 + x^6 - x^8 + \cdots,$$

and letting $x \to 1$

$$1 - 1 + 1 - 1 + 1 - 1 + \cdots = \tfrac{2}{3}.$$

Likewise, we generate a list of possibilities

$$(1 - x^k)/(1 - x^{k+1}) \to k/(k+1) = 1 - 1 + 1 - 1 + 1 - 1 + \cdots.$$

Alternatively, we could simply decide once and for all to pick an infinite $\omega \in {}^*\mathbf{N}_\infty$ and take the $*$-finite sum $\sum_{k=0}^{\omega} (-1)^k$ as our answer. Unfortunately, when ω is even,

$$1 - 1 + 1 - 1 + \cdots + 1 = 1$$

and when ω is odd,

$$1 - 1 + 1 - 1 + \cdots - 1 = 0.$$

The point is that there might be many "reasonable" answers.

We now give the abstract essentials of a **generalized limit** for bounded sequences. These limits must be applied to the sequence of partial sums $s_n = \sum_{k=0}^{n} a_k$ in order to obtain the generalized sum. In the example above, $s_n = \{1, 0, 1, 0, \ldots\}$. We will show how to construct generalized limits infinitesimally so that they appear similar to classical summability methods.

Robinson [1964] points out the result of Steinhaus that summability methods are never defined on all bounded sequences. The nonstandard ones are. Robinson also proves that each generalized limit can be represented as an infinitesimal summability method. We will take this matter up in greater generality in Part 2 when we study the second dual space.

Robinson's lemma will play a key role in this theory of generalized limits.

(6.4.3) DEFINITION *Let $\ell^\infty(\mathbf{N})$ denote the space of bounded sequences. A generalized (Banach) limit is a function* $\mathrm{GL}: \ell^\infty(\mathbf{N}) \to \mathbf{R}$ *satisfying the following conditions:*

(a) $\mathrm{GL}(as_n + bt_n) = a\,\mathrm{GL}(s_n) + b\,\mathrm{GL}(t_n)$ (linearity).
(b) *If* $s_n \geq 0$ *for all n, then* $\mathrm{GL}(s_n) \geq 0$ (positivity).
(c) $\liminf(s_n) \leq \mathrm{GL}(s_n) \leq \limsup(s_n)$, *in particular*, $\mathrm{GL}(s_n)$ *must agree with the usual limit when it exists* (continuity).

Robinson begins his construction of generalized limits by taking an internal sequence $a = (a_n : n \in \mathbf{N})$ with the property that $\sum_{k=0}^{n} |a_k|$ is bounded by a finite constant for all $n \in {}^*\mathbf{N}$ (finite as well as infinite n). This is the same as requiring that $({}^*\sum_{k=0}^{\infty})(|a_k|)$ is finite (or that a is a finite point of ${}^*\ell^1(\mathbf{N})$).

(6.4.4) DEFINITION [a-limit for an internal sequence $a \in \mathrm{fin}({}^*\ell^1)$] *For each standard bounded sequence* $s \in {}^\sigma\ell^\infty(\mathbf{N})$,

$$a\text{-}\lim(s_n) = \mathrm{st}\left(\sum_{k=0}^{\infty} a_k s_k\right).$$

This operation could be defined on all finitely bounded $*$-sequences since $|\sum_{k=0}^{\infty} a_k s_k| \leq \sup_k (s_k) \sum_{k=0}^{\infty} |a_k|$ (or, in other words, a-lim is S-continuous on ${}^*\ell^\infty$).

(6.4.5) THEOREM *For standard bounded sequences s,* a-$\lim(s_n)$ *agrees with* $\lim_{n\to\infty}(s_n)$ *when the limit exists if and only if*

$$(\mathrm{f}): a_n \approx 0 \qquad \text{for all finite } n,$$

and

$$(\mathrm{u}): \sum_{k=1}^{\infty} a_k \approx 1.$$

PROOF: ($\Rightarrow$): The standard sequences

$$s_k = \delta_k{}^n = \begin{cases} 0, & k \neq n \\ 1, & k = n \end{cases}$$

have limit zero and $\sum a_k s_k = a_n$, so (f): $a_n \approx 0$ when n is finite. The constant sequence $s_n = 1$ for all n, whose limit is 1, shows $(\mathrm{u}): \sum a_k \approx 1$.

($\Leftarrow$): Let s_n be a standard sequence with bound B. The fact that $a_n \approx 0$ for finite n makes the partial sums $|\sum_{k=0}^{n} a_k s_k| \le B \sum_{k=0}^{n} |a_k|$ infinitesimal for all finite n, and by Robinson's lemma there is an infinite ω so that $\sum_{k=0}^{\omega} a_k s_k \approx 0$ and $\sum_{k=0}^{\omega} a_k \approx 0$.

By the $*$-transform of the Weierstrassian definition of $\sum_{k=0}^{\infty} a_k$, taking $\varepsilon \approx 0$, we know there is an infinite Ω so that

$$\left({}^*\sum_{k=0}^{\infty}\right)(a_k s_k) \approx \sum_{k=0}^{\Omega} a_k s_k$$

and

$$\left({}^*\sum_{k=0}^{\infty}\right)(a_k) \approx \sum_{k=0}^{\Omega} a_k \approx \sum_{k=\omega}^{\Omega} a_k.$$

Suppose now that s has a limit. By the infinitesimal characterization of $\lim s_n = L$ we know that $s_k \approx L$ for $\omega \le k \le \Omega$. Let $\delta = \max[|s_k - L|: \omega \le k \le \Omega]$, an infinitesimal by the preceding remark. Finally,

$$\left|\sum_{k=\omega}^{\Omega} a_k s_k - \sum_{k=\omega}^{\Omega} a_k L\right| \le \delta \sum_{k=\omega}^{\Omega} |a_k| \approx 0$$

and

$$\sum_{k=\omega}^{\Omega} a_k L = L \sum_{k=\omega}^{\Omega} a_k \approx L \qquad \text{by (u).}$$

EXAMPLES OF *a*'s

(6.4.6) Multiplicative functionals can be obtained by taking

$$a_n = \delta_n{}^{\omega} = \begin{cases} 0, & n \ne \omega \\ 1, & n = \omega, \end{cases}$$

where ω is infinite. Multiplicativity simply says $\text{st}(s_\omega \cdot t_\omega) = \text{st}(s_\omega) \cdot \text{st}(t_\omega)$ when $s, t \in \ell^\infty$. (We can represent all the multiplicative linear functionals this way in a saturated model.)

(6.4.7) Cesaro $(C, 1)$ limits are defined by averaging and taking the limit

$$(C, 1) - \lim s_n = \lim_{n \to \infty} \left((1/n) \sum_{k=0}^{n} s_k\right),$$

provided the latter limit exists. These can be subsumed by taking $a = c^\omega$, where

$$c_k{}^{\omega} = \begin{cases} 1/\omega, & 0 \le k < \omega \\ 0, & \omega \le k \end{cases}$$

with ω infinite. By the infinitesimal characterization of $\lim_{n\to\infty}$ we see that $(C, 1) - \lim(s_n) = L$ if and only if $c^{\omega} - \lim(s_n) = L$ for all infinite ω.

(6.4.8) A regular a-limit which violates positivity is given by taking

$$a_n = \begin{cases} 1/\omega, & n \text{ even and } 0 \le n < 2\omega \\ -1/\omega, & n \text{ odd and } 0 \le n < 2\omega \\ 1/\omega, & 2\omega \le n \le 3\omega \\ 0, & \text{otherwise.} \end{cases}$$

We still have $a_n \approx 0$ for finite n, $\sum_{k=0}^{\infty} |a_k| \le 3$ and $\sum_{k=0}^{\infty} |a_k| \approx 1$, but the sequence $s = \{0, 1, 0, 1, 0, \ldots\}$ has a-$\lim(s_n) = -1$.

EXERCISE: Why are the above examples internal? What are the corresponding a-limits on the series $1 - 1 + 1 - 1 + \cdots$? Find an internal a so that the series $1 - 1 + 1 - 1 + \cdots$ has a-limit $\frac{2}{3}$.

(6.4.9) THEOREM *If in addition to* (f) *and* (u) *of* (6.4.5), (S): $\sum_{k=0}^{\infty} |a_{k+1} - a_k| \approx 0$, *then* a-$\lim(s_n) = a$-$\lim(s_{n+1})$ (*that is,* a-lim *is shift invariant*) *and agrees with* $\lim_{n\to\infty}$ *where the limit exists.* (Shift-invariant limits are called *Banach–Mazur limits.*)

PROOF: Left as an exercise. Also show which of the above examples satisfy the additional hypothesis (S). (*Hint*: Write the difference and use the triangle inequality.)

(6.4.10) THEOREM *If a is an internal sequence such that* $(p): a_k \ge 0$ *for all* $k \in {}^*\mathbf{N}$, (u) : $\sum_{k=0}^{\infty} a_k \approx 1$, *and* (f) : $a_n \approx 0$, *for finite n, then* a-lim( ) *is a generalized limit in the sense of* (6.4.3).

PROOF: Linearity and positivity are clear from the definition of a-limit when one adds that $a_k \ge 0$.

Let s_n be a bounded sequence with $I = \lim\inf(s_n)$ and $S = \lim\sup(s_n)$. Let ω and Ω be as in the proof of the converse of (6.4.5). We claim that $I \lessapprox s_k \lessapprox S$ for $\omega \le k \le \Omega$ and this proves the theorem since $\sum_{k=\omega}^{\Omega} a_k s_k \approx \sum_{k=0}^{\infty} a_k s_k$ and $I \approx I \sum_{k=\omega}^{\Omega} a_k \lessapprox \sum_{k=\omega}^{\Omega} a_k s_k \lessapprox S \sum_{k=\omega}^{\Omega} a_k \approx S$. To establish the claim, assume for example that $s_k - S$ is finite with k infinite. Then for every finite n, there exists an $m \ge n$ so that $s_m - S \ge \frac{1}{2}\,\mathrm{st}(s_k - S)$. Applying Leibniz' principle we see that $S \ne \lim\sup(s_n)$ in the standard model.

To come full circle on our beginning example of a divergent series, we give the definition of Abel summability:

$$(A) - \sum_{k=0}^{\infty} a_k = \lim_{x\uparrow 1}\left(\sum_{k=0}^{\infty} a_k x^k\right),$$

provided the latter limits exist. Notice that this distinguishes between $1 - 1 + 1 - 1 + \cdots$ and $1 + 0 - 1 + 1 + 0 - 1 + 1 + 0 - 1 + 1 + \cdots$; however, if $(A) - (a_0 + a_1 + a_2 + \cdots) = L$, then $(A) - (0 + a_0 + 0 + a_1 + 0 + a_2 + \cdots) = L$. Our infinitesimal generalized limits apply to the sequence of partial sums, so a-lim$(1 - 1 + 1 - 1 + \cdots)$ becomes a-lim$(\{1, 0, 1, 0, 1, \ldots\})$ while a-lim$(1 + 0 - 1 + 1 + 0 - 1 + \cdots) = a$-lim$(\{1, 1, 0, 1, 1, 0, \ldots\})$. We close the section by mentioning two classic results for the reader to investigate: First, $(C, 1) - \sum_{k=0}^{\infty} a_k = L$ implies $(A) - \sum_{k=0}^{\infty} a_k = L$. Second (Tauber's theorem), if $(A) - \sum_{k=0}^{\infty} a_k = L$ and $\lim_{k\to\infty} ka_k = 0$, then $\sum_{k=0}^{\infty} a_k = L$ (see Goldberg [1964]).

6.5 DYNAMICAL SYSTEMS

In this section we give a new proof of one of the celebrated theorems of Birkhoff's [1927] classic monograph. This is one simple example of how infinitesimal techniques apply to the theory of dynamical systems. Other work along these lines and many reformulation theorems can be found in the articles by Hurd [1971a, b]. We believe infinitesimal analysis could be useful in open problems related to families of flows (via (8.4.42), for example) and in the study of related questions associated with partial differential equations. The reader can no doubt also have a lot of fun translating portions of Bhatia and Szegö [1970], Nemytskii and Stepanov [1960], and Gottschalk and Hedlund [1955] into the language of infinitesimals. We feel that infinite time and infinitesimal perturbation make many of those results more intuitive.

We will use the modern axiomatic definition of a dynamical system. Integrating a vector field on a manifold to obtain a flow (as in Birkhoff's monograph) is an important elementary problem, but we shall not take that up here despite the fact that such flows constitute the primary examples (consult Nemytskii and Stepanov [1960] or Hirsch and Smale [1974], for example).

We will also use some topological facts from the general theory of monads. Perhaps these metric applications will serve as problems for the interested reader and as motivation even for the casual reader. In any case we reference the exact theorems.

(6.5.1) *A* (real) *flow on a space* X *is a* (time) *action on the points of* X *satisfying the group property*: $x \in X$, $t, s \in \mathbf{R}$ imply that $x_{(t+s)} = (x_t)_s$ (that is, if you throw a stick in the water and follow it for time $t + s$ you end up in the same place as if you followed for time s beginning from the position after time t). This means $(x, t) \to x_t$ is a map from $X \times \mathbf{R} \to X$, and we shall require that the map is continuous. But this requires a "closeness" on X. If X were a submanifold of $\mathbf{R}^n$, $^*X$ would inherit a notion of infinitesimal from $^*\mathbf{R}^n$, and we could use that. In keeping with the axiomatic spirit, we shall assume (X, d) is a metric space and $x \overset{d}{\approx} y$ or simply $x \approx y$ in $^*X$ shall mean $^*d(x, y) \approx 0$ in $^*\mathbf{R}$. Now continuity means that if $a \in {}^\sigma X$ and $t \in {}^\sigma\mathbf{R}$ and if $x \approx a$ and $s \approx t$, then $x_s \approx a_t$.

For simplicity, we shall assume that every point of $^*X$ is infinitesimally near a standard point in ${}^\sigma X$. This is equivalent to compactness of X by (8.3.7) and (8.4.32); we saw the real-line version in (5.1.8). (You may wish to prove the equivalence for a metric space yourself.) Thus each point $x \in {}^*X$ has a *standard part* $\operatorname{st}(x) = y \in {}^\sigma X$ with $y \approx x$. If $x \approx y$, then $x_t \approx y_t$ for all finite t in $\mathcal{O}$. This need not remain true for infinite time (think of the points near the peak of a dome in a rainstorm—the drops there eventually spread over the whole thing).

(6.5.2) ***REMARK:*** The *limit sets* of a point are:

$$\alpha(x) = \{\operatorname{st}(x_T) : T \in {}^*\mathbf{R}_\infty{}^-, \text{ negative infinite time}\}$$

and

$$\omega(x) = \{\operatorname{st}(x_T) : T \in {}^*\mathbf{R}_\infty{}^+, \text{ positive infinite time}\}.$$

Limit sets are closed and invariant under the flow, since if y is standard and $y \approx x_T$, then $y_t \approx x_{T+t}$ for $t \in {}^\sigma\mathbf{R}$ by continuity. Since the limit sets are standard (although externally defined) and invariant for standard values, they are invariant. They are closed since they are an intersection of $\operatorname{st}\{x_t : t > n,\ n \in {}^\sigma\mathbf{N}\}$ and the standard part of an internal set is closed. (Prove this in the metric case or see (8.3.9).)

(6.5.3) ***REMARK:*** A point x is *periodic* if it repeats its trajectory. That is, if there is a $p \in {}^*\mathbf{R}$ so that $x_t = x_{(t+p)}$ for all $t \in {}^*\mathbf{R}$. If x is standard, some p must be standard (by Leibniz' principle). Integer multiples of periods are also periods of course, so infinite ones exist.

(6.5.4) ***REMARK:*** A point x is *almost periodic* if there is a P in each infinitely long interval of $^*\mathbf{R}$ satisfying $x_t \approx x_{(t+P)}$ for all $t \in {}^*\mathbf{R}$.

(6.5.5) EXERCISE: Give a standard formulation of "almost periodic" (see above references for hints). Show that P must be infinite when x is standard and not periodic. Give the connection between periodic, almost periodic, and recurrent points.

(6.5.6) ***REMARK:*** A point x is *recurrent* if every infinite interval of $^*\mathbf{R}$ contains a T such that $x \approx x_T$. Naturally, the continuity implies that for finite t, $x_t \approx x_{(T+t)}$; however, this need not remain true for infinite time (see Nemytskii and Steponov [1960, Example 7.12]). If x is *Lyapunov stable*, that is, if whenever $x \approx y$, then $x_t \approx y_t$ for *all* $t \in {}^*\mathbf{R}$, clearly a recurrent point is almost periodic. *When x is a standard recurrent point, the complete orbit $x_{*\mathbf{R}}$ is contained in the infinitesimal tube around any infinite time arc of x*, that is, each point of $\{x_t : t \in {}^*\mathbf{R}\}$ is infinitesimally near a point of $\{x_r : s \le r \le S\}$, so long as $(S - s)$ is infinite. To prove this, first notice that there is an R in $[s, S]$ such that $x \approx x_R$. Second, one of $[s, R]$ or $[R, S]$ is infinite, say $[s, R]$, and there is an r in $[s, R - ((R - s)/2)]$, with $x \approx x_r$ and $(R - r)$ infinite. *The set* $A = \mathrm{st}\{x_t : r \le t \le R\}$ *is compact, invariant, and contains* x. Then A is compact because the standard part of an internal set of near-standard points is always compact. [Prove this or see (8.3.11).] Thus A is invariant because if y and u are standard and $y \approx x_t$, then $y_u \approx x_{t+u}$, where if $t + u > R$ you switch over to x_r at time R or if $t + u < r$, you switch over to x_R at time r. Now ${}^*A = \mathrm{ns}({}^*A) = \{x_t : r \le t \le R\}$ and $A \supseteq \alpha(x) \cup x_{\mathbf{R}} \cup \omega(x)$.

(6.5.7) Suppose on the other hand that x *is NOT recurrent*, so that there is an infinite interval (s, S) with $x \not\approx x_t$ for all t in (s, S). We already saw that $\alpha(x)$ and $\omega(x)$ are invariant and naturally $\mathrm{st}\{x_t : s \le t \le S\}$ is contained in one of these two sets. *The set* $\lambda(x) = \mathrm{st}\{x_t : t - s$ *and* $S - t$ *are infinite*$\}$ *is an invariant closed set not containing* x. It clearly does not contain x and if a standard $y \approx x_t$, then $y_r \approx x_{t+r}$ for standard r shows that this standard set is invariant. The fact that this set is the intersection of a family of closed sets $\bigcap\, [\mathrm{st}\{x_t : s + 1/n < t < S - 1/n\} : n \in {}^\sigma\mathbf{N}]$ shows that it is closed.

(6.5.8) THEOREM *X is a compact dynamical system. A compact invariant subset $M \subseteq X$ has no proper closed subsets that are invariant precisely when each $m \in M$ is recurrent under the flow and such that $\alpha(m) = \omega(m) = M$. Such sets M are called minimal.*

PROOF: If M is minimally closed and invariant, then the invariant sets $\alpha(m)$ and $\omega(m)$ must equal M. Moreover, by (6.5.7), m must be recurrent, since otherwise $\lambda(m) \subset \omega(m)$ violates the minimality.

Conversely, the closed invariant set generated by a point m is $\alpha(m) \cup m_{\mathbf{R}} \cup \omega(m)$ and each of these is M.

(6.5.9) A point x is said to *wander* if in infinite time the infinitesimal neighborhood of x, $\mu_0 = \{y : y \approx x\}$, moves completely away from its initial position under the flow, $\mu_T \perp \mu_0$ for $T \in {}^*\mathbf{R}_\infty$. An equilibrium point ($x_t = x$ for all t) is *nonwandering* and so are the periodic, almost periodic, and recurrent points.

(6.5.10) *The monad* μ_0 *of a standard wandering point does not overlap its image in any finite time*, because if $\mu_t = \{y_t : y \approx x\} \cap \mu_0 \neq \phi$, then $x \approx z = y_t \approx x_t \approx x_p$, where $p = \mathrm{st}(t)$ and therefore x is periodic. (An infinite multiple of the period violates the wandering condition.) Birkhoff's own description of this was $\cdots$ x "never recurs to the infinitesimal neighborhood of any points once passed."

(6.5.11) EXERCISE: Show that it would suffice to know that $\mu_T \perp \mu_0$ for only positive infinite T or for only negative infinite T in order to know that x wanders.

Now if $x_T \approx x$ for some infinite T, then x is nonwandering, but the wandering condition is an open condition so it may happen that $x_T \not\approx x$ but some $y \approx x$ satisfies $y_T \approx y \approx x$ for infinite T. An example where this happens can be constructed as follows. Start with a circle with one rest point and circulation around the remainder (see Fig. 6.5.1). Let the center of the circle be a rest point and fill in the disk with spirals (see Nemytskii and Stepanov [1960, Example 5.06] for the differential equations). If x is on the circle, x_T goes near the rest point in infinite time, but a point infinitesimally inside the disk near x spirals back around near x in large enough infinite time. (Of course, things slow down near the inside of the rest point, but the spirals continue on around by $*$-transform of the statement that finite ones do.)

This example also shows that the notion of wandering changes if one restricts the flow to a subspace. Here x wanders to the rest point when the flow is restricted to the circle, but is nonwandering for the flow on the disk. Except for the center, the standard points inside the disk wander out to the boundary and if we then restrict our attention to the circle everything wanders to the rest point. (On the other hand, if one permits infinitesimal jumps after infinite time under the flow, then the points on the circle recur.

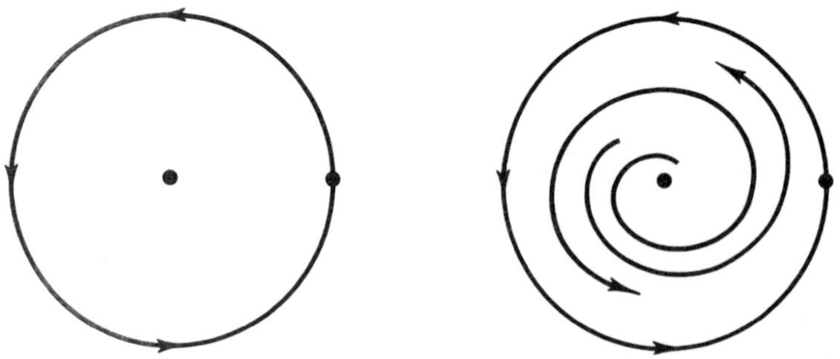

Fig. 6.5.1

In some respects generalized recurrence is more compatible with the geometry of the flow, while Birkhoff's recurrence is more central to the question of where points probably end up—all points spend most of the infinite time near recurrent points.)

(6.5.12) EXERCISE: Let x be a standard wandering point and T an infinite time. Show that $\mathrm{st}(x_T)$ is nonwandering. Show that there are infinite times S such that $\mathrm{st}(x_S)$ is recurrent.

(6.5.13) EXERCISE: We parametrize the torus with real numbers φ, ψ by taking the point $x(\varphi, \psi) = (\exp(i2\pi(\varphi)), \exp(i2\pi(\psi)))$. We define a flow in terms of these coordinates as follows: $(x, t) \mapsto (\exp(i2\pi(\varphi + pt)), \exp(i2\pi(\psi + qt)))$. Describe this when p/q is rational and when p/q is irrational. Give near-periods in both cases.

6.6 GEOMETRY OF THE UNIT BALL AND BOUNDARY BEHAVIOR

Let $B = \{z \in \mathbf{R}^3 : |z| < 1\}$. When a Euclidean plane or a 2-sphere intersects the unit sphere orthogonally we shall call the portion inside B a "*plane*." Two planes intersect in a "*line*" which is either a Euclidean line through 0 or a circle intersecting the unit sphere orthogonally; in the plane of the centers this appears as shown in Fig. 6.6.1. Lines intersect in "points" which are just the points of B. Two points determine a line and three points a plane.

Reflection across a plane is a mapping of the form $z \to \tilde{z}$ if the plane is Euclidean, z and $\tilde{z}$ lie equidistant on a Euclidean line perpendicular to the plane but on opoosite sides. If the plane is a sphere $\{z : |z - c| = r\}$, then

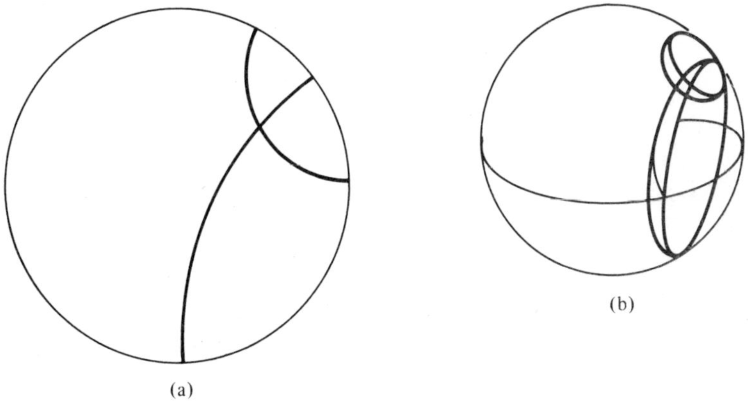

Fig. 6.6.1 (a) In E-plane of three centers; (b) three dimensions.

z and $\tilde{z}$ lie on the same ray from c with $|z - c||\tilde{z} - c| = r^2$. In a plane through c and z this appears as shown in Fig. 6.6.2. Reflections reverse orientation, so we single out the mappings consisting of an even number of reflections and call them the "*motions*" of the geometry of the unit ball. Both reflections and motions are bijections of B onto itself. In addition, motions are conformal.

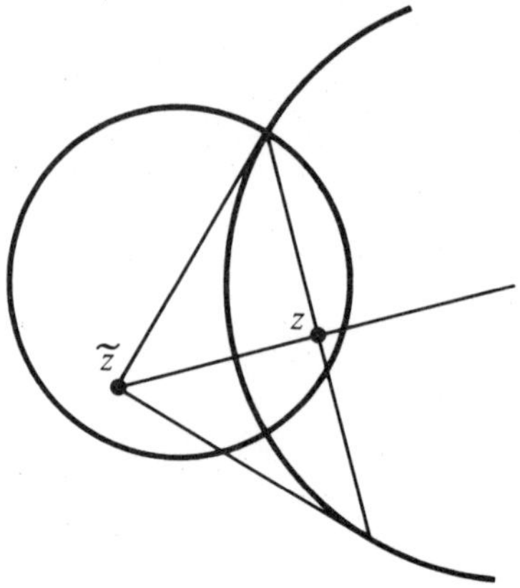

Fig. 6.6.2 $\tilde{z} = c + r^2(z - c)/|z - c|$.

The geometry is homogeneous under the motions, that is, given an arbitrary point z, a plane P containing z, and a line L in P through z, we can move

$$\begin{aligned} &z \to 0, \\ &P \to \{z : z \perp k\} = P_0, \qquad \text{the } x\text{–}y \text{ plane}, \\ &L \to (i\text{-axis in } P_0) = L_0, \end{aligned}$$

all by a single motion. To see this, first observe that the plane P goes onto a Euclidean plane through 0 under the reflection shown through P_1 followed by an arbitrary reflection across that Euclidean plane (see Fig. 6.6.3). Next we can rotate around 0 by reflections through planes through zero. Finally we move z in the x–y plane to zero similarly and rotate the final product so the x-axis is the image of L.

Now we ask for a metric $\eta(x, y)$ on B that is invariant under the motions, satisfies the triangle equality along lines [$\eta(a, b) + \eta(b, c) = \eta(a, c)$ when a, b, and c are in the right order on a line], and is asymptotic to the plane metric at zero. By homogeneity it suffices to give the metric on the positive x-axis L_0^+ then move it around to the other lines. This way the motions are η-rigid.

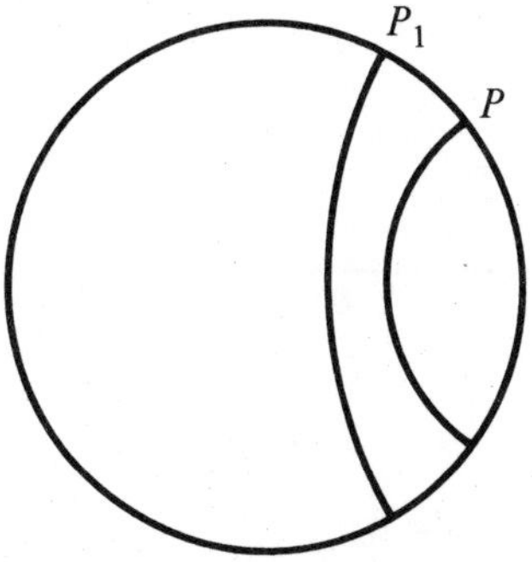

Fig. 6.6.3

Reflecting $r \in L_0{}^+$ to zero through a plane perpendicular to $L_0{}^+$ we know $(c - x)(c - \tilde{x}) = (1 - r^2 + (c - r)^2)^{1/2}$, where c is the center of the plane. Following that with reflection through the y–z plane we obtain the motion $x \to y$, where $y = (x - r)/(1 - rx)$ for x on L_0.

Now consider three points 0, r, $r + \delta$ with $0 < \delta \approx 0$. The motion sends $r + \delta$ to $\delta/(1 - r\delta - r^2)$. Additivity on $^*L_0$ means

$$\eta(0, r) + \eta(r, r + \delta) = \eta(0, r + \delta)$$

hence

$$\eta(0, r + \delta) - \eta(0, r) = \eta(0, \delta/(1 - \delta r - r^2)).$$

The asymptotic condition means $\eta(0, \varepsilon)/|\varepsilon| \approx 1$ when $|\varepsilon| \approx 0$. When r is standard, therefore,

$$[\eta(0, r + \delta) - \eta(0, r)]/\delta = \frac{1}{1 - \delta r - r^2} \frac{1 - \delta r - r^2}{\delta} \eta\left(0, \frac{\delta}{1 - \delta r - r^2}\right)$$

$$\approx \frac{1}{1 - r^2}.$$

Hence the right derivative of $\eta(0, r)$ is $1/(1 - r^2)$. The integral is $\eta(0, r) = \frac{1}{2}\log(1 + r)/(1 - r)$, and thus the metric exists uniquely on $L_0{}^+$. For $z, w \in B$, $\eta(z, w) = \eta(Tz, Tw)$ where T is the rigid motion taking z to zero and the ray $\overline{zw}$ into $L_0{}^+$.

(6.6.1) DEFINITION *The hyperbolic geometry of the unit ball is the above collection of objects; points, lines, planes, rigid motions, and the hyperbolic metric.* For specificity it is sometimes convenient to call them "hyperbolic lines," etc.

Consider the following entities:

$$
\begin{aligned}
B &= \{z : |z| < 1\},\\
Li &= \{L : L \text{ is an } h\text{-line}\},\\
Pl &= \{P : P \text{ is an } h\text{-plane}\},\\
\eta &: B \times B \to \mathbf{R}^+,\\
R &= \text{the set of reflections},\\
M &= \text{the set of rigid motions}.
\end{aligned}
$$

All of these have nonstandard extensions and the $*$-transform of the standard descriptions apply; for example, an arc of a circle orthogonal to the unit sphere is a $*$-line (of course the radius could be infinite or infinitesimal) and an internal rigid motion is an η-isometry of $^*B$ onto itself which takes internal lines and planes to internal lines and planes.

Each hyperbolic plane looks alike as far as the geometry of its lines is concerned. When the x–y plane, say, is thought of as complex numbers $x + iy$, the motions are $z \to \exp(i\theta)(z - a)/(1 - \bar{a}z)$ in that plane (the reader should consult Carathéodory [1932] for proof). This gives us a convenient way to compute the action of a motion that fixes a particular plane in that plane.

Let δ be an infinitesimal vector and z a standard point in B. The distance from z to $z + \delta$ can be computed by first rotating z around zero to the positive real axis, say $z \to r$. Then $z + \delta \to r + \varepsilon$ with $|\delta|/|\varepsilon| = 1$, since rotations are rigid in both η and the plane metric. Three points $0, r, r + \varepsilon$ determine a plane in which the motion $z \to (z - r)/(1 - rz)$ sends r to zero and $r + \delta$ to $\delta/(1 - \delta r - r^2)$ with complex coordinates put on the plane. Computing as above

$$\eta(z, z + \delta) = \frac{|\varepsilon|}{|1 - \varepsilon r - r^2|} \cdot k = \frac{|\delta|}{|1 - r^2|} \cdot h,$$

where $h, k \approx 1$. A more suggestive version would be to let $\delta = dz$; then

$$\eta(z, z + dz) = \frac{|dz|}{1 - |z|^2} \cdot h,$$

that is, the ratio of these is near one for infinitesimal dz.

There are many strong connections between the hyperbolic plane and holomorphic mappings, for example, Pick's version of Schwarz' lemma states that a holomorphic mapping of the hyperbolic plane into itself is either a strict contraction $\eta(f(z), f(w)) < \eta(z, w)$ or one of the rigid motions. Most of these connections are tied up with the differential form of η. Poincaré initiated its use in the study of automorphic functions.

Important contributions have been made even quite recently, although we shall not pursue this line. Instead we look at some simple and more general connections with harmonic functions.

The standard part mapping extends to $\mathcal{O}^n$ the finite vectors of $\mathbf{R}^n$ by $\mathrm{st}(x) = (\mathrm{st}(x_1), \mathrm{st}(x_2), \ldots, \mathrm{st}(x_n))$. We will call this the *Euclidean standard part* st_{E}.

(6.6.2) ***REMARK:*** In the hyperball we define $z \overset{\eta}{\approx} w$ if and only if $\eta(z, w) \approx 0$ and also

$$o_\eta[z] = \{w \in {}^*B : w \overset{\eta}{\approx} z\},$$

or, in general, if $A \subseteq {}^*B$,

$$o_\eta[A] = \{z \in {}^*B : \text{there exists } a \in A \text{ with } z \overset{\eta}{\approx} a\}.$$

This gives us an intrinsic notion of "infinitesimal" or *hyperbolic infinitesimal.* We also define the *hyperbolic galaxies*:

$$\mathrm{Gal}_\eta(z) = \{w \in {}^*B : \eta(z, w) \in \mathcal{O}\}.$$

This gives us an intrinsic notion of "finite." Since the hyperbolic distance is unbounded in the standard model there are lots of different galaxies. Each galaxy "looks" like the *principal galaxy* Gal(0) by the $*$-transform of the fact that B is homogeneous under the action of the motions M, specifically, there is an η-isometry in ${}^*M$ taking $\alpha \to 0$ so in particular $\mathrm{Gal}(\alpha) \to \mathrm{Gal}(0)$. The principal galaxy "looks" like the open unit ball once one identifies mod $\overset{\eta}{\approx}$. The Euclidean standard part restricted to Gal(0) is onto the open unit ball and 1–1 up to η-infinitesimals on Gal (0). We refer to points of Gal(0) as *hyperbolically finite.*

If $P \in {}^*Pl$ and $P \cap \mathrm{Gal}(0) \neq \varnothing$, then the Euclidean standard part of P in Gal(0) is a standard plane.

(6.6.3) ***REMARKS:*** The motions form a group under composition and there are natural notions of finite and infinitesimal in ${}^*M$ as follows.

The group of *finite motions* is $FM = \{W \in {}^*M : \eta(W(z), z)$ is finite for each finite $z\}$.

The group of *infinitesimal motions* is $IM = \{W \in {}^*M : \eta(W(z), z)$ is infinitesimal for each finite $z\}$.

(6.6.4) EXERCISE: *The standard part of a finite motion is a standard motion:* $B \xrightarrow{*} {}^\sigma B \xrightarrow{f} \mathrm{Gal}(0) \xrightarrow{\mathrm{st}} B$. *IM is a normal subgroup of FM and* $FM/IM \simeq M$ *with standard part as the canonical homomorphism.*

In order to describe translations and rotations, it is convenient to have three families of surfaces associated with a line L: first, the *latitude planes* of L, consisting of all h-planes perpendicular to L, second, the *longitude planes* of L, consisting of all h-planes through L. We ask the reader to sketch these with their associated ends on the unit sphere **S** (which they all meet perpendicularly; when the line L is through the north and south poles on **S** the ends are the usual latitude and longitude lines). The third family is the family of *equidistant tubes* around L. We ask the reader to show that the intersection of a longitude plane and an equidistant tube is an arc of a circle through the ends of L on **S**. This *equidistant curve* is not an h-line since it does not intersect **S** orthogonally. Also show that an h-sphere is a sphere (the center can be different). This means that the intersection of a latitude plane and an equidistant tube is a circle.

A *translation* along L is a motion that fixes L as a line, fixes longitude planes, and shifts the family of latitude planes among themselves. We could parametrize translations along L by the hyperbolic distance which they move along the equidistant curves and thus think of the family along L as a flow.

Translations are made up by reflections in planes which are ultraparallel (see Fig. 6.6.4). One must be careful here to distinguish this from parallel reflections where the planes are tangent on **S** even though they do not intersect inside B.

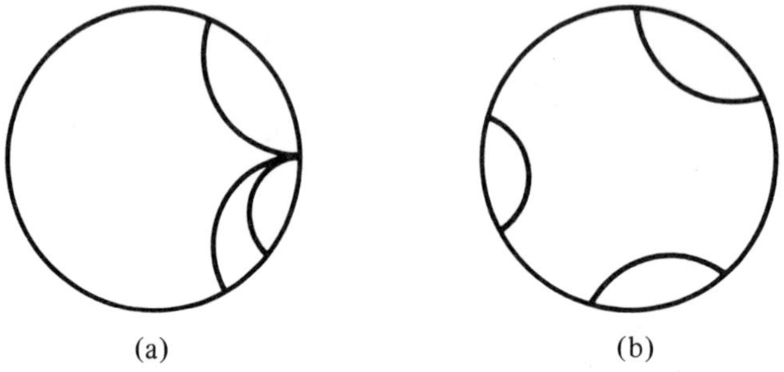

Fig. 6.6.4 (a) Parallel; (b) ultraparallel.

A *rotation* around L is a motion that fixes every point of L and fixes the latitude planes and the equidistant tubes while shifting the longitude planes. We could parametrize these by the hyperbolic distance moved around an orthogonal equidistant circle—that is, the intersection curve of a latitude plane and an equidistant tube—and thus obtain a flow.

Rotations are made of reflections through intersecting planes.

Extrinsically at hyperbolic infinity on **S** the action of translations and rotations take the respective forms

(6.6.5) T: $(w - a)(z - b) = r(w - b)(z - a)$, r real,

(6.6.5) R: $(w - a)(z - b) = e^{i\theta}(w - b)(z - a)$, θ real,

in complex coordinates (after stereographic projection, see Section 4.7). Both translations and rotations have two fixed points on **S** written here as a and b. Another form arises when a and b coincide:

(6.6.5) $$\text{L:} \quad \frac{1}{w - a} = \frac{1}{z - a} + c.$$

The reader should be aware of the fact that the action of a motion on **S** completely determines the motion. This is because the action on **S** is through inversions in circles which can be extended into B by reflections through the corresponding planes. As a result of this we can coordinatize M by the matrices

$$\begin{pmatrix} \alpha & \beta \\ \gamma & \delta \end{pmatrix},$$

where the extended motion sends $z \to w$ on **S** with $(\alpha z + \beta)/(\gamma z + \delta) = w$ after stereographic projection. The α, β, γ, δ can be computed directly from the forms above. Composition in M corresponds to matrix product, as the reader can verify by direct calculation. The normalization $\alpha\delta = \beta\gamma = 1$ eliminates the identity transformation written in the form $w = \alpha z/\alpha$ whose matrix is

$$\begin{pmatrix} \alpha & 0 \\ 0 & \alpha \end{pmatrix}$$

The group of 2×2 complex matrices with determinant one is referred to as SL(2, **C**)—the special linear group of dimension 2 over the complex field. We have just shown that it is equivalent to our group of motions.

In Euclidean geometry all motions can be decomposed into rotations and translations; hyperbolically this is only nearly true. Consider a finite translation (or rotation) along a $*$-line Lwhich lies completely outside the principal galaxy. The standard part of this map has only one fixed point on **S** and the standard parts of the ends of the associated family of latitude planes become a family of circles tangent at the fixed point on **S**; these might be called *horospheres*. The latitude planes have a family of parallel planes as their standard parts and the orthogonal family of parallel planes comes from standard parts of longitude planes. The standard parts of the equidistant tubes are the family of spheres tangent to the single fixed point on **S**; naturally only infinite equidistant tubes around L meet the principal galaxy. These motions and their standard parts are called *limit motions.*

(6.6.6) EXERCISE: *Show that every standard motion is nearly a product of finite $*$-translations and finite $*$-rotations.* (*Hint*: Consider the cases needed to prove this for a product of two reflections. Successive reflections in two nonintersecting planes is the hardest when they have no common perpendicular line. This happens when they are tangent at **S**.) *Derive the form* (6.6.5)L *from the two forms* (6.6.5)T *and* R *giving conditions on r and θ in order that* $\mathrm{st}(c) \neq 0$. (*Hint*: Combine T and R by letting r be complex, let $a - b = \varepsilon$, separate z and w by changing $(w - a + \varepsilon)/w - a$ into $1 + (\varepsilon/(w - a))$, rearrange, and divide by ε.)

Now we study the infinitesimal group from a finite point of view. An appropriate name for the Lie algebra we are about to construct might be the hyperbolic virtual motions to distinguish it from the nonstandard infinitesimal group. Exercise (6.6.4) shows that we cannot just take standard parts of the transformations of the infinitesimal group—that would collapse completely.

(6.6.7) DEFINITION *Let δ be a positive infinitesimal. The δ-virtual motions* $\delta VM = \{W \in {}^*M : \eta(z, W(z))/\delta$ *is finite for finite* $z\}$.

Since the motions are all analytic maps of B, these are δ-infinitesimal transformations in the sense of Section 5.9. (Note that $z \to z/(1 - |z|^2)$ is a standard chart.) The results of that section specialize to δVM, in particular, virtual motions act as derivations on SU-differentiable functions by

$$(Wf)(z) = (1/\delta)[f(Wz) - f(z)]$$

with δ-equivalent virtual motions being nearly the same derivation, that is, $W \overset{\delta}{\approx} V$ implies $\mathrm{st}(Wf) = \mathrm{st}(Vf)$, where $W \overset{\delta}{\approx} V$ means W and V are δ-equivalent at all finite points.

The brackets $[V, W]$ are associated with the commutator $W^{-1}V^{-1}WV$ in ${}^*M$ in that the commutator is a δ^2-virtual motion satisfying:

$$(1/\delta^2)\{f[W^{-1}V^{-1}WV(z)] - f[z]\} \approx V_z(Wf) - W_z(Vf) \approx [V, W]_z f.$$

The *infinitesimal hull* $VM = \delta VM/\overset{\delta}{\approx}$ can be thought of as the tangent space to the manifold of motions at the identity $T_E M$ or as the space of M-invariant *standard* derivations. The latter interpretation is even compatible with the factorization $FM/IM \simeq M$.

First, $\delta VM \subseteq \delta T_E M$ (in the notation of Section 5.9). This can be shown by relating the hyperbolic distance on the sphere, or better, in the complex plane after stereographic projection and then by coordinatizing M with the matrices of the associated Möbius functions. Each point of $T_E M$ has a representative in δVM given by the endpoint in $\delta T_E M$.

The "translation" map $\mathscr{J}: G \mapsto F \circ G$ in M has the derivative $\mathscr{J}_\vee(G, W) \to (F \circ G, F \circ W)^\wedge$ when $(G, W) \in \delta TM$. At the identity this gives $(F, F \circ V)$. When g is a function on M this means

$$\mathscr{J}_\wedge[G, VG]g \approx V_G(g \circ F),$$

which is a way to express invariance.

Examples of virtual motions are easy to construct from the geometric form of the group of motions. For example, all the rotations about a fixed line can be parametrized by the distance moved around a unit perpendicular equidistant circle. Taking the δ-rotation from the extended family of rotations about that line we obtain a δ-virtual motion. Going the other direction from VM to M is more difficult; this map is called the *exponential function.* One way to proceed is by solving the standard differential equation associated with a virtual motion for the local flow; we saw in Section 5.9 that this was equivalent to the infinitesimal transformation when the flow is evaluated at δ. Robinson [1966] gives another account of one-parameter groups and a basic treatment of Lie groups and algebras. We hope that the interaction between infinitesimal analysis and Lie groups will develop in more profound ways in the future.

We now consider boundary behavior of functions that are uniformly continuous in the hyperbolic metric. First we reformulate the standard condition. The *hyperbolically near-standard points* of $*B$ are those which are within a hyperbolic infinitesimal of a standard point; the non-near-standard points are called *remote.* In terms of the Euclidean metric, remote points satisfy $|z| \approx 1$. (X, d) is a metric space.

(6.6.8) THEOREM *The following are equivalent for a continuous standard function $f: (B, \eta) \to (X, d)$:*

(a) *f is uniformly continuous on B.*
(b) *$z \overset{\eta}{\approx} w$ implies $f(z) \overset{d}{\approx} f(w)$ for all remote points z, w in $*B$.*

PROOF: Left as an exercise. Write the formal sentence of uniform continuity and apply the ideas of Section 5.2.

(6.6.9) THEOREM *Let $f: (*B, \eta) \to (*X, d)$ be an internal function and let $a \in *B$. Then* (a) *and* (b) *are equivalent as are* (c) *and* (d).

(a) *For each* $z \in {}^*B$, $z \overset{\eta}{\approx} a$ *implies* $f(z) \overset{d}{\approx} f(a)$.

(b) *f is S-continuous at a, that is, for each* $\varepsilon \in {}^\sigma\mathbf{R}^+$ *there is a* $\delta \in {}^\sigma\mathbf{R}^+$ *such that* $\eta(z, a) < \delta$ *implies* $d(f(z), f(a)) < \varepsilon$.

(c) *For each* $z, w \in {}^*B$, $z \overset{\eta}{\approx} w$ *implies* $f(z) \overset{d}{\approx} f(w)$.

(d) *f is uniformly S-continuous on* ${}^*B$, *that is, for each* $\varepsilon \in {}^\sigma\mathbf{R}^+$, *there is a* $\delta \in {}^\sigma\mathbf{R}^+$ *such that for each* $z, w \in {}^*B$ *with* $\eta(z, w) < \delta$ *we have* $d(f(z), f(w)) < \varepsilon$.

Proof is omitted; we consider the notion of *S*-continuity more generally below. It has been included for comparison with (6.6.8). Notice that (c) appears to be local while (d) is global.

A continuous arc $C(t)$: $[0, 1) \to B$ is called a *boundary arc* if $|C(t)| \to 1$ as $t \to 1$. A boundary arc $C(t)$: $[0, 1) \to B$ with $C(t) \to a \in \mathbf{S}$ (the unit sphere) approaches a nontangentially provided $a \cdot (a - C(t))/|a - C(t)| \geq M > 0$.

APPLICATION OF (6.6.8): *Suppose a uniformly continuous standard function* $f \to b$ *along a boundary arc* C: $[0, 1) \to B$ *and that* D: $[0, 1) \to B$ *is a boundary arc for which* $\eta(C(t), D(t)) \to 0$, *as* $t \to 1$. *Then* $f \to b$ *along D also*, since when $t \approx 1$, $C(t) \overset{\eta}{\approx} D(t)$ and by the theorem $f(C(t)) \overset{d}{\approx} f(D(t)) \overset{d}{\approx} b$. (We are applying the infinitesimal characterization of $\lim g(t) = 1$.)

A point $z \in {}^*B$ *is tangential to* $a \in {}^*\mathbf{S}$ *if* $a \cdot (a - z)/|a - z| \approx 0$. A boundary arc $C(t) \to a$ approaches a nontangentially provided none of the points $C(t)$, $t \in {}^*[0, 1)$, are tangential to a.

A subset D of B is nontangential provided none of the points of ${}^*D$ are tangential to points in the closure of D on $\mathbf{S}$. In this case ${}^*D$ is contained in a standard cone near each of the closure points $a \in \mathbf{S}$ since $\{\alpha : z \in {}^*D$ and $a \cdot (a - z)/|a - z| > \alpha > 0\}$ contains all positive infinitesimals and is standard.

(6.6.10) THEOREM *Let* $f: B \to X$ *be a standard function. Then* $f(z) \to b$ *uniformly inside cones* $\mathrm{Cn}(a, \alpha) = \{z \in B : a \cdot (a - z)/|a - z| > \alpha > 0\}$ *as* $z \to a \in \mathbf{S}$ *if and only if for every* $z \in {}^*B$ *with* $z \overset{e}{\approx} a$ *and z nontangential to a*, $f(z) \overset{d}{\approx} b$.

PROOF: We give the "if" part, the rest is left to the reader. Let $\mathrm{Cn}(a, \alpha)$ be a standard cone at a inside B. Let $\varepsilon \in {}^\sigma\mathbf{R}^+$ be given. Since the points of ${}^*\mathrm{Cn}(a, \alpha)$ are nontangential to a, "there exists a δ such that $|z - a| < \delta$ and $z \in {}^*\mathrm{Cn}(a, \alpha)$ imply $d(b, f(z)) < \varepsilon$" holds, taking $\delta \approx 0$. By Leibniz' principle it holds in the standard model and $f(z) \to b$ uniformly inside the cone.

(6.6.11) *The hyperbolic distance from a point* $z \in {}^*B$ *near* $a \in {}^*\mathbf{S}$ *but nontangential to a to the radius terminating at a is finite. Thus the radial galaxies out to a contain all the points nontangential to a.* To see this, we may rotate the plane of z and the radius to a into the x–y plane sending a to 1; this is rigid in both the Euclidean and hyperbolic geometries. Thus there is no loss in generality in assuming $z = x + iy$ a complex number with $x \approx 1$, $y \approx 0$, and $(1-x)^2/[(1-x)^2 + y^2] \not\approx 0$ the nontangentiality. Now $\eta(x, z) \geq \eta(z, {}^*\mathbf{R}^+)$ since the shortest hyperbolic distance is along the hyperbolic line from z to ${}^*\mathbf{R}^+$:

$$\begin{aligned}\eta(z, x) &= \eta(0, (z-x)/(1-xz)) \\ &= \tfrac{1}{2}\log[1 + (|y|/|1 - x^2 - ixy|)]/[1 - (|y|/|1 - x^2 - ixy|)],\end{aligned}$$

which is finite. This requires that the reader verify that $(1-x)^2/[(1-x)^2 + y^2] \not\approx 0$ if and only if $|y|/|1 - x^2 - ixy| \not\approx 1$.

The meaning of S-continuity for internal functions on Gal(0) has more than just a local significance. The standard part of an S-continuous function externally defines a standard continuous function and the nonstandard extension differs from the internal function by a uniform infinitesimal on any standard compact set ${}^*K$. We will prove this below (or the reader can supply a proof), but now we give the following application:

(6.6.12) THEOREM *Let h be a $*$-harmonic function defined on the near-standard points of ${}^*B$. If h is S-continuous and finite on the near-standard points, then its standard part $k(z) = \mathrm{st}(h(z))$ is a standard harmonic function.*

PROOF: Note that if h is complex valued, $\mathrm{st}(h(z)) = \mathrm{st}(u(z)) + i\,\mathrm{st}(v(z))$. Since k is continuous by the $\varepsilon - \delta$ version of S-continuity applied to h, it suffices to show that k has the average value property. Let $c \in B$ and $B(c, r)$ be a ball of radius r and center c whose closure is contained in B. Then by the remarks about approximation on compact sets, $h(z) \approx k(z)$ for all $z \in \bar{B}(c, r)$. By the internal definition principle $|h(z) - k(z)| < \delta \approx 0$. Now, the average of h over the small ball is $h(c)$ by $*$-harmonicity and $k(c) \approx \mathrm{Ave}(k, B(c, r))$ by the δ-estimate. Since $k(c)$ and the average are standard, they are equal and this proves the theorem.

S-continuity on Gal(0), the near-standard points, is the same for either the Euclidean metric or the hyperbolic metric. This is not the case at the hyperbolically remote points.

The positive real numbers have two natural metrics: $e(x, y) = |x - y|$, the Euclidean metric, and $p(x, y) = |\log x/y|$, the product metric. Our next result concerns S-continuity of positive harmonic functions as mappings from (B, η) into $(\mathbf{R}^+, e)$ and $\mathbf{R}^+, p)$.

(6.6.13) THEOREM *Let h be a positive $*$-harmonic function defined on $^*B$ with $h(0)$ finite. Then*

(a) *the map $h: (^*B, \eta) \to (^*\mathbf{R}^+, e)$ is S-continuous on* Gal(0), *and*

(b) *the map $h: (^*B, \eta) \to (^*\mathbf{R}^+, p)$ is S-continuous everywhere. Moreover, part* (a) *extends to functions $h = h_1 - h_2$ where h_1 and h_2 satisfy the hypotheses above.*

PROOF: In both parts we shall need the Poisson integral representation for a function harmonic in a neighborhood of the closed ball $\{z : |z - y| \le r\}$. If h is harmonic and $|z - y| < r$, then

$$h(z) = \frac{1}{4\pi r} \int_{|x-y|=r} \frac{r^2 - |y - z|^2}{|x - z|^3} h(x)\, da(x),$$

the area integral over the sphere of radius r.

Next we need to say what $z \overset{\eta}{\approx} w$ means in terms of Euclidean distances. Take $z \overset{\eta}{\approx} w$. It is enough to work in the plane of 0, z, and w with complex coordinates where we find that $|z - w|/(1 - |w|) \approx |z - w|/(1 - |z|) \approx 0$.

To show part (a) we assume further that z and w are near-standard. Let r be a number, $0 < r < 1$, with $r \approx 1$ and apply the Poisson integral formula with $y = 0$ over $|x| = r$. The difference between the Poisson kernels for z and w with $|x| = r$ is

$$\Delta P(x) = \frac{1 - |z|^2}{|x - z|^3} - \frac{1 - |w|^2}{|x - w|^3} \approx 0.$$

Hence, $\sup(|\Delta P(x)| : |x| = r) = \varepsilon \approx 0$ and

$$|h(z) - h(w)| \le (\varepsilon/4\pi r) \int_{|x|=r} h(x)\, da(x) = \varepsilon \cdot h(w)/4\pi r \approx 0.$$

To show part (b) we no longer assume that z and w are near-standard. Now we need to apply Harnack's inequality to the ball of radius $r = \min(1 - |z|, 1 - |w|)$ centered at the hyperbolic midpoint y of $\overline{zw}$. We include the proof of this inequality. In effect, (b) is just a geometric form of the inequality.

First, the closed ball $|x - y| \le r$ is contained in the unit ball since one of z or w is closer (in Euclidean distance) to the unit sphere than y. Hence we may apply the Poisson integral formula to h on $|x - y| = r$.

Now z and w lie inside the ball $|x - y| = \delta$, where $\delta = |z - w|$, so with $u = z$ or w we have

$$r - \delta \leq |x - u| \leq r + \delta$$

and

$$\frac{r^2 - |y - u|^2}{(r + \delta)^2} \leq \frac{r^2 - |y - u|^2}{|x - u|^3} \leq \frac{r^2 - |y - u|^2}{(r - \delta)^3}.$$

Therefore,

$$\frac{r^2 - |y - u|^2}{|x - z|^3} \leq \left(\frac{r^2}{r^2 - \delta^2}\left(\frac{r + \delta}{r - \delta}\right)^3\right)\frac{r^2 - |y - w|^2}{|x - w|^3},$$

so by taking Poisson integrals

(H) $$\frac{h(z)}{h(w)} \leq \frac{1}{1 - \varepsilon^2}\left(\frac{1 + \varepsilon}{1 - \varepsilon}\right)^3 \qquad \text{with } \varepsilon = \frac{\delta}{r},$$

and also by interchanging z and w,

(H) $$\frac{h(w)}{h(z)} \leq \frac{1}{1 - \varepsilon^2}\left(\frac{1 + \varepsilon}{1 - \varepsilon}\right)^3.$$

We already observed above that $z \overset{\eta}{\approx} w$ implies $\varepsilon \approx 0$, since $\delta/r = |z - w|/(1 - |w|) \approx 0$. Now by Harnack's inequality (H) we have $h(z)/h(w) \approx 1$, whence $\log(h(z)/h(w)) \approx 0$ or $h(z) \overset{p}{\approx} h(w)$.

We now give an application to standard positive harmonic functions.

(6.6.14) THEOREM *Let h be a standard positive harmonic function on B and suppose $h \to 0$ along a nontangential boundary arc $C(t) \to a \in S$. Then $h \to 0$ uniformly inside cones in B at a as $z \to a$.*

PROOF: It is sufficient to show that if $z \overset{e}{\approx} a$ and z is nontangential, that $h(z) \approx 0$. We know for $w = C(t)$ with $C(t) \overset{e}{\approx} a$ that $h(w) \approx 0$ by the limit on the arc. Since C is nontangential and z is nontangential, z is a finite hyperbolic distance from some point $w = C(t) \overset{e}{\approx} a$. Now $p(h(z), h(w))$ is finite by S-continuity, and since $h(w) \approx 0$ we must have $h(z)/h(w)$ finite so $h(z) \approx 0$.

We close with a construction that previews things in the general theory of monads and infinitesimals. In order to develop that theory we need the additional model theory that is outlined in the next chapter.

Let $A = \{z \in {}^*B : |z| \leq r\}$, where $r \approx 1$ and $r < 1$. Of course we can form Poisson integrals over $|z| = r$ in the internal sense. Let $h^+(B)$ denote the set of all (standard) positive harmonic functions on B. Define a relation on A by $z \overset{h^+}{\approx} w$ if and only if for each $h \in {}^\sigma h^+(B)$ either $h(z)$ and $h(w)$ are both infinite or $h(z) \approx h(w)$. Then in particular, $z \overset{\eta}{\approx} w$ implies $z \overset{h^+}{\approx} w$. We can consider each $h \in {}^\sigma h^+(B)$ extended to $(A/\overset{h^+}{\approx}) = \hat{A}$ via $\text{st}({}^*h(z))$ (or ∞). Endow $\hat{A}$ with the weakest topology that makes all the extensions continuous. What sort of space is $\hat{A}$? Does it describe all possible boundary behavior of positive harmonic functions? What is the connection between $\overset{\eta}{\approx}$ and $\overset{h}{\approx}$? What is the standard description on $\hat{A}$ of the Poisson integrals on $|x| = r$?

INFINITESIMALS IN FUNCTIONAL ANALYSIS PART 2

MORE TOOLS FROM MODEL THEORY CHAPTER 7

In this chapter we shall discuss some very general external properties that a nonstandard model could have. We will point out where some of the literature departs from each of these additional properties, but our aim is a fairly inclusive kind of model called a *polysaturated* model. Explicit external knowledge of a model is relatively hard to obtain, but also extremely important in applications since the main power of this technique involves shifting between internal and external concepts.

Our choice of polysaturated models is motivated by applications that we have in mind (especially in functional analysis). Other choices may be advantageous for other applications.

7.1 COUNTABLE ULTRAPOWERS

To demonstrate the kind of question that might arise in applications consider the following. Let J be a countable set ($J = \{0, 1, 2, \ldots\}$ for convenience). Consider an ultrapower model of $\mathcal{N}$, ${}^*\mathcal{N} = \mathcal{N}^J/\mathcal{U}$. Suppose $\lambda \in {}^*\mathbf{N}$ is an infinite natural number. When can we say there is a set $J_\infty \in \mathcal{U}$ such that

$$\lim_{j \in J_\infty, j \to \infty} \lambda(j) = \infty ?$$

In a different setting Rudin [1956] and Choquet [1968] have shown that the continuum hypothesis implies that there are free ultrafilters which are *δ-stable* in the sense that if $(J_n : n \in \mathbf{N})$ is any sequence of elements of $\mathcal{U}$, there is a set $J_\infty \in \mathcal{U}$ *almost contained in each* J_n, precisely, $J \backslash J_n$ *is finite* for each n. (Of course we cannot generally have $J_\infty \subseteq \bigcap J_n$ because of the δ-incompleteness). A proof of the existence of these ultrafilters is also contained in the work of Puritz [1971], where many other external properties of ultrapowers are considered. We recommend this paper to the interested reader.

(7.1.1) THEOREM *If ${}^*\mathcal{N} = \mathcal{N}^J/\mathcal{U}$ is a δ-stable free ultrapower with $J = \{0, 1, 2, \ldots\}$, and $\lambda \in {}^*\mathbf{N}$ is infinite, then there exists a set $J_\infty \in \mathcal{U}$ so that*

$$\lim_{j \to \infty, j \in J_\infty} \lambda(j) = \infty$$

as an ordinary limit.

PROOF: Since λ is infinite, $J_n = \{j : \lambda(j) > n\} \in \mathscr{U}$ for each $n = 0, 1, 2, \ldots$. By δ-stability of $\mathscr{U}$, there is a set $J_\infty \in \mathscr{U}$ so that $J_\infty \backslash J_n$ is finite for each n. Let $M > 0$ be given. Take $n = [M] + 1$ and $m = \max(1, [(J_\infty \backslash J_n)] + 1)$. If $m \leq j \in J_\infty$, then $j \in J_n$ so $\lambda(j) > n > M$, which proves the theorem.

Puritz [1971] introduces some interesting relations, which we summarize here. The reader can find some of the simpler properties for himself.

(7.1.2) *First a notion we have already discussed: for $a, b \in {}^*\mathbf{N}$ we write*

$$a \sim b$$

if and only if $a - b$ is finite. This is an equivalence relation whose equivalence classes are called *galaxies*. (Of course, $a \approx b$ if and only if $a = b$ in $^*\mathbf{N}$, so infinitesimals do not appear.)

Let $\mathscr{J} = \mathscr{J} : \mathbf{N} \to \mathbf{N}$ be the standard set of standard functions $f: \mathbf{N} \to \mathbf{N}$. We are interested in what extent *standard* functions act on $^*\mathbf{N}$.

(7.1.3) *$a \uparrow b$ means there is a standard $f \in {}^\sigma\mathscr{J}$ so that $f(a) \geq b$.*
(7.1.4) *$a \ll b$ means that $a \uparrow b$ does not hold.*
(7.1.5) *$a \uparrow\downarrow b$ means $a \uparrow b$ and $b \uparrow a$.*
(7.1.6) *$a \to b$ means there exists a standard $f \in {}^\sigma\mathscr{J}$ so that $f(a) = b$.*
(7.1.7) *$a \leftrightarrow b$ means $a \to b$ and $b \to a$.*

Some simple observations whose proofs are left as exercises are given below. Property (7.1.4) will have applications in analysis—we will want models with infinite natural numbers $\lambda, \omega \in {}^*\mathbf{N}_\infty$ satisfying $\lambda \ll \omega$. These pairs (λ, ω) are called *random* pairs.

(7.1.8) *$(\uparrow\downarrow)$ and $(\leftrightarrow)$ are equivalence relations with $(\leftrightarrow)$ stronger than $(\uparrow\downarrow)$, that is, smaller equivalence classes.*

(7.1.9) *If $a \sim b$, then $a \uparrow\downarrow b$ and $a \leftrightarrow b$, that is, the equivalence classes of $(\uparrow\downarrow)$ and $(\leftrightarrow)$ contain the galaxies. Puritz calls $(\leftrightarrow)$-classes* **constellations** *and $(\uparrow\downarrow)$-classes* **skies**.

We will return to some of these relations in our stronger models. Similar properties of action by standard functions in continuous situations are important in analysis and will be looked into below.

7.2 ENLARGEMENTS

In this section we describe the kind of models most frequently encountered in the literature on nonstandard analysis. In particular, Robinson [1966] considers only enlargements from the beginning of his applications. We start with some basic concepts.

(7.2.1) DEFINITION *A binary relation Φ is called concurrent on* $\mathrm{dom}(\Phi)$ *provided that for any finite set of elements* $x_1, \ldots, x_n \in \mathrm{dom}(\Phi)$ *there is a* $y \in \mathrm{rng}(\Phi)$ *such that*

$$(x_j, y) \in \Phi \qquad \text{for each} \quad j = 1, \ldots, n.$$

In this case we say y satisfies the x_j*'s,* $j = 1, 2, \ldots, n$.

(7.2.2) DEFINITION *Let $\mathscr{A}$ be a standard entity of $\mathscr{X}$ which consists of a family of entities* $\{A : A \in \mathscr{A}\}$. *The intersection monad of $\mathscr{A}$ is*

$$\bigcap^{\sigma} \mathscr{A} = \bigcap [{}^*A : A \in \mathscr{A}] = \mu(\mathscr{A}).$$

The union monad of $\mathscr{A}$ is

$$\bigcup^{\sigma} \mathscr{A} = \bigcup [{}^*A : A \in \mathscr{A}] = \nu(\mathscr{A}).$$

We remark that the set of infinitesimals $o \subseteq {}^*\mathbf{R}$ is the monad of the standard family of standard intervals $\{(-a, a) : a > 0 \text{ in } \mathbf{R}\}$. The set of infinitesimals around an infinite number $\lambda \in {}^*\mathbf{R}$ is *not a monad* as it turns out because no standard family of standard sets has $\lambda + o$ as its monad.

We will say a monad is *principal* if

$$\mu(\mathscr{A}) = {}^*\bigcap \mathscr{A} = \bigcap [B : B \in {}^*\mathscr{A}] = {}^*(\bigcap [A : A \in \mathscr{A}])$$

or

$$\nu(\mathscr{A}) = {}^*\bigcup \mathscr{A} = \bigcup [B : B \in {}^*\mathscr{A}] = {}^*(\bigcup [A : A \in \mathscr{A}]).$$

The finite intersection property for a family $\mathscr{A}$ means that if $A, B \in \mathscr{A}$, then $A \cap B \neq \varnothing$. The finite union property for $\mathscr{A}$ means that if $A, B \in \mathscr{A}$, then $A \cup B \neq \bigcup \mathscr{A}$.

Monads form an extension of the family of standards sets; they are either principal or external, that is, if internal, they are standard. (At least in sufficiently saturated models.)

Monads have both model-theoretical and topological significance as we shall see.

The $*$-transform of the standard notion of finiteness is what we consider next. The following properties all characterize finiteness in "standard analysis" [AC].

(7.2.3) *An entity A is finite if there is a bijection entity from an initial segment* $\{0, 1, \ldots, n\}$ *of* $\mathbf{N}$ *onto A. The $*$-transform reads:* "An internal entity A is $*$-finite if there is an internal bijection from an initial segment of ${}^*\mathbf{N}$ onto A." Here of course an initial segment can be externally infinite, $\{n \in {}^*\mathbf{N} : o \leq n \leq \omega\}$ with $\omega \in {}^*\mathbf{N}_\infty$, for example.

(7.2.4) *An entity A is finite if every injection of A into itself is a bijection.* The $*$-transform reads: "An internal set A is $*$-finite if every internal injection of A into itself is a bijection." The reader should observe here that having infinite $*$-finite sets implies that not all mappings between internal sets are internal.

(7.2.5) The following is due to A. Tarski: *An entity A is finite if every nonempty subset of its power set $\mathscr{P}(A)$, ordered by inclusion, has a maximal element.* The $*$-transform here is: "An internal entity A is $*$-finite if every nonempty internal subset of the set of internal subsets of A has an inclusion-maximal element." Perhaps this should be formalized a little more than the preceding characterizations.

A is finite means:

$$(\forall \mathscr{B})[\mathscr{B} \in \mathscr{P}(\mathscr{P}(A)) \Rightarrow [(\exists B)B \in \mathscr{B} \; \& \; [\forall C[C \in \mathscr{B} \Rightarrow [C \subset B]]]]];$$

the $*$-transform is:

$$(\forall \mathscr{B})[\mathscr{B} \in {}^*\mathscr{P}(\mathscr{P}(A)) \Rightarrow [(\exists B)B \in \mathscr{B} \; \& \; [(\forall C)[C \in \mathscr{B} \Rightarrow [C \subset B]]]]].$$

The reader should observe that this statement could be made bounded and that ${}^*\mathscr{P}(\mathscr{P}(A))$ is the set of internal subsets of ${}^*\mathscr{P}(A)$, which is the set of internal subsets of ${}^*A$.

We remark that $*$-finite sets have the same internal formal properties as finite sets in the standard model. We already used this in $*$-finite summation above.

EXERCISE: Let ${}^*\mathscr{X}$ be an ultrapower model $\mathscr{X}^J/\mathscr{U}$. If A is a $*$-finite set of ${}^*\mathscr{X}$, show that $\text{card}({}^\circ A) = \text{card}(\{a : {}^*a \in A\}) \leq \text{card}(J)$.

(7.2.6) THEOREM *The following are equivalent properties of a nonstandard model ${}^*\mathscr{X}$ of the superstructure $\mathscr{X}$.*

(a) *Every standard binary relation entity $\Phi \in \mathscr{X}$ that is concurrent on* $\text{dom}(\Phi)$ *is satisfied in ${}^*\mathscr{X}$ in the sense that there exists a $z \in$* $\text{rng}({}^*\Phi)$ *so that for each $x \in$* $\text{dom}(\Phi)$, *$({}^*x, z) \in {}^*\Phi$, that is,* ${}^\sigma(\text{dom}(\Phi))$ *is satisfied by z.*

(b) *Every standard entity $E \in \mathscr{X}$ can be embedded in a $*$-finite entity $F \in {}^*\mathscr{X}$, that is, there is a $*$-finite $F \in {}^*\mathscr{P}(E)$ so that $F \supseteq {}^\sigma E = \{{}^*e : e \in E\}$.* (The embedding is external when E is infinite.)

(c) *The intersection monad of every nonempty standard family of entities $\mathscr{A} \in \mathscr{X}$ that has the finite intersection property is nonempty,* i.e., $\mu(\mathscr{A}) \neq \varnothing$.

(d) *The union monad of every standard family of entities $\mathscr{A} \in \mathscr{X}$ that has the finite union property is submaximal,* i.e., $\nu(\mathscr{A}) \subset {}^*(\bigcup \mathscr{A})$.

Such models are called **enlargements**.

PROOF: (a)$\Rightarrow$(b) The relation $\Phi = \{(x, y) : x \in y$ and y is a finite subset of $E\}$ is concurrent on E. Hence there is a $*$-finite subset of $^*E$ satisfying $(^*e, F) \in \Phi$, i.e., $^*e \in F$, for each $e \in E$.

(b)$\Rightarrow$(c) Embed $^\sigma\mathscr{A}$ in the $*$-finite set $\mathscr{B} \subseteq {}^*\mathscr{A}$. Then $\bigcap \mathscr{B} \neq \varnothing$ by the $*$-finite intersection property of $^*\mathscr{A}$ and $\mu(\mathscr{A}) \supseteq \bigcap \mathscr{B}$.

(c)$\Leftrightarrow$(d) These are dual notions.

(c)$\Rightarrow$(a) Define a family as follows:
for $x_1, \ldots, x_n \in \operatorname{dom} \Phi$, $A(x_1, \ldots, x_n) = \{y : (x_j, y) \in \Phi,\ j = 1, \ldots, n\}$, $\mathscr{A} = \{A(x_1, \ldots, x_n)$: over all finite subsets of dom $\Phi\}$. Let $z \in \mu(\mathscr{A})$; then $(^*x, y) \in {}^*\Phi$ for all $x \in \operatorname{dom}(\Phi)$.

This concludes the proof.

(7.2.7) If $\mathscr{A} \subseteq \mathscr{P}(X)$ has the finite intersection property, then the set $\operatorname{Fil}(\mathscr{A})$ of all sets B, $X \supseteq B \supseteq A_1 \cap \cdots \cap A_n$ forms a filter, i.e., $\varnothing \notin \operatorname{Fil}(\mathscr{A})$; $B_1, B_2 \in \operatorname{Fil}(\mathscr{A})$ implies $B_1 \cap B_2 \in \operatorname{Fil}(\mathscr{A})$, and $X \supseteq C \supseteq B \in \operatorname{Fil}(\mathscr{A})$ implies $C \in \operatorname{Fil}(\mathscr{A})$. A filter $\mathscr{F}_1$ is *finer* than $\mathscr{F}_2$ provided $\mathscr{F}_1 \supseteq \mathscr{F}_2$ and $\mathscr{F}_2$ is *coarser* than $\mathscr{F}_1$.

(7.2.8) THEOREM *Let $^*\mathscr{X}$ be an enlargement of $\mathscr{X}$. Let $\mathscr{A}$ and $\mathscr{B}$ be standard families of subsets of an entity $X \in \mathscr{X}$. Then*

(a) *$\mu(\mathscr{A}) \neq \varnothing$ if and only if $\mathscr{A}$ has the finite intersection property.*

When $\mathscr{A}$ and $\mathscr{B}$ have the finite intersection property,

(b) $\operatorname{Fil}(\mathscr{A}) \supseteq \operatorname{Fil}(\mathscr{B})$ *if and only if $\mu(\mathscr{A}) \subseteq \mu(\mathscr{B})$, in particular, $\mu(\operatorname{Fil}(\mathscr{A})) = \mu(\mathscr{A})$.* (We say a monad μ_1 with $\mu_1 \subseteq \mu_2$ is *finer* than the monad μ_2, respectively, *coarser*.)

PROOF: Proving (a) is easy; one only needs to show the "only if" part. (b) follows by embedding $^\sigma\mathscr{A}$ in a $*$-finite set $\mathscr{E} \subseteq {}^*\mathscr{A}$ and observing that $\varnothing \neq \bigcap\mathscr{E} \in {}^*\operatorname{Fil}(\mathscr{A})$ by the $*$-finite intersection property of $^*\mathscr{A}$. Now if A is an internal subset of $\mu(\mathscr{A})$, $\mathscr{E}(A) = \{E \in \mathscr{E} : E \supseteq A\}$ is still a $*$-finite set containing $^\sigma\mathscr{A}$ and satisfying

$$A \subseteq \bigcap\mathscr{E}(A) \subseteq \mu(\mathscr{A}).$$

This means in particular that $\mu(\mathscr{A}) = \bigcup[E : E$ is an internal subset of $\mu(\mathscr{A})]$. Take $B_1, \ldots, B_n \in \mathscr{B}$; then $B_1 \cap \cdots \cap B_n \supseteq \mu(\mathscr{A}) \supseteq \bigcap\mathscr{E}$, a nonempty $*$-finite intersection of elements of $^*\mathscr{A}$. Apply the converse of Leibniz' principle in the standard model to see that $\operatorname{Fil}(\mathscr{A}) \supseteq \operatorname{Fil}(\mathscr{B})$. The converse of (b) is easily shown.

(7.2.9) *As a result of this theorem*

$$\operatorname{Fil}(\mathscr{A}) = \{B \in \mathscr{P}(X) : {}^*B \supseteq \mu(\mathscr{A})\}.$$

We will discuss the relation between monads and filters (or ideals) in the next chapter.

(7.2.10) *If $\mathscr{F}$ is a standard filter entity, we call the following the* **filter of sections of** $\mathscr{F}$,

$$\mathscr{F}(S) = \{\mathscr{E}(F) = \{E \in \mathscr{F} : E \supseteq F\} : F \in \mathscr{F}\}.$$

The reader will observe that the proof of (7.2.8(b)) uses the fact that the elements of $\mu(\mathscr{F}(S))$ determine $\mu(\mathscr{F})$, that is, $F \in (\mathscr{F}(S))$ *implies* $F \in {}^*\mathscr{F}$ *and* $F \subseteq \mu(\mathscr{F})$. We leave it to the reader to show that ${}^*\mathscr{X}$ is an enlargement if and only if $\mu(\mathscr{F}(S)) = \varnothing$ for each filter entity. An ultrapower enlargement is constructed in (7.5.3) below.

7.3 COMPREHENSIVE MODELS

We begin with the following simple question. Suppose that for each standard $n \in {}^\sigma\mathbf{N}$ a number $a_n \in {}^*\mathbf{R}$ is given. When can we find a standard sequence $(b_n : n \in \mathbf{N})$ so that ${}^*b_n = a_n$ when $n \in {}^\sigma\mathbf{N}$? It is clear that we cannot do this if any $a_n \notin {}^\sigma\mathbf{R}$, because ${}^*b_n \in {}^\sigma\mathbf{R}$ when $n \in {}^\sigma\mathbf{N}$. We then ask if there is an internal sequence $(c_n : n \in {}^*\mathbf{N})$ such that $a_n = c_n$ when $n \in {}^\sigma\mathbf{N}$. For example, if $a_n = n + \omega$ with $\omega \in {}^*\mathbf{N}_\infty$, then $c_n = n + \omega$ for all $n \in {}^*\mathbf{N}$ is internal but not standard.

We say a nonstandard model ${}^*\mathscr{X}$ is *comprehensive* with respect to $\mathscr{X}$ provided that for each entity $A \in \mathscr{X}$ any function f: ${}^\sigma A \to {}^*B$, where ${}^*B$ is a standard entity, can be extended to an internal function g: ${}^*A \to {}^*B$, $f(a) = g(a)$ for $a \in {}^\sigma A$.

(7.3.1) THEOREM *Ultrapower models are comprehensive with respect to the standard model.*

PROOF: Viewed in the ultrapower an internal mapping g is a function from J, the index set, into the set of mappings of A into B. Now for a particular $j, f(a(j)) = b(j) \in B^J$ and taking $g(a)(j) = b(j) = f(a(j))$ we obtain g as a map from J into functions from $A \to B$. Let $a(j) \equiv a$; then $\{j : g(a)(j) = f(a)\} \in \mathscr{U}$ and $f = g$ on ${}^\sigma A$.

Other questions of extending external functions are considered below.

7.4 SATURATED MODELS

Now we consider the external notion of concurrence on an internal binary relation in the nonstandard model. The internal relation Ψ is *concurrent* on a (possibly external) subset $A \subseteq \mathrm{dom}(\Psi)$ if whenever a finite set $a_1, \ldots, a_n \in A$

of elements of A is given, there exists a $y \subseteq \text{rng}(\Psi)$ so that $(a_j, y) \in \Psi$, $j = 1, \ldots, n$. (The finite set $a_1, \ldots, a_n$ is finite in the absolute sense, not $*$-finite and infinite.) As above, Ψ is *satisfied* for A if there exists a $z \in \text{rng}(\Psi)$ so that $(a, z) \in \Psi$ for each $a \in A$.

We introduce the following convenient notation. The terminology arises from topological considerations.

(7.4.1) ***NOTATION:*** Let $A \subseteq {}^*X$ be a subset of a standard entity ${}^*X$. The *discrete standard part* of A is the set

$$°A = \{a \in X : {}^*a \in A\} \in \mathscr{X}.$$

We abuse our previous notation slightly by defining

$$^{\sigma}A = {}^{\sigma}(°A) = \{{}^*a \in A\} \qquad \text{and} \qquad {}^{\#}A = A \backslash {}^{\sigma}A.$$

This is just *notation for the set of standard elements* of A in $\mathscr{X}$ and in ${}^*\mathscr{X}$ and the set of *nonstandard elements* of A in ${}^*\mathscr{X}$.

Observe that if $A \in \mathscr{X}$, $°({}^*A) = A$ and ${}^{\sigma}({}^*A) = {}^{\sigma}A$. Also, $°({}^*(°B)) = °B$.

Now we consider properties of nonstandard models which fail in enlargements, although we shall not present the counterexamples here.

Compact enlargements are characterized in the next theorem and one is constructed in (7.5.5).

(7.4.2) THEOREM *The following properties of an enlargement of $\mathscr{X}$ are equivalent.*

(a) *Every standard binary relation $\Psi = {}^*\Phi$ that is concurrent on $A \subseteq$ dom(Ψ) where card$({}^{\#}A) < \aleph_0$ is satisfied for A.*

(sa) *Every standard binary relation Ψ that is S-concurrent on $A \subseteq$ dom(Ψ) [$a_1, \ldots, a_n \in A$ implies there is $y \in °\text{rng}(\Psi)$ with $(a_i, y) \in \Psi : i = 1, \ldots n$], where card$({}^{\#}A) < \aleph_0$ is satisfied for A.*

(b) *Every (possible external) set $\mathscr{B} \subseteq {}^*\mathscr{P}(X)$ of internal subsets of a standard entity ${}^*X$ that has the finite intersection property and for which* card$({}^{\#}\mathscr{B}) < \aleph_0$ *satisfies* $\bigcap \mathscr{B} = \bigcap [B : B \in \mathscr{B}] \neq \varnothing$.

(sb) *Every set $\mathscr{B} \subseteq {}^*\mathscr{P}(X)$ of internal subsets of a standard entity ${}^*X$ that has S-finite intersection property, $°(B_1 \cap \cdots \cap B_n) \neq \varnothing$, and for which* card$({}^{\#}\mathscr{B}) < \aleph_0$ *satisfies* $\bigcap \mathscr{B} \neq \varnothing$.

(c) *If $\mathscr{F}$ is a filter entity of $\mathscr{X}$ and $\mathscr{B} \in {}^*\mathscr{P}(\mathscr{F})$ is an internal subset of ${}^*\mathscr{F}$ that satisfies $°\mathscr{B} = \mathscr{F}$, then there exists $B \in \mathscr{B}$ with $B \subseteq \mu(\mathscr{F})$, in other words, $\mathscr{B} \cap \mu(\mathscr{F}(S)) \neq \varnothing$.*

(d) *For every $\mathscr{F}$, a filter entity of $\mathscr{X}$, if $\mathscr{E} \in {}^*\mathscr{P}(\mathscr{F})$ is an internal subset of ${}^*\mathscr{F}$ with the property that "$\mu(\mathscr{F}) \supseteq E \in {}^*\mathscr{F}$ implies $E \in \mathscr{E}$," then $\mathscr{E}$ contains a standard element of $\mathscr{F}$, $°\mathscr{E} \neq \varnothing$ or ${}^{\sigma}\mathscr{F} \cap \mathscr{E} \neq \varnothing$ or there is a ${}^*F \in \mathscr{E}$ with $F \in \mathscr{F}$.*

REMARK: Since $^{\#}\mathscr{B}$ is finite, (b) can be phrased "if A is an internal set for which $^*\mathscr{F} \cap A \neq \varnothing$ for each $F \in \mathscr{F}$, a standard filter entity, then $A \cap \mu(\mathscr{F}) \neq \varnothing$."

PROOF: Left for the reader as an exercise.

(7.4.3) THEOREM *$^*\mathscr{X}$ is a compact enlargement. Let $\mathscr{F}$ be a filter entity of subsets of X in $\mathscr{X}$. Let f: $X \to Y$ be a standard function entity. Let $f(\mathscr{F})$ be the filter generated by the sets $f(F)$ for $f \in \mathscr{F}$. Then $\mu(f(\mathscr{F})) = f(\mu(\mathscr{F}))$.*

PROOF: Take $y \in \mu(f(\mathscr{F}))$, which is nonempty since the model is an enlargement. Now let $A = f^{-1}(y)$. Then $A \bigcap {^*F} \neq \varnothing$ for $F \in \mathscr{F}$, hence $A \bigcap \mu(\mathscr{F}) \neq \varnothing$ since the model is a compact enlargement. Therefore $\mu(f(\mathscr{F}))$ is contained in $f(\mu(\mathscr{F}))$ and the opposite containment is trivial.

(7.4.4) THEOREM *$^*\mathscr{X}$ is a compact enlargement.*

(a) *Let $X \in \mathscr{X}$ and let $\mathscr{B} \in {^*\mathscr{P}}(\mathrm{FP}(X))$ be an internal set of $*$-finite subsets of $^*X$ that has the property that whenever $E \in {^*\mathrm{FP}}(X)$ is a $*$-finite subset of $^*X$ with $^\circ E = X$, then $E \in \mathscr{B}$.* (Here FP(X) stands for finite power set of X, the set of all finite subsets of X.) *Then $^\circ\mathscr{B} \neq \varnothing$, that is, there exist $x_1, \ldots, x_n \in X$ such that*

$$\{^*x_1, \ldots, {^*x_n}\} \in \mathscr{B}.$$

(b) *If A is internal and $A \subseteq \bigcup [^*A_\lambda : \lambda \in \Lambda]$, where $A_\lambda \subseteq X$ for each λ, then there exist $\lambda_1, \ldots, \lambda_n$ such that $A \subseteq {^*A_{\lambda_1}} \cup \cdots \cup {^*A_{\lambda_n}}$, that is, an internal subset of a union monad is contained in a finite union.*

PROOF: Case (a) follows from (7.4.2) by consider the Frechet filter of X. Case (b) follows from (a) by embedding Λ in a $*$-finite set. We shall summarize this fundamental result in **Cauchy's principle** (7.6.4) and (8.1.3).

(7.4.5) COROLLARY *In a compact enlargement, a monad is either principle or external; it cannot be internal and nonstandard.*

PROOF: If $\mu(\mathscr{F})$ is internal and not principal, then $\mathscr{B} = {^\sigma\mathscr{F}} \cap \{^*X \backslash \mu(\mathscr{F})\}$ has the finite intersection property and $\bigcap \mathscr{B} = \varnothing$, contradicting the above result. We leave the dual case $\nu(\mathscr{G})$ to the reader.

(7.4.6) DEFINITION *Let κ denote an infinite cardinal number. A nonstandard model is κ-**saturated** if whenever an internal binary relation Ψ is concurrent on a set A with card$(A) < \kappa$, then Ψ is satisfied for A. A nonstandard model of $\mathscr{X}$ is polysaturated if it is κ-saturated with card$(\mathscr{X}) \leq \kappa$.*

In model theory the term "saturated" is conventionally applied to models saturated to their own cardinality; we do not need this restriction. The cardinal number in the definition of polysaturated is so chosen because constructions of analysis will be smaller than it.

Luxemburg [1969] introduced saturated models in infinitesimal analysis and they have gradually become the most widely used framework. Roughly, this is because they treat internal objects in the same way that they treat standard objects (a theme we have been emphasizing since Chapter 5). This is evident in much of the recent literature, at least to the countable level. In Section 7.7 we will even prove a metatheorem of Henson that says (a form of) this.

Keisler showed that δ-incomplete ultrapowers, as above, are $\aleph_1$-saturated and he constructed κ-saturated ultrapowers using the generalized continuum hypothesis. Kunen later removed the use of the continuum hypothesis. The reader should consult the work of Chang and Keisler [1973]. We shall use ultralimits to produce polysaturated models.

EXERCISE: *Show that an enlargement of $\mathscr{X}$ is $\aleph_1$-saturated if and only if it is countably comprehensive.*

7.5 ULTRALIMITS

This section has one objective: the construction of a polysaturated model by means of an ultralimit of successive ultrapowers. This is probably the easiest construction of polysaturated models. We also choose this construction because of the stronger properties examined in Section 7.7 and because we obtain ultrapower enlargements as fallout. (Robinson [1966] applies the compactness principle of model theory to produce enlargements.)

We begin with the following question: *When does $\mathscr{X}^J/\mathscr{U}$ produce an enlargement*? Let Φ be a concurrent standard binary relation entity on a domain D with range R. We saw above that the family of sets $A_d = \{y : (d, y) \in \Phi\}$ then has the finite intersection property. The relation $*\Phi$ comes from the pointwise extension: $(x, y) \in *\Phi$ if and only if $(x(j), y(j)) \in \Phi$ for j on an element of $\mathscr{U}$. We need to construct J and $\mathscr{U}$ so that there is a sequence $r(j)$ with values in R satisfying "$\{j : (d, r(j)) \in \Phi\}$ belongs to $\mathscr{U}$ for each constant d in $\mathscr{D}$."

(7.5.1) DEFINITION *A filter $\mathscr{F}$ over an index set J is called λ-adequate if to every nonempty family $\mathscr{B}$ of subsets of λ with the finite intersection property, there is a map $s: J \to \lambda$ that maps $\mathscr{F}$ so that $s(\mathscr{F}) \supseteq \mathscr{B}$, that is, to each $B \in \mathscr{B}$, there is an $F \in \mathscr{F}$ with $s(F) \subseteq B$.*

Suppose for the moment that we had an index set J and a $\lambda = \text{card}(\mathscr{X})$-*adequate* ultrafilter $\mathscr{U}$. We let $\mathscr{B}$ of the definition be the family $\{A_d : d \in D\}$ above, and then $r(j) = s(j)$ satisfies the relation Φ over D.

EXERCISE: Prove that

(1) If $\mathscr{F}$ is λ-adequate and $\mathscr{U}$ is finer than $\mathscr{F}$, show that $\mathscr{U}$ is λ-adequate.
(2) If $\mathscr{F}$ is μ-adequate and $\mu > \lambda$, show that $\mathscr{F}$ is λ-adequate.

(7.5.2) A λ-ADEQUATE FILTER Let λ be an infinite cardinal, take J, the index set, equal to the set of finite families of subsets of λ, that is, $J = \text{FP}(\mathscr{P}(\lambda))$, where FP denotes "finite power" set. Take $\mathscr{F}$ to be the filter generated by the sets

$$\{F_A : A \subseteq \lambda\}, \qquad \text{where} \quad F_A = \{j \in J : A \in j\}.$$

This has the finite intersection property and therefore generates a filter, because

$$F_A \cap F_B \quad \text{contains} \quad j = \{A, B\}.$$

$\mathscr{F}$ is λ-adequate, since if $\mathscr{B}$ has the finite intersection property, then

$$A_j = \bigcap [B \in \mathscr{B} : B \in j] \neq \varnothing.$$

Let $s\colon J \to \{A_j : j \in J\}$ be a choice function

$$s(j) \in A_j.$$

Finally, since $s(F_B) \subseteq B$, we see that $\mathscr{F}$ is λ-adequate. In summary, so far we have

(7.5.3) *ASIDE:* *If we apply Definition* (3.10.1) *to a* $\text{card}(\mathscr{X})$*-adequate ultrapower* $\mathscr{X}^J/\mathscr{U}$, *the resulting factored model is an enlargement.*

(7.5.4) SEQUENTIAL ULTRALIMITS We are going to construct a direct limit of a chain of models; for the unfamiliar we begin with a sequential limit. We start the sequence by letting $\mathscr{M}_0 = \mathscr{X}$ and $I^0 = I$, the interpretation of Chapter 3. Let J_1 be an index set as is (7.5.2) with $\lambda = \text{card}(\mathscr{M}_0)$ and take $\mathscr{U}_1$ to be an ultrafilter refinement of the λ-adequate filter on J_1 constructed there. By the exercise above, $\mathscr{U}_1$ is λ-adequate. Let $\mathscr{M}_1 = \mathscr{M}_0{}^{J_1}$ and $\eta_0{}^1 : \mathscr{M}_0 \to \mathscr{M}_1$ be the constant sequence extension map. $(\mathscr{M}_1, \in_{\mathscr{U}_1}, =_{\mathscr{U}_1})$ is a model of analysis under the $I^{\mathscr{U}}$-interpretation of Section 3.8, $I^{\mathscr{U}} = I^1$. Next, apply (7.5.2) with $\lambda = \text{card}(\mathscr{M}_1)$ obtaining J_2 and an adequate ultrafilter $\mathscr{U}_2$. Let $\mathscr{M}_2 = \mathscr{M}_1{}^{J_2}$ and let $\eta_1{}^2 : \mathscr{M}_1 \to \mathscr{M}_2$ denote the constant sequence embedding. Again $(\mathscr{M}_2, \in_2, =_2)$ is a model of analysis and the map $\eta_0{}^2 = \eta_1{}^2 \circ \eta_0{}^1$ is the constant–constant sequence embedding. (We shorten $\in_{\mathscr{U}_n}$ to $\in_n$ and $=_{\mathscr{U}_n}$ to $=_n$.) The interpretation of $\mathscr{L}$ in $\mathscr{M}_2$ is defined to be an extension of $\eta_1{}^2 \circ I^1$, that is, the interpretation of constants of $\mathscr{L}$ which map to points of $\mathscr{M}_1$ is the constant sequence of that interpretation in $\mathscr{M}_2$ (see Fig. 7.5.1).

Now, continue this procedure inductively: given $\mathscr{M}_n$, let $\lambda = \text{card}(\mathscr{M}_n)$ and apply (7.5.2), obtaining J_{n+1} and an adequate ultrafilter $\mathscr{U}_{n+1}$. Take $\mathscr{M}_{n+1} = \mathscr{M}_n^{J_{n+1}}$ and $\eta_n^{n+1} : \mathscr{M}_n \to \mathscr{M}_{n+1}$, the constant sequence extension, and $\eta_m{}^n = \eta_{n-1}^n \circ \cdots \circ \eta_m^{m+1}$ when $m < n$. By successive application of the fundamental theorem on ultrapowers (3.8.3) we know each $(\mathscr{M}_n, \in_n, =_n)$ is a model of analysis where we inductively extend the interpretation from $\mathscr{M}_n$ to $\mathscr{M}_{n+1}$ so that when a constant of $\mathscr{L}$ is interpretable in both, $\eta_n^{n+1}(I^n(\alpha))$ equals $I^{n+1}(\alpha)$. We actually know $\mathscr{M}_{n+1}$ is a model of (the internal theory) $\mathscr{M}_n$ by the proof of (3.8.3).

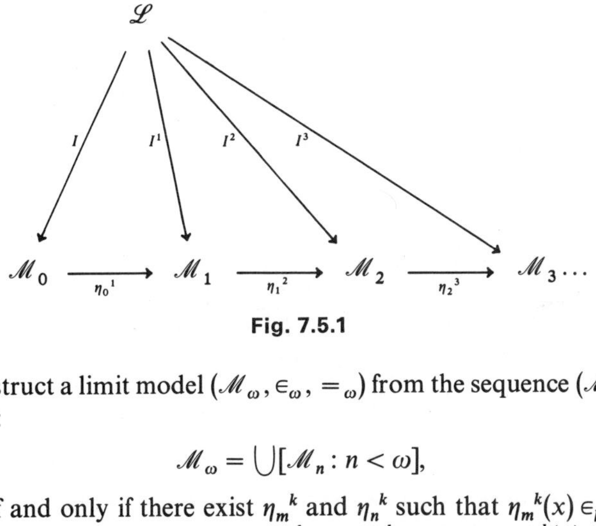

Fig. 7.5.1

We construct a limit model $(\mathscr{M}_\omega, \in_\omega, =_\omega)$ from the sequence $(\mathscr{M}_n, \in_n, =_n)$ as follows:

$$\mathscr{M}_\omega = \bigcup [\mathscr{M}_n : n < \omega],$$

$x \in_\omega y$ if and only if there exist $\eta_m{}^k$ and $\eta_n{}^k$ such that $\eta_m{}^k(x) \in_k \eta_n{}^k(y)$,
$x =_\omega y$ if and only if there exist $\eta_m{}^k$ and $\eta_n{}^k$ such that $\eta_m{}^k(x) =_k \eta_n{}^k(y)$.

The interpretation I^ω of $\mathscr{M}_\omega$ is given by taking $I^n(\alpha)$ for the minimum n at which it is defined and $\text{dom}(I^\omega) = \bigcup [\text{dom}(I^n) : n < \omega]$. Since we have $I^n = \eta_m{}^n I^m$, constants continue to be interpreted the same way mod $=_\omega$ once they enter the limit process. The map $\eta_n{}^\omega : \mathscr{M}_n \to \mathscr{M}_\omega$ is the inclusion into the limit.

This is a special case of a direct limit, and we claim now that $(\mathscr{M}_\omega, \in_\omega, =_\omega)$ also is a model of bounded formal analysis $K_0(\mathscr{X})$. If W is a bounded formal sentence that is true in $\mathscr{X} = \mathscr{M}_0$, that is, so that ${}^I W$ holds where $W = (q_1 x_1) \cdots (q_n x_n) V(x_1, \ldots, x_n)$, then the associated n-ary relation (see Section 3.9) $\Phi = \{(x_1, \ldots, x_n) : x_1 \in A_1 \ \& \cdots \& \ x_n \in A_n \ \& \ {}^I V(x_1, \ldots, x_n)\}$ is defined on all of $A_{i_1} \times \cdots \times A_{i_m}$, where $(i_i, \ldots, i_m)$ are the indices of the universal quantifiers. Since all the constants of W are interpreted at every level consistently, Φ is true exactly on $A_{i_1} \times \cdots \times A_{i_m}$ in the limit and W is true.

(7.5.5) ***ASIDE:*** *Factorization of $\mathscr{M}_\omega$ produces a compact enlargement of bounded formal analysis.*

PROOF: Left as an exercise for the reader.

(7.5.6) **κ-ULTRALIMITS** By transfinite induction,

$$\mathscr{M}_0 = \mathscr{X}, \qquad \text{“}\in_0\text{”} = \text{“}\in,\text{”} \qquad \text{“}=_0\text{”} = \text{“}=,\text{”} \qquad I^0 = I.$$

For $\alpha < \kappa$ and α not a limit ordinal $(\mathscr{M}_{\alpha+1}, \in_{\alpha+1}, =_{\alpha+1})$ is a card$(\mathscr{M}_\alpha)$-adequate ultrapower of $\mathscr{M}_\alpha$,

$$\mathscr{M}_{\alpha+1} = \mathscr{M}_\alpha{}^J, \qquad \text{“}\in_{\alpha+1}\text{”} = (\in_\alpha)_{\mathscr{U}_{\alpha+1}}, \qquad \text{“}=_{\alpha+1}\text{”} = (=_\alpha)_{\mathscr{U}_{\alpha+1}},$$

$$\eta^{\alpha+1} = \text{constant sequence extension} \quad \text{and} \quad \eta_\beta^{\alpha+1} = \eta_\alpha^{\alpha+1} \circ \eta_\beta{}^\alpha,$$

$$I^{\alpha+1} \quad \text{is an extension of} \quad I^\alpha;$$

and if β is a limit ordinal,

$$\begin{aligned}
&\mathscr{M}_\beta = \bigcup [\mathscr{M}_\alpha : \alpha < \beta], \\
&a \in_\beta A \qquad \text{if and only if} \qquad \eta_\alpha{}^\delta(a) \in_\delta \eta_\gamma{}^\delta(A), \\
&a =_\beta b \qquad \text{if and only if} \qquad \eta_\alpha{}^J(a) \in_\delta \eta_\gamma{}^\delta(b), \qquad \text{for} \quad \alpha, \gamma < \delta < \beta, \\
&\quad \eta_\alpha{}^\beta \qquad \text{is the inclusion into} \quad \mathscr{M}_\beta, \\
&I^\beta = I^\alpha \qquad \text{on the subdomain of minimum} \quad \alpha.
\end{aligned}$$

Finally, $(\mathscr{M}_\kappa, \in_\kappa, =_\kappa; I^\kappa)$ is the direct limit of the chain (see Fig. 7.5.2).

(7.5.7) **THEOREM** *Let $\kappa = 2^{\text{card}(\mathscr{X})}$. The set factorization of the bounded portion of $(\mathscr{M}_\kappa, \in_\kappa, =_\kappa; I^\kappa)$ into a superstructure $(\mathscr{Y}, \in, =)$ is a poly-saturated model of bounded formal analysis.*

PROOF: An internal binary relation comes from some level $\mathscr{M}_\alpha$ in the direct limit. The set of concurrence contains fewer than card$(\mathscr{X})$ internal constants and thus lies in some $\mathscr{M}_\beta$, $\beta < \kappa$. The model $\mathscr{M}_{\beta+1}$ is an adequate extension and therefore satisfies the relation. Satisfaction is maintained in $\mathscr{Y}$.

(7.5.8) ***BLANKET ASSUMPTION:*** We shall assume throughout the remainder of the text that we are working with a factored $2^{\text{card}(\mathscr{X})}$-ultralimit model $(\mathscr{Y}, \in, =)$.

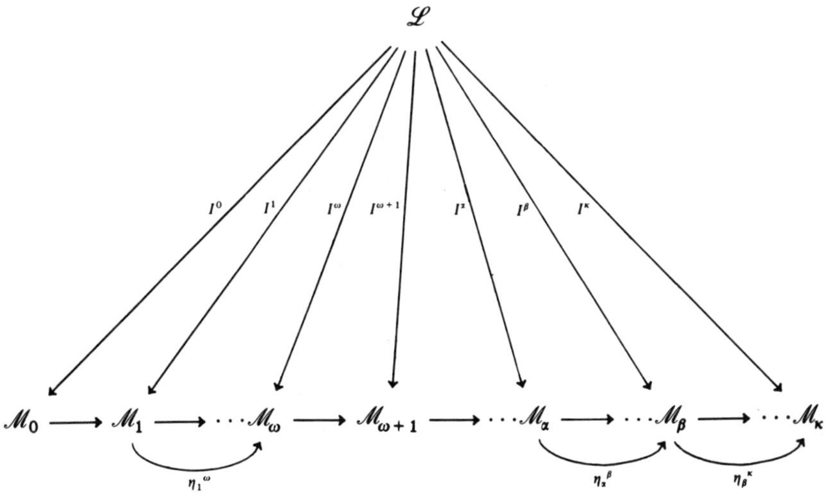

Fig. 7.5.2

The λ-adequate filters were first introduced by Luxemburg [1969] for (7.5.2) and (7.5.3).

7.6 PROPERTIES OF POLYSATURATED MODELS

All the properties in Sections 7.2–7.4 hold in polysaturated models (with $\kappa = \text{card}(\mathcal{X})$ where applicable).

(7.6.1) THEOREM *The following is equivalent to the definition of polysaturated model: If $\mathcal{B} \subseteq {}^*\mathcal{P}(X)$, for $X \in \mathcal{X}$, has the finite intersection property and* $\text{card}(\mathcal{B}) < \text{card}(\mathcal{X})$, *then* $\bigcap \mathcal{B} \neq \varnothing$.

Also, in a polysaturated model, if $B \subseteq {}^*X$ for $X \in \mathcal{X}$ and $\text{card}(B) < \text{card}(\mathcal{X})$, then there is a $*$-finite subset C of ${}^*X$ with $B \subseteq C$. (This condition is weaker than saturation. See Keisler [1967] and Luxemburg [1969].)

We omit the proof since it is similar to the above discussion for enlargements.

(7.6.2) THEOREM *A polysaturated model is comprehensive in the sense that if $A \subseteq {}^*X$ has* $\text{card}(A) < \text{card}(\mathcal{X})$ *and if $f: A \to B$, $B \in {}^*\mathcal{P}(Y)$, is given, then f can be extended to an internal map into B.*

PROOF: "f_1 and f_2 are internal maps with values in B and f_2 extends f_1" is a concurrent internal binary relation on A since $f: \{a\} \to \{f(a)\}$ is internal.

(7.6.3) THEOREM *Let* $\mathbf{N} \in \mathscr{X}$. *Then* $^*\mathbf{N}$ *contains an infinite random pair* $\lambda, \omega \in {}^*\mathbf{N}_\infty$ *with* $\lambda \ll \omega$; *that is, if* $f : \mathbf{N} \to \mathbf{N}$ *is a standard function* $^*f(\lambda) < \omega$.

PROOF: Take $\lambda \in {}^*\mathbf{N}_\infty$. Let $^\sigma\mathscr{F}$ be the set of standard functions of $^*\mathbf{N}$. The set $\{\lambda' : \lambda' = f(\lambda), f \in {}^\sigma\mathscr{F}\}$ has cardinality less than card($\mathscr{X}$) and hence lies in a $*$-finite subset of $^*\mathbf{N}$. Take ω greater than the maximum of that $*$-finite set.

(7.6.4) CAUCHY'S PRINCIPLE (also see (8.1.3)) *If* $P(x)$ *is a bounded formal internal property depending on the free variable x, and if* $P(m)$ *holds for all* $m \in \mu(\mathscr{F})$, *then* $P(f)$ *holds for all* $f \in {}^*F$, *a standard element of the filter entity* $\mathscr{F}$.

For example, in $\mathbf{R}$ if $|f(x) - b| < \delta$ for all $x \approx 0$, then there is a standard $\varepsilon > 0$ so that $|x| < \varepsilon$ implies $|f(x) - b| < \delta$, so long as f is internal.

PROOF: Let $P = \{x \in {}^*X : P(x)\}$. Then P is internal and $P \supseteq \mu(\mathscr{F})$. If $\mu(\mathscr{F}) = {}^*F$ for some set F, we are done, otherwise apply (7.4.4(a)) to the $*$-finite subsets of $^*\mathscr{F}$ that satisfy $F_1 \cap \cdots \cap F_\lambda \subseteq P$.

It is worthwhile to keep (7.4.4(a)) in mind as a separate principle.

(7.6.5) LUXEMBURG'S $*$-FINITE LEMMA *If an internal family of $*$-finite sets contains all the $*$-finite extensions of a standard set, then that family must contain a properly finite standard set.*

7.7 THE ISOMORPHISM PROPERTY OF ULTRALIMITS

Henson [1974] recently pointed out a new model-theoretical property of ultralimit nonstandard models. Since this property can be used to answer various natural internal–external questions and since it shows promise for analysis applications (beginning with Henson's proof of a theorem of Amir and Lindenstrauss on weakly compactly generated Banach spaces), we shall describe it here.

For simplicity we make all our cardinality restrictions relative to card($\mathscr{X}$) and a fixed κ-ultralimit as in Section 7.5. This should present no essential restriction in the study of analysis, since $\kappa >$ card($\mathscr{X}$) $>$ card(E) whenever E is an entity. In other words we shall adhere to our blanket assumption (7.5.8). The reader is refered to Henson's paper for more refined treatment of cardinal numbers.

(7.7.1) 'SMALL' FIRST-ORDER THEORIES We want to study first-order substructures of $\mathscr{X}$ and $^*\mathscr{X}$ that involve only internal relations and whose cardinality is smaller than card($\mathscr{X}$). For this purpose we introduce other formal languages, $L(\rho_k{}^n : k \in K,\ n \in \mathbf{N}) = L$ which have the *logical connectives, variables, quantifiers,* and *brackets* of Section 3.5, but which have a *list* $\{\rho_k{}^n : k \in K,\ n \in \mathbf{N}\}$ *of basic n-ary predicates* with $\text{card}(K) < \text{card}(\mathscr{X})$. The list is permitted to be empty for some values of n. The *extralogical constants* of L are the nullary relations of the list $\{\rho_k{}^0\}$. This restriction means in particular that the theory can only refer to a limited number of specific points of the base space when interpreted (and this makes the "isomorphism class" of models larger).

Well-formed formulas are defined in the same way starting from *atomic formulas* as in Section 3.5, but *atomic formulas* now take the forms:

$$\rho_1{}^2(x, \rho_2{}^0), \quad \rho_1{}^2(x, y), \quad \rho_1{}^2(x, \rho_3{}^3(y, \rho_2{}^0, z)), \quad \text{etc.,}$$

where $\rho_1{}^2$ is binary, $\rho_2{}^0$ a constant, and $\rho_3{}^3$ ternary. An atomic formula consists of one of the n-ary relations with its n-places filled with constants and/or variables and with finite compositions of relations with their spaces filled.

The idea here is not very difficult, but it is very abstract so we cement it with the following example. Suppose we want to discuss group theory. We need to discuss products, inverses, and formulas like "every x has an inverse," or "for every x in G, $x \cdot x^{-1} = 1$." Thus, L needs the constant $1 = \rho_1{}^0$, the binary inverse relation $y = x^{-1}$, $\rho_2{}^2$, and the ternary product relation $z = x \cdot y$, $\rho_3{}^3$. The axioms of group theory can now be built in this language where *one thinks of quantifiers running over the points of a particular group*—this is the meaning of "first order." Naturally, this is the usual informal interpretation in abstract algebra; we must be specific about this restriction on the interpretation of quantifiers.

Besides the axioms, one might ask which formal sentences are part of "all group theory," that is, which sentences are true in every interpretation. Naturally, a sound method of formal proof will supply some of these. The conventional notation is

$$\mathscr{G} \vdash V$$

for "the axioms $\mathscr{G}$ formally prove V," while if G is a particular group in which V holds,

$$G \models V$$

stands for "G satisfies V." Symbolically, *soundness* of formal proof is the theorem:

If $\mathscr{G} \vdash V$, then $G \models V$ for each model of $\mathscr{G}$.

Completeness is the theorem:

If $G \models V$ in every model G of $\mathscr{G}$, then $\mathscr{G} \vdash V$.

[Incompleteness is a theorem that says, for example, "not every formal property of number theory that is true in **N** has a formal proof," consequently, there are nonstandard models of arithmetic.]

Model theory deals with a broader kind of "isomorphism" (elementary equivalence) than in abstract algebra; we hope these remarks will help the reader to understand the difference. More examples are given below as well.

(7.7.2) STANDARD AND INTERNAL SUBINTERPRETATIONS OF 'SMALL' FIRST-ORDER THEORIES *If* $L(\rho_k{}^n : k \in K,\ n \in N)$ *is a "small"* $(<\text{card}(\mathscr{X}))$ *first-order language, then a set of sentences of L,* $T = T(L(\rho_k{}^n))$ *will be referred to as a theory or* **"small" first-order theory**. We are primarily interested in consistent theories and the *compactness theorem* tells us that it suffices for all finite subsets of T to be consistent.

A **standard subinterpretation** *of* $T(L(\rho_k{}^n))$ *is a one-to-one map i from* $\{\rho_k{}^n : k \in K,\ n \in \mathbf{N}\}$ *into* $\mathscr{X}$ *such that* ${}^i\rho_k{}^n = i(\rho_k{}^n)$ *is an n-ary relation entity on a fixed* **entity** *A*; ${}^i\rho_k{}^0$ is a point in A for each k, ${}^i\rho_k{}^2$ is a subset of $A \times A$, and so on. The list

$$\mathscr{A} = (A, R_k) = (A, {}^i\rho_k{}^n)$$

is a model of T provided the i-interpretation ${}^iV$ is true in $\mathscr{A}$ for each V in T where quantifiers are interpreted as bounded by A. If a standard subinterpretation exists, the cardinality restriction $\text{card}(K) < \text{card}(\mathscr{X})$ is automatically fulfilled; $\mathscr{A} = (A, {}^i\rho_k{}^n)$ is also referred to as a *standard L-substructure.*

An **internal subinterpretation** *of* $T(L(\rho_k{}^n))$ *is an injection* i' *of* $\{\rho_k{}^n\}$ *into the internal n-ary relations on a fixed internal set B.* For example, $i'(\rho_k{}^0) = {}^{i'}\rho_k{}^0$ is a point of B, ${}^{i'}\rho_k{}^2$ is an internal subset of $B \times B$, and so on. The list

$$\mathscr{B} = (B, S_k) = (B, {}^{i'}\rho_k{}^n)$$

is a model of T provided the interpretation ${}^{i'}V$ is true in B for each V in T where quantifiers are interpreted as bounded by B. Notice that the family $\{{}^{i'}\rho_k{}^n : k \in K,\ n \in \mathbf{N}\}$ need **not** be an internal set. $\mathscr{B}$ is referred to as an *internal L-substructure.*

(7.7.3) DEFINITION *A first-order subtheory of analysis* (that is, our fixed $\mathscr{X} \xrightarrow{*} \mathscr{Y}$) *is a first-order theory* $T(L(\rho_k{}^n : k \in k,\ n \in \mathbf{N}))$ *with* $\text{card}(K) <$ $\text{card}(\mathscr{X})$ *which has an internal subinterpretation.*

(7.7.4) EXERCISE: If T is a theory that has a standard subinterpretation, show that it has an internal subinterpretation and is of "small" cardinality, thus is a subtheory. (*Hint*: $i' = * \circ i$; rewrite formulas of L as bounded formulas of $\mathscr{L}$; apply Leibniz' principle.)

The notion of elementary equivalence is a model-theoretical notion that depends on the language L (not a particular theory). An *interpretation* of $L(\rho_k{}^n : k \in K,\ n \in \mathbf{N})$ is simply a one-to-one map taking the $\rho_k{}^n$ to n-ary relations on some fixed set $\mathscr{C}$ (or one could say an interpretation of the theory $(\exists x)[x = x]$ if equality was one of the $\rho_k{}^2$'s.) Let ${}^C\rho_k{}^n$ be the interpretation of $\rho_k{}^n$ over $\mathscr{C}$.

(7.7.5) DEFINITION *Two L-structures $(C, R_k{}^n) = \mathscr{C}$ and $(D, s_k{}^n) = \mathscr{D}$ are L-elementarily equivalent provided each sentence V of L satisfies*

$$\mathscr{C} \models ({}^C V) \text{ if and only if } \mathscr{D} \models ({}^D V),$$

that is, they have the same true sentences in L.

(7.7.6) EXERCISE: Let L be the first-order language of only the order relation $<$ and equality and no other extralogical constants. Show that $(\mathbf{Q}, {}^{\mathbf{Q}}<, =)$ and $(\mathbf{R}, {}^{\mathbf{R}}<, =)$ are L-elementarily equivalent.

(7.7.7) DEFINITION *An L-elementary monomorphism between two L-structures $\mathscr{A} = (A, R_k)$ and $\mathscr{B} = (B, S_k)$ is an injection $h\colon A \to B$ such that*

$$\mathscr{A} \models {}^i V[a_1, \ldots, a_n] \text{ if and only if } \mathscr{B} \models {}^{i'} V[ha_1, \ldots, ha_n]$$

whenever $a_1, \ldots, a_n \in A$ and $V[x_1, \ldots, x_n]$ is a predicate of L with $x_1, \ldots, x_n$ free. Here ${}^i V[a_1, \ldots, a_n]$ and ${}^{i'} V[ha_1, \ldots, ha_n]$ denote the predicates interpreted in the respective structures with the free variables replaced with the constants indicated. We include the possibility that $n = 0$, that V is a sentence.

Notice that this definition goes outside the realm of L or specifically of $\{\rho_k{}^0\}$, in fact, another way to define an L-monomorphism is to extend the collection of constants to include all the points of A and require that $(A, R_k, a : a \in A)$ is elementarily equivalent to $(B, S_k, ha : a \in A)$ in the enriched language.

An L-isomorphism is an L-monomorphism that is onto.

(7.7.8) DEFINITION *$\mathscr{A}$ is an L-elementary substructure of $\mathscr{B}$ and $\mathscr{B}$ is an L-elementary extension of $\mathscr{A}$ in case the inclusion map of $A \subseteq B$ is an L-monomorphism. In this case we write $\mathscr{A} \prec \mathscr{B}$.*

(7.7.9) EXERCISE: If (A, R_k) is a standard submodel, show that $*$ restricted to A is an L-monomorphism. We denote the list $({}^*A, {}^*R_k)$ by ${}^\sigma\mathscr{A}$.

EXERCISE: Is $(\mathbf{Q}, {}^{\mathbf{Q}}<, =)$ L-isomorphic to $(\mathbf{R}, {}^{\mathbf{R}}<, =)$ in the language of "less than"?

An intuitively plausible example of equivalent nonisomorphic structures is as follows. Let L be the language of algebraically closed fields of characteristic zero; specifically, L has the constants 0 and 1, the relation for equality, and the ternary relations of multiplication and addition. The theory consists of the axioms for a field of characteristic zero and the infinite list saying "polynomials have roots":

$$\{\text{degree } 2\} \qquad (\forall x)(\forall y)(\forall z)[x \neq 0 \Rightarrow [(\exists w)[x \cdot w \cdot w + y \cdot w + z = 0]]],$$

$$\{\text{degree } 3\} \qquad (\forall x)(\forall y)(\forall z)(\forall w)[x \neq 0 \Rightarrow [(\exists u)[x \cdot u \cdot u \cdot u + y \cdot u \cdot u + z \cdot u + w = 0]]],$$

and so on for each degree.

All the models of this theory are elementarily equivalent, but the isomorphism class depends on the cardinal number of a transcendence base; see Chang and Keisler [1973]. Transcendental elements are "indiscernible," so isomorphism is just a matter of shuffling them around. On the other hand, since constants are not available, the theory cannot detect how many there are.

(7.7.10) HENSON'S LEMMA ($\mathscr{X} \xrightarrow[\mathscr{L}]{*} \mathscr{Y}$ is an ultralimit extension as in Section 7.5.) *Elementarily equivalent internal L-substructures of $\mathscr{Y}$ are (externally) L-isomorphic.*

This means that if $L(\rho_k{}^n)$ is a first-order language with fewer than card$(\mathscr{X})$ constants and relations $\{\rho_k{}^n\}$, and if $\mathscr{A} = (A, R_k)$ and $\mathscr{B} = (B, S_k)$ are internal subinterpretations with the same true sentences in L, then there is a bijection $h: A \to B$ that preserves the formulas of L. For example, Henson [1974] shows that the atomic "internal subsets" Boolean algebra on an infinite Z is equivalent to the Boolean algebra of "$*$-finite subsets and their complements" in the language of Boolean algebras, consequently *there is a map (on atoms) that satisfies "A is an internal subset of Z if and only if $h(A)$ or its complement is a $*$-finite subset of Z."*

(7.7.11) COROLLARY *If $\mathscr{A} = (A, R_k)$ and $\mathscr{B} = (B, S_k)$ are L-elementarily equivalent standard substructures, then ${}^\sigma\mathscr{A} = ({}^*A, {}^*R_k)$ and ${}^\sigma\mathscr{B} = ({}^*B, {}^*S_k)$ are (externally) L-isomorphic.* ($\mathscr{A}$ and $\mathscr{B}$ need not be isomorphic; see (7.7.6) for example.)

PROOF: See Henson [1974].

(7.7.12) THEOREM *If $\mathscr{B} = (B, S_k)$ is an internal subtheory, then there exists a standard subtheory $\mathscr{A} = (A, R_k)$ such that ${}^\sigma\mathscr{A} = ({}^*A, {}^*R_k)$ is L-isomorphic to $\mathscr{B}$.*

PROOF: By the downward Löwenheim–Skolem theorem there is an elementarily equivalent model of $\mathscr{B}$ with small cardinality. Embed this model in $\mathscr{X}$ and apply Leibniz' principle to see that ${}^\sigma\mathscr{A} \equiv \mathscr{B}$. Finally apply Henson's lemma.

The difficulty in applying Henson's lemma is that one must select the language and prove L-equivalence. This is generally difficult since it involves "all properties expressible in L." The most typical examples of elementary equivalence from model theory are focused on algebraic structures, so that they do not seem to be directly applicable in analysis.

Perhaps the **back-and-forth argument** used in the proof of Henson's lemma could be applied directly in certain analytical settings.

The article by Barwise [1974] is a nice introduction to back-and-forth arguments and is related to the following alternative to elementary equivalence in Henson's lemma. We are most grateful to Ward Henson for privately communicating this formulation of the isomorphism property in terms of "concurrent" families of partial isomorphisms.

(7.7.13) DEFINITION *Let $\mathscr{A} = (A, R_k)$ and $\mathscr{B} = (B, S_k)$ be $L(\rho_k)$-structures. A family $\Phi = \{\varphi\}$ of injections partially defined on A into B has the back-and-forth property provided that given (φ, a, b) with $\varphi \in \Phi$, $a \in A$, and $b \in B$, there exist φ_a and φ^b in Φ such that $a \in \mathrm{dom}(\varphi_a)$ and φ_a extends φ while $b \in \mathrm{rng}(\varphi^b)$ and φ^b extends φ. If each $\varphi \in \Phi$ is an L-isomorphism on its domain (that is, preserves the interpretation of L-sentences that only involve constants from $\mathrm{dom}(\varphi)$ and where the relations are restricted to $\mathrm{dom}(\varphi)$), then we say Φ is a partial L-isomorphism, Φ :*

$$\mathscr{A} \overset{p}{\underset{\mathscr{L}}{\cong}} \mathscr{B}.$$

REMARK: See Problem 1.3.15 in the work of Chang and Keisler [1973].

Let us suppose that $\mathscr{A}$ and $\mathscr{B}$ are L-equivalent structures. If L contains constants, and equality, then $\varphi({}^i\rho_0) = {}^{i'}\rho_0$ for a monomorphism φ since $(\exists 1x)[x = \rho]$ is a formula that must be preserved. In this case we let $\Phi = \{\varphi : \varphi$ is an L-monomorphism defined on a finite extension of the set of constants$\}$.

(7.7.14) LEMMA *If $\mathscr{A}$ and $\mathscr{B}$ are internal L-equivalent substructures in a polysaturated model, then Φ has the back-and-forth property.*

PROOF: Left as an exercise for the reader.

(7.7.15) HENSON'S LEMMA (Alternate form) *Let $^*\mathscr{X}$ be a polysaturated ultralimit as in Section 7.5 and let $\mathscr{A}$ and $\mathscr{B}$ be internal L-substructures. If $\mathscr{A}$ and $\mathscr{B}$ are partially L-isomorphic, then $\mathscr{A}$ and $\mathscr{B}$ are (externally) L-isomorphic.*

PROOF: By Karp's theorem (see Barwise [1974]) $\mathscr{A}$ and $\mathscr{B}$ are elementarily equivalent for $L_{\infty,\omega}$ and thus for L.

ALGEBRAIC APPLICATION: Let δ and ε be two positive infinitesimals. There is an ordered field automorphism taking $(^*\mathbf{R}, \delta)$ onto $(^*\mathbf{R}, \varepsilon)$.

Let $L(0, 1, +, \cdot, <, =, dx)$ be the language of an ordered field with a positive infinitesimal and take $\mathscr{A} = (^*\mathbf{R}, \delta)$, $\mathscr{B} = (^*\mathbf{R}, \varepsilon)$ with the usual interpretations of $0, 1, +, \cdot, <, =$ and ${}^{i}dx = \delta$ while ${}^{i'}dx = \varepsilon$. Both $\mathscr{A}$ and $\mathscr{B}$ satisfy the axioms in L of an ordered field with a positive infinitesimal. We will show that $\mathscr{A}$ and $\mathscr{B}$ are partially isomorphic. The standard rationals ${}^\sigma\mathbf{Q}$ are pointwise fixed since there are formulas in L describing them in terms of 0, 1, and algebra $(2 : (\exists 1x)[1 + 1 = x], \ldots)$. We want to send $\delta \mapsto \varepsilon$ and thus the field generated by δ and ${}^\sigma\mathbf{Q}$ is mapped onto that generated by ε ard ${}^\sigma\mathbf{Q}$. Since δ and ε are positive infinitesimal transcendentals, the order isomorphism can be built by sending the rational polynomial $p(\delta) \mapsto p(\varepsilon)$. A $*$-real a, which is algebraic over the field of δ and ${}^\sigma\mathbf{Q}$, satisfies a standard integer polynomial in two variables:

$$P(\delta, a) = 0.$$

Moreover, the $*$-real roots of $p(x) = P(\delta, x)$ are ordered and we can express in L the statement that a is the kth. We take the image of $a \mapsto b$ to be the kth $*$-real root of

$$P(\varepsilon, b) = 0.$$

EXERCISE: Show that the polynomial $q(x) = P(\varepsilon, x)$ has the same number of $*$-real roots as $p(x) = P(\delta, x)$. Now let D be the field of reals that are algebraic in δ and $\mathbf{Q}$ and let E be its order isomorph of reals algebraic in ε and $\mathbf{Q}$. Call the isomorphism φ_0. Our partial isomorphism

$$\Phi = \{\varphi : \varphi \text{ is an algebraically closed ordered field extension of } \varphi_0\}$$

has the back-and-forth property. For example, suppose t is transcendental over D. Consider all the formulas $I(t) = \{d_1 < t < d_2 : d_1, d_2, \in D\}$. Since $\varphi_0(d_1) = e_1$ and $\varphi_0(d_2) = e_2$ preserves order, the family of sets $\{x : e_1 < x < e_2\}$ has the finite intersection property, and since it is only countable and $\mathscr{B}$ is polysaturated, there is an s in the intersection. Send $t \mapsto s$ and extend to the algebraic closure $D(t) \to E(s)$.

Henson has also extended this idea to include the valued fields considered by Robinson [1973b].

THE GENERAL THEORY OF MONADS AND INFINITESIMALS

CHAPTER 8

8.1 MONADS WITH RESPECT TO A RING OF SETS

A set $\mathscr{E}$ of subsets of an entity X in $\mathscr{X}$ is called a *ring of sets* ($\mathscr{E} \subseteq \mathscr{P}(X)$) provided $\varnothing, X \in \mathscr{E}$ and if $E, F \in \mathscr{E}$, then $E \cup F \in \mathscr{E}$ and $E \cap F \in \mathscr{E}$. $\mathscr{E}$ itself is a standard entity.

(8.1.1) ***EXAMPLES:*** (a) the *discrete ring of all subsets of* X, $\mathscr{D} = \mathscr{P}(X)$,

(b) the family τ of *open sets* of a topological space (X, τ),

(c) the family σ of *closed sets* of a topological space (X, τ), $\sigma = \{X \backslash U : U \in \tau\}$,

(d) the *zero sets* of continuous real-valued functions on a topological space $\mathscr{Z} = \{f^{-1}(0) : F \in C(X)\}$,

(e) the *cozero sets* $\mathscr{C} = \{X \backslash Z : Z \in \mathscr{Z}\}$,

(f) if $\mathscr{E}$ is a ring over X, $\mathscr{G} = \{X \backslash E : E \in \mathscr{E}\}$ is the *dual ring* to $\mathscr{E}$ in $\mathscr{D}$,

(g) the *macles* $\mathscr{M}$ in a Cartesian product which is the ring generated by sets of the form $E \times F$ where $X = Y \times Z$ and $E \subseteq Y$ and $F \subseteq Z$.

(8.1.2) ***NOTATION:*** When $A \subseteq {}^*X$ we introduce the operators $\mu_{\mathscr{E}}$, $\nu_{\mathscr{E}}$, $\mathrm{Fil}_{\mathscr{E}}$, and $\mathrm{Idl}_{\mathscr{E}}$ as follows:

$$\mu_{\mathscr{E}}(A) = \bigcap\,[{}^*E : E \in \mathscr{E} \text{ and } {}^*E \supseteq A],$$

$$\mathrm{Fil}_{\mathscr{E}}(A) = \{E : E \in \mathscr{E} \text{ and } {}^*E \supseteq A\},$$

$$\nu_{\mathscr{E}}(A) = \bigcup\,[{}^*E : E \in \mathscr{E} \text{ and } {}^*E \subseteq A],$$

$$\mathrm{Idl}_{\mathscr{E}}(A) = \{E : E \in \mathscr{E} \text{ and } {}^*E \subseteq A\}.$$

With these conventions $\mathrm{Fil}_{\mathscr{E}}(A)$ (respectively, $\mathrm{Idl}_{\mathscr{E}}(A)$) is a standard filter in the ring $\mathscr{E}$ (respectively, ideal in $\mathscr{E}$) and the point of (7.2.8) is that the monads determine the filters (respectively, ideals) and conversely. In fact, the correspondence given here is a lattice isomorphism.

(8.1.3) DEFINITION The monad μ_0 of any subset of $\mathscr{E}$ will be called an $\mathscr{E}$-*monad* and by (7.2.8) $\mu_{\mathscr{E}}(\mu_0) = \mu_0$. In other words $\mu_{\mathscr{E}}(A)$ runs through all the $\mathscr{E}$-monads. Cauchy's principle and its covering dual hold for $\mathscr{E}$-monads with the stronger conclusion that the set is in $\mathscr{E}$.

(8.1.4) CAUCHY'S PRINCIPLE (for a ring of sets) *Let $\mathscr{E}$ be a ring of sets over $X \in \mathscr{X}$ and let B be an internal subset of $^*X$. Then*

(a) *if $\mu_0 \subseteq B$ and μ_0 is an $\mathscr{E}$-$\bigcap$-monad, there is an $E \in \mathscr{E}$ with*

$$\mu_0 \subseteq {}^*E \subseteq B;$$

(b) *if $B \subseteq \nu_0$ and ν_0 is an $\mathscr{E}$-$\bigcup$-monad, there is an $E \in \mathscr{E}$ with*

$$B \subseteq {}^*E \subseteq \nu_0;$$

(c) *if $\mu_0 \subseteq \nu_0$ where μ_0 is an $\bigcap$-monad, ν_0 a $\bigcup$-monad, and one of them is an $\mathscr{E}$-monad, there is an $E \in \mathscr{E}$ with*

$$\mu_0 \subseteq {}^*E \subseteq \nu_0.$$

PROOF: Apply (7.4.4) to obtain B in a finite intersection or union of $\mathscr{E}$-sets and observe that the ring property of $\mathscr{E}$ means the intersection or union is in $\mathscr{E}$ itself. Case (c) follows from (b) by taking a $*$-finite intersection inside μ_0 for example.

REMARK: We could rephrase (b) to say "a monadic cover of an internal set can be replaced by a finite subcover."

We already used the following very important result which we restate here as another continuity principle.

(6.4.1) ROBINSON'S SEQUENTIAL LEMMA *An internal sequence $(a_k : k \in {}^*\mathbf{N})$ which is infinitesimal for standard subscripts, is in fact infinitesimal out to some infinite subscript, that is, if $a_k \approx 0$ for $k \in {}^\sigma\mathbf{N}$, then there is an infinite λ for which $a_k \approx 0$ when $1 \le k \le \lambda$.*

PROOF: From Section 6.4 we use the internal definition principle on the set $\{n \in {}^*\mathbf{N} : \text{for all } k < n,\ k|a_k| < 1\}$.

A generalization of Robinson's lemma for polysaturated models suggested by M. Wolff is as follows. A similar idea underlies the completeness of the nonstandard hull in Section 8.4.

(8.1.5) LEMMA *Let φ: $^*A \to B^*$ be an internal map between standard sets and let $\mathscr{F}$ and $\mathscr{G}$ be filters on A and B, respectively. If $\varphi(^*F) \bigcap \mu(\mathscr{G}) \neq \varnothing$ for $F \in \mathscr{F}$*, then $\varphi(\mu(\mathscr{F})) \bigcap \mu(\mathscr{G}) \neq \varnothing$.

PROOF: The family $\{\varphi(^*F) \bigcap {}^*G : F \in \mathscr{F},\ G \in \mathscr{G}\}$ has the finite intersection property and small cardinality (relative to $\mathscr{X}$), therefore it has nonempty intersection.

EXERCISE: Let $A = \mathbf{N}, B = \mathbf{C}$, and $\varphi(n) = a_n$. Which filters in Wolff's lemma produce the result of Robinson's sequential lemma?

In case $\mathscr{E} = \mathscr{D}$ we use the terminology *discrete monad* or when $\mathscr{E} = \tau$, *open monad*, etc. Discrete monads have the following characterization:

(8.1.6) THEOREM *Let $a \in {}^*A$, $A \in \mathscr{X}$, be an internal entity. The discrete monad of a, $\mu_{\mathscr{D}}(\{a\})$, is the set of all entities that have all the same standard bounded formal properties as a. In particular, $\mu_{\mathscr{D}}\{a\} = \{a\}$ if and only if a is standard. If $b \in \mu_{\mathscr{D}}\{a\}$, there is a bijection m: ${}^*A \to {}^*A$ that leaves every standard relation on A invariant and $m(a) = b$. If B is any subset of ${}^*A$, $\mu_{\mathscr{D}}(B)$ is the set of elements of ${}^*A$ with all the same standard bounded formal properties shared by all the elements of B.*

PROOF: Suppose $P(x)$ is a standard bounded formal predicate with x free and suppose "for all $b \in B$, ${}^{I'}P(b)$" holds. Then ${}^*P = \{x \in {}^*A : P(x)\}$ is standard and contains B.

Conversely, ${}^{I'}P(x) =$ "$x \in D$" is a bounded (interpreted) formal predicate with only standard constants. If ${}^*D \supseteq B$, then "for all $b \in B$, $P(b)$" holds. We can apply this to $B = \{a\}$.

The first-order language L with one constant ρ and a relation symbol for each standard relation on A can be interpreted as $({}^*A, a, {}^*R_k)$ and $({}^*A, b, {}^*R_k)$. There are $\aleph_0 \cdot 2^{\text{card}(A)}$ such relations so these are internal substructures. The above remarks specialize to L to yield the fact that these are elementarily equivalent. Henson's lemma provides m.

(8.1.7) COROLLARY *Let $\omega \in {}^*\mathbf{N}_\infty$ be an infinite natural number. Then $\mu_{\mathscr{D}}(\{\omega\})$ contains points different from ω but none within a finite distance of ω, that is, $\lambda \in \mu_{\mathscr{D}}(\{\omega\})$ implies either $\lambda = \omega$ or $|\lambda - \omega|$ is infinite.*

PROOF: $\lambda - \omega$ is divisible by each finite n without a remainder because "$\omega = n \cdot k + m$ with $m < n$" is a standard formula about ω.

We shall return to the question of "infinitesimals" around nonstandard points below. While $\mu_\tau(\{x\})$, with τ a topology and x standard, is an infinitesimal neighborhood, this corollary serves as a warning against identifying monads and infinitesimals.

As an operator on all subsets of ${}^*X$, $\mu_{\mathscr{D}}$ is a closure operator; the topology it produces is called the *discrete S-topology* on ${}^*X$. It is a compact weakly Hausdorff topology and when closures of points are identified, the result is the discrete Čech–Stone compactification of X. Internal sets are compact in the discrete S-topology. Some of these remarks are contained in the next result, which is of general interest. (We will give analysis applications later in Chapter 9.)

(8.1.8) THEOREM *Let M be a monadic closure operator on ${}^*X$, that is, a closure that is always $\bigcap$-monad valued. Then ${}^\sigma X$ is dense in ${}^*X$ and ${}^*X$ is* (non-Hausdorff or quasi-) *compact.*

PROOF: We remind the reader that a closure operator satisfies

(1) $M(\varnothing) = \varnothing$ and $M(*X) = *X$,
(2) $M(M(A)) = M(A)$,
(3) $A \subseteq B$ implies $A \subseteq M(A) \subseteq M(B)$,
(4) $M(A \cup B) = M(A) \cup M(B)$.

${}^\sigma X$ is dense because $\mu_{\mathscr{D}}({}^\sigma X) = *X$ is the finest monad containing ${}^\sigma X$, thus $M({}^\sigma X) = *X$.

We will show that $*X$ has the *finite intersection property*. Let $(M_\lambda : \lambda \in \Lambda)$ be a family of closed sets and let $F_\lambda = \mathrm{Fil}_{\mathscr{D}}(M_\lambda)$. If $M_{\lambda_1} \cap \cdots \cap M_{\lambda_n} \neq \varnothing$ for any finite collection of λ_i's, then the filter generated by $f_1 \cap \cdots \cap f_n$, $f_i \in F_{\lambda_i}$, over all finite collections of λ's is a proper filter $\mathscr{F}$. This is because

$$*f_1 \cap \cdots \cap *f_n \supseteq M_{\lambda_1} \cap \cdots \cap M_{\lambda_n} \neq \varnothing.$$

Now $\mu(\mathscr{F}) = \bigcap (M_\lambda : \lambda \in \Lambda) \neq \varnothing$ since the monad of a proper filter is never empty. This proves that M induces a compact topology.

REMARK: We could rephrase this result to say, *any topology with a closed base of a standard sets is* (quasi-) *compact.*

We refer the reader to the work of Luxemburg [1969], Machover and Hirschfeld [1969], Robinson [1969], Haddad [1972], and Stroyan [1972b] for additional results and details in the theory of monads. We will restrict ourselves to some basic general theorems and applications.

(8.1.9) EXERCISE: Show that $\mu_{\mathscr{E}}$ is a monadic closure on $*X$. What can you say about the spaces resulting from Examples (8.1.1(a–g))?

8.2 CHROMATIC SETS

This section deals with the questions of how monadic a given set is at a point and to what extent we can say that being a monad is a "local" property.

(8.2.1) DEFINITION *$\mathscr{E}$ is a ring over $X \in \mathscr{X}$. A subset $A \subseteq *X$ is $\mathscr{E}$-* **chromatic** (just chromatic if $\mathscr{E} = \mathscr{D}$) *provided $\mu_{\mathscr{E}}(\{a\}) \subseteq A$ for each $a \in A$.*

We remark that both kinds of $\mathscr{E}$-monads are automatically $\mathscr{E}$-chromatic since $\mu_0 = \mu_{\mathscr{E}}(\mu_0)$ and $\mu_{\mathscr{E}}(\{a\}) \subseteq \mu_{\mathscr{E}}(A)$ and $\nu_0 = \bigcup *E$ with each $*E$ $\mathscr{E}$-chromatic. An $\bigcap$-monad $\mu_0 \subseteq *X$ that is $\mathscr{E}$-chromatic is an $\mathscr{E}$-monad when $\mathscr{E}$ is closed under arbitrary unions of its elements, as is the case when $\mathscr{E} = \tau$, a topology. Thus a monad is a τ-monad if and only if it is τ-chromatic, in particular $*A$ is τ-open if and only if it is τ-chromatic.

Haddad [1972] points out that an essential mechanism in the proof of the next result is the fact that internal sets are compact in the discrete S-topology. Compare this result with (7.4.5).

(8.2.2) THEOREM *$\mathscr{E}$ is a ring over $X \in \mathfrak{X}$, B is an internal subset of $^*X$. The $\mathscr{E}$-monad of B is the smallest $\mathscr{E}$-chromatic set containing B, $\mu_{\mathscr{E}}(B) = \bigcup [\mu_{\mathscr{E}}(\{b\}) : b \in B]$. In particular, the smallest chromatic set containing B is its discrete monad. An internal set is standard if and only if it is chromatic. In other words, internal nonstandard sets are pale.*

(8.2.3) ***REMARK:*** An application of **Cauchy's principle** somewhat like the last part of this theorem is that if B is internal and $\mathscr{E}$-chromatic at some point $b \in B$, $\mu_{\mathscr{E}}(\{b\}) \subseteq B$, then B contains a standard set $^*E$ containing b.

PROOF: Since $\mu_{\mathscr{E}}(B)$ is $\mathscr{E}$-chromatic we have $\mu_{\mathscr{E}}(B) \supseteq C = \bigcup [\mu_{\mathscr{E}}(\{b\}) : b \in B]$. If the inclusion is strict, there is an $m \notin C$ so that for each $b \in B$ there is an $E(b) \in \mathscr{E}$ with $b \in {}^*E(b)$ and $m \notin {}^*E(b)$. Now $B \subseteq \bigcup [{}^*E(b) : b \in B]$ and by Cauchy's principle B is contained in a finite union of $^*E(b)$'s, say, $b_1, \ldots, b_n$. Since the reader will easily verify that a finite union of monads is a monad, $\mu_{\mathscr{E}}(B) = \mu_{\mathscr{E}}(\{b\}_1) \cup \cdots \cup \mu_{\mathscr{E}}(\{b\}_n)$.

(8.2.4) EXERCISE: Suppose $A = \bigcup B$, where B is internal. Show that

$$\mu_{\mathscr{D}}(A) = \bigcup \mu_{\mathscr{D}}(B).$$

8.3 TOPOLOGICAL ASPECTS OF MONAD THEORY

In this section we summarize the applications of monads to topology. Let X be a topological space with τ the open sets, σ the closed sets, $\mathscr{Z} = \{f^{-1}(0) : f$ is a continuous real-valued function on $X\}$ the zero sets, and $\mathscr{C} = \{X \backslash Z : Z \in \mathscr{Z}\}$ the cozero sets. Since we are interested in applications in analysis we shall only consider the weakly Tychonoff spaces. This means essentially that real-valued functions can be used to "measure" the neighborhoods. First, we assume the weak Hausdorff property: if x, y are standard points, either $\mu_\tau\{x\} = \mu_\tau\{y\}$ or $\mu_\tau\{x\} \perp \mu_\tau\{y\}$. This means that either standard points are indistinguishable topologically (like L^1-functions which differ on a set of measure zero) or they can be separated by disjoint open neighborhoods. (To prove this from the monad property use Cauchy's principle on a monad and its complement or apply Leibniz' principle and the fact that there are $*$-neighborhoods inside the disjoining monads.) Besides this we need to assume that if $x \notin S$, a closed set, that there is a real-valued continuous function that is zero on S and nonzero on x. It turns out that it is sufficient to assume $\mu_{\mathscr{Z}}\{x\} \subseteq \mu_{\mathscr{C}}\{x\} = \mu_\tau\{x\}$ for each standard $x \in {}^\sigma X$. In this case $\mu_{\mathscr{Z}}$ and $\mu_{\mathscr{C}}$ as closures on $^*X$ are both related to the Čech–Stone compactification (see Robinson [1969] and Stroyan [1972b]).

We remark that Tychonoff spaces are regular, which means that points and closed sets can be separated by open sets. This follows from the condition $\mu_{\mathscr{Z}}\{x\} \subseteq \mu_{\mathscr{C}}\{x\} = \mu_\tau\{x\}$, since then $\mu_{\mathscr{C}}\{x\} \perp \mu_{\mathscr{C}}(^*S)$ and consequently $\mu_\tau\{x\} \perp \mu_\tau(^*S)$. By two applications of Cauchy's principle $x \in U \perp V \supseteq S$ and the point and closed set are separated by open sets.

(8.3.1) THEOREM *A standard map $f\colon (X, \tau) \to (X', \tau')$ is continuous at x if and only if $f(\mu_\tau\{x\}) \subseteq \mu_{\tau'}\{f(x)\}$.*

PROOF: That conventional open neighborhood continuity implies the nonstandard condition we leave to the reader (see Robinson [1966]). If U is a neighborhood of $f(x)$, $f^{-1}(^*U) \supseteq \mu_\tau\{x\}$. Applying Cauchy's principle, there is a τ-set V so that $f^{-1}(^*U) \supseteq {}^*V \supseteq \mu_\tau\{x\}$ and f is continuous.

(8.3.2) DEFINITION The *τ-near-standard points* of $^*X$ are defined by $\text{ns}_\tau(^*X) = \{y \in {}^*X : \text{there is an } x \in {}^\sigma X \text{ with } y \in \mu_\tau\{x\}\}$.

(8.3.3) DEFINITION The *τ-standard-part* map $\text{st}_\tau : \text{ns}(^*X) \to X$ is given by $\text{st}_\tau(A) = \{x \in X : \text{for some } a \in A,\ a \in \mu_\tau\{^*X\}\}$, where $A \subseteq {}^*X$.

The map st is singleton valued on near-standard points in a proper Hausdorff space, in which case we usually take the point as the value.

(8.3.4) DEFINITION The non-near-standard points are called *τ-remote points* $\text{rmt}_\tau(^*X) = {}^*X \backslash \text{ns}_\tau(^*X)$.

The near-standard points form the smallest τ-chromatic set containing the standard points.

By (8.2.2), the smallest τ-chromatic set containing an internal set is its τ-monad. Since τ-monads are either standard τ-sets or external, we have the following theorem as a result of Cauchy's principle:

(8.3.5) THEOREM *An internal τ-chromatic set is necessarily a standard open set. In particular, if $\text{ns}_\tau(^*X)$ is internal, it is all of $^*X$.*

(8.3.6) *OBSERVATION:* *If Λ is a standard open cover for X, then* $\text{ns}_\tau(^*X) \subseteq \nu(\Lambda)$.

Now suppose $\text{ns}_\tau(^*X) = {}^*X$. By Cauchy's principle (b), Λ can be replaced by a finite subcover and X is compact. This proves

(8.3.7) THEOREM *X is compact if and only if $\text{ns}_\tau(^*X)$ is internal and thus equal to $^*X$.*

Since the near-standard points form a chromatic set, $x \in \text{rmt}_\tau(^*X)$ is remote if and only if $\mu_{\mathscr{D}}\{x\}$ consists only of remote points.

The standard-part map is connected with the closure on X in several ways, the first of which is:

(8.3.8) THEOREM *If M is any $\bigcap$-monad, in particular, if $M = {}^*A$, for $A \subseteq X$, then* $\mathrm{st}_\tau(M)$ *is* τ-closed, *in particular,* $\mathrm{cl}_\tau(A) = \mathrm{st}_\tau({}^*A)$.

PROOF: To see this let $x \in \mathrm{st}_\tau(M)$. This means $\mu_\tau\{x\} \cap M \neq \varnothing$ and every standard neighborhood of x intersects M. If y is an adherent point of $\mathrm{st}_\tau(M)$, then for each standard open neighborhood V of y there is an $x \in \mathrm{st}_\tau(M)$ so that $x \in V$. Then ${}^*V$ intersects M and it follows that $\mu_\tau\{y\} \cap M \neq \varnothing$ so $y \in \mathrm{st}_\tau(M)$. In the particular case that $M = {}^*A$, take y adherent to A in the above discussion. This says the standard part of a standard set is closed.

(8.3.9) THEOREM *The standard part of an internal set is closed.*

PROOF: Let B be an internal subset of ${}^*X$. The τ-monad of B is the smallest τ-chromatic set containing B and has the same standard part as B.

(8.3.10) THEOREM *Let (X, τ) be a Tychonoff space. Consider ${}^*X$ with μ_τ as a closure and* $\mathrm{ns}({}^*X)$ *as a subspace in the μ_τ topology. The map* $\mathrm{st}\colon \mathrm{ns}({}^*X) \to X$ *is closed and continuous.*

In effect this supplements the knowledge that τ is contained in ${}^*X$ locally inside $\mu_\tau\{x\}$ for $x \in {}^\sigma X$.

PROOF: $\mathrm{st}(\mu_\tau(A))$ is closed by (8.3.8). If S is closed and $r \notin S$, the regularity of X assures us that $\mu_\tau\{r\} \perp \mu_\tau({}^*S)$. Thus, if $\mathrm{st}(t) = r \notin S$, $t \notin \mu_\tau({}^*S)$. On the other hand, if $t \in \mu_\tau({}^*S) \cap \mathrm{ns}_\tau({}^*X)$, then $t \in \mu_\tau\{s\}$ for some $s \in {}^*S$ by (8.2.2) and $\mathrm{st}(t) \in S$ by (8.3.8). Together this shows $\mathrm{st}^{-1}(S) = \mu_\tau({}^*S) \cap \mathrm{ns}_\tau({}^*X)$.

As a corollary, $(\mathrm{ns}({}^*X), \mu_\tau)$ modulo equal standard parts is homeomorphic to X.

The connection between near-standard points and compactness is further revealed in the next result.

(8.3.11) THEOREM *Suppose B is an internal set of near-standard points $B \subseteq \mathrm{ns}({}^*X)$. Then* $\mathrm{st}_\tau(B)$ *is a compact subset of X. In particular, $A \subseteq X$ is relatively compact if and only if ${}^*A \subseteq \mathrm{ns}({}^*X)$.*

PROOF: We remind the reader of our assumption that X is regular. Let Λ be an open cover of $\mathrm{st}_\tau(B)$. Then $B \subseteq \nu(\Lambda)$ and by Cauchy's principle, B is contained in a finite union $B \subseteq {}^*V_1 \cup \cdots \cup {}^*V_n$. Since the finite union is τ-chromatic it contains the τ-monad of B and the closures of the V's cover $\mathrm{st}(B)$. This proves compactness in a regular space.

Let $(X_k : k \in K)$ be a standard family of topological spaces. The product $\prod [X_k : k \in K] = \{f\colon K \to \bigcup_k X_k \mid f(k) \in X_k\}$ may be endowed with the coarsest topology that makes the projections $f \overset{\pi_k}{\mapsto} f(k)$ continuous. Infinitesimally, this means that *if f is standard,*

$$g \in \mu_\pi(f) \qquad \text{if and only if} \qquad g(k) \in \mu_k(f(k)), \quad k \in {}^\sigma K.$$

(8.3.12) TYCHONOFF'S THEOREM *A product of (quasi-) compact spaces is (quasi-) compact in the weak product topology given above.*

PROOF (Robinson): Let $f\colon {}^*K \to {}^*\bigcup [X_k : k \in K]$ be a given internal map with $f(k) \in {}^*X_k$. Since each ${}^*X_k$, $k \in {}^\sigma K$, consists entirely of near-standard points, $G(k) = \mathrm{st}_k(f(k))$, $k \in {}^\sigma K$, defines a standard family of sets. If the X_k are Hausdorff, these are singleton sets $G(k) = \{g(k)\}$ and $g(k)$ is a standard point of ${}^*\Pi$ for which $f \in \mu_\pi(\{g\})$. **Notice that this case does not require the axiom of choice**, but exploits the amount of choice built into our model. Otherwise, when the X_k are non-Hausdorff, let $g(k) \in G(k)$ be a (standard) choice function and still $f \in \mu_\pi(\{{}^*g\})$ in ${}^*\Pi$.

The interested reader will recall that Kelley showed that the axiom of choice is equivalent to the non-Hausdorff Tychonoff theorem. This shows that the prime ideal theorem implies the Hausdorff Tychonoff theorem.

A natural monad to consider when studying compactness is as follows. Let $\mathscr{K} = \{K : K \subset\subset X\}$, the standard family of compact subsets of X. Then $K \subset\subset X$ means K is compactly contained in X.

(8.3.13) DEFINITION The *compact points of ${}^*X$ are given by*

$$\mathrm{cpt}_\tau({}^*X) = \nu(\mathscr{K}).$$

In light of the last result each ${}^*K \subseteq \mathrm{ns}({}^*X)$ for $K \in \mathscr{K}$, so $\mathrm{cpt}({}^*X) \subseteq \mathrm{ns}({}^*X)$. The set of compact points is chromatic but not necessarily τ-chromatic, in fact,

(8.3.14) THEOREM *X is locally compact at $x \in X$ if and only if* $\mathrm{cpt}({}^*X) \supseteq \mu_\tau\{x\}$, *in particular X is locally compact if and only if* $\mathrm{cpt}({}^*X)$ *is τ-chromatic, or,* $\mathrm{cpt}({}^*X) = \mathrm{ns}({}^*X)$.

The proof of this uses Cauchy's principle: if $\nu(\mathscr{K}) \supseteq \mu_\tau\{x\}$, then x has a compact neighborhood.

EXAMPLE OF A NONCOMPACT NEAR-STANDARD POINT: We begin with a basic simple example from analysis, the noncompactness of the closed unit ball in the Hilbert space $l^2(\mathbf{N}) = \{(a_k\colon \mathbf{N} \to \mathbf{R}) : \sum_{k=1}^\infty |a_k|^2 < \infty\}$. The sequences

$$a_k = \delta_k{}^n = \begin{cases} 1, & n = k \\ 0, & n \neq k \end{cases}$$

are unit vectors. The balls

$$B(\pm\delta^n; (1.1)) = \left\{a_k\colon \sum_{k=1}^{\infty} |a_k \mp \delta_k{}^n|^2 < (1.1)\right\}$$

cover the unit ball

$$\bar{B}(0; 1) = \left\{a_k\colon \sum_{k=1}^{\infty} |a_k|^2 \le 1\right\}$$

but cannot reduce to a finite subcover since each $B(\delta^n; (1.1))$ contains only one of the δ^m's. This is simply because the distance between δ^m and δ^n is $\sqrt{2}$ when $m \ne n$. In fact, this shows that a set containing infinitely many of the δ^n vectors is noncompact and we may scale down by a fixed scalar λ, so any set containing infinitely many vectors $\lambda\delta^n$ is noncompact.

Now select an infinite random pair $\omega \ll \Omega$ as in (7.6.3). *The vector* $(1/\omega)\delta^\Omega$ *is infinitesimally close to zero and not an element of any standard compact set.* Suppose that $(1/\omega)\delta^\Omega \in {}^*K$, but for each $m \in {}^\sigma\mathbf{N}$, there is a maximum $n \in \mathbf{N}$ for which $(1/m)\delta^n \in K$, that is, suppose only finitely many δ^m's satisfy $(1/m)\delta^n \in K$ for each standard m. The standard map $f(m) = \max(n: (1/m)\delta^n \in K)$ necessarily satisfies $f(\omega) < \Omega$ in the extension, contradicting the hypothesis that $(1/\omega)\delta^\Omega \in^* K$. Therefore, K contains $(1/m)\delta^n$ for infinitely many n's and a fixed m and is noncompact. Also see (8.4.44).

REMARK: The "Hilbert cube" $\{a \mid |a_k| \le 1/k \text{ for all } k\}$ *is* compact. First, take $b_k = \operatorname{st}(a_k)$ for finite k, then compare $^*b_k$ and a_k. By Leibniz' principle, b is in the cube since "$b_k \le 1/k$ for all $k \in \mathbf{N}$" holds. By Robinson's sequential lemma,

$$\sum_{k=1}^{n} |a_k - b_k|^2 \approx 0 \qquad \text{for} \quad 1 \le n \le \Omega \in {}^*\mathbf{N}_\infty.$$

By the computation in the examples of Section 5.1, $\sum_{n=\Omega}^{\infty} (1/n^2) \approx 0$, therefore,

$$\|a - b\|^2 \le \sum_{k=1}^{\Omega} |a_k - b_k|^2 + 2\sum_{n=\Omega}^{\infty} (1/n^2) \approx 0.$$

Hence a is near standard and the cube is compact.

The same remarks as above for $\operatorname{ns}({}^*X)$ can now be made about the case where $\operatorname{cpt}({}^*X)$ is internal, namely, since it is chromatic it must be standard, therefore ${}^*X$ forcing $\operatorname{ns}({}^*X) = {}^*X$ and X to be compact. This proves

(8.3.15) THEOREM *X is compact if and only if* $\operatorname{cpt}({}^*X)$ *is internal and hence equal to* ${}^*X$.

Next we view paracompactness in a nonstandard extension. A space is called *paracompact* if each open cover has a locally finite open refinement. A nonstandard characterization in terms of what happens at "infinity" is

(8.3.16) THEOREM *The space X is paracompact if and only if for each internal set $K \subseteq \mathrm{rmt}({}^*X)$ that consists only of remote points there is a standard locally finite open cover $\Lambda(K)$ of X so that $K \perp \nu(\Lambda(K))$.*

PROOF: Note that $A \perp B$ for sets means $A \cap B = \varnothing$, hence $K \perp \nu(\Lambda(K))$ means ${}^*X \backslash K \supseteq \nu(\Lambda(K))$. If X is paracompact and K is a remote internal set, we will show that there is a cover $\Lambda(K)$ with the desired properties. First, given $x \in {}^\sigma X$, there is a neighborhood $N(x)$ of x so that $N(x) \perp K$. This is because $X \backslash K \supseteq \mathrm{ns}({}^*X) \supseteq \mu_\tau\{x\}$. Since $X \backslash K$ is internal, Cauchy's principle says there is a ${}^*N(x)$ satisfying

$$X \backslash K \supseteq {}^*N \supseteq \mu_\tau\{x\}.$$

Now the cover $\Lambda = \{N(x) : x \in X\}$ satisfies

$$K \perp \nu(\Lambda).$$

Since X is paracompact, Λ has a locally finite open refinement $\Lambda(K)$ and $\nu(\Lambda) \supseteq \nu(\Lambda(K))$ so

$$K \perp \nu(\Lambda(K)).$$

Conversely, suppose the nonstandard condition is fulfilled and Λ is an open cover of X. Let L be a $*$-finite subset of ${}^*\Lambda$ that contains ${}^\sigma\Lambda$ and define $K = {}^*X \backslash (\bigcup L)$. Take $\Lambda(K)$ a locally finite open cover with $K \perp \nu(\Lambda(K))$; if $\mathcal{O} \in \Lambda(K)$, then ${}^*\mathcal{O} \subseteq (\bigcup L)$, and by Leibniz' principle $\mathcal{O}$ is a subset of a finite union of elements of Λ, say,

$$\mathcal{O} \subseteq \mathcal{O}_1 \cup \cdots \cup \mathcal{O}_n, \qquad \mathcal{O}_j \in \Lambda.$$

Now finitely refine $\Lambda(K)$ to a refinement of Λ, proving that X is paracompact in the process.

(8.3.17) THEOREM *Let $f\colon (X, \tau) \to (X', \tau')$ be a standard map. Then f is open if and only if f maps τ-chromatic sets to τ'-chromatic sets.*

PROOF: Apply (8.3.5).

(8.3.18) THEOREM *Let $f\colon (X, \tau) \to (X', \tau')$ be a continuous standard map. Then f is proper if and only if $f^{-1}(\mathrm{ns}({}^*X')) \subseteq \mathrm{ns}({}^*X)$.*

PROOF: Left as an exercise for the reader.

The point of some of the preceding results can be summarized by saying that the τ-monad of a standard point is an infinitesimal neighborhood of the point. We ask the reader to recall (8.1.6), which says that the points in the discrete monad of an infinite point are infinitely far apart. The next exercise is another result of this kind.

EXERCISE: *Show that the relation $x \in \mu_\tau\{y\}$ for x, $y \in {}^*X$ is not symmetric unless τ is the discrete topology on the closures of points.* (*Hint*: Use the Hausdorff axiom internally.)

Thus y is close to x but x is not close to y. The next section concerns infinitesimal relations that will always be symmetric, in fact, equivalence relations, but "infinitesimal" is not a topological notion.

8.4 UNIFORM INFINITESIMAL RELATIONS AND FINITE POINTS

This section gives the basic infinitesimal reformulations of the standard theory of metric and uniform spaces from Robinson [1966], Luxemburg [1969], Machover and Hirschfeld [1969], Haddad [1972], and Henson [1972]. Some generalizations of these ideas are possible, for example, the study of internal spaces and metrics, and we will introduce those features in the linear theory of the next chapter; in fact, we already have done this in the tangent bundles of Chapters 5 and 6. We feel that infinitesimal analysis will be most useful where it works in special ways that arise "naturally" in examples with abundant structure, rather than where it reformulates abstract standard theories. Nonetheless, it is convenient to see the reformulations so that infinitesimal methods can assimilate past knowledge as well as shed more light on it.

(8.4.1) DEFINITION *A (standard) uniform infinitesimal relation on an entity ${}^*X$ is an equivalence relation described by a family Λ of pseudometrics on X as follows:*

$$x \overset{\Lambda}{\approx} y \qquad \text{if and only if} \qquad \lambda(x, y) \approx 0 \qquad \text{for all} \quad \lambda \in {}^\sigma\Lambda.$$

Since Λ is standard its cardinality is "small" for saturation properties, $\text{card}(\Lambda) < \text{card}(\mathscr{X})$. Many of the ideas here carry over to the case where ${}^*X$ is replaced by an internal set and ${}^\sigma\Lambda$ is replaced by an external family of internal pseudometrics with cardinality less than $\text{card}(\mathscr{X})$.

(8.4.2) DEFINITION The *finite points* of ${}^*X$ with respect to Λ and some preferred $x \in X$ are given by

$$\text{fin}_\Lambda({}^*X, x) = \{y \in {}^*X : \lambda(x, y) \text{ is finite for every } \lambda \in {}^\sigma\Lambda\}.$$

(8.4.3) DEFINITION A function f is Λ-*S-continuous* at x provided $x \overset{\Lambda}{\approx} y$ implies $f(x) \approx f(y)$. The space of *internal* functions that are S-continuous everywhere is denoted $SC_\Lambda({}^*X)$.

A subset $A \subseteq {}^*X$ is $\overset{\Lambda}{\approx}$-*saturated* if whenever $x \in A$ and $y \overset{\Lambda}{\approx} x$, then $y \in A$. The set $\text{fin}_\Lambda({}^*X)$ is saturated.

(8.4.4) DEFINITION The *uniform infinitesimal hull* of X with respect to Λ (and x), $\hat{X}_\Lambda = \hat{X}$ is the set of finite points with the uniformity described by the family of semimetrics

$$\hat{\lambda}(\hat{x}, \hat{y}) = \text{st}(\lambda(x, y))$$

wherever λ is an S-continuous $*$-pseudometric on an internal set containing the finite points $\text{fin}_\Lambda({}^*X, x)$. If f is S-continuous and finite on the finite points, the hull of f is

$$\hat{f}(\hat{x}) = \widehat{f(x)} = \text{st}(f(x)),$$

where $\hat{x}$ represents any $y \overset{\Lambda}{\approx} x$ in $\hat{x}$.

(8.4.5) ***NOTATION:*** One more convenient notation for *infinitesimal neighborhoods*:

$$o_\Lambda[x] = \{y \in {}^*X : y \overset{\Lambda}{\approx} x\}.$$

The following examples indicate how natural the idea of infinitesimals really is. Perhaps the reader can supply further examples and even generalizations.

(8.4.6) ***EXAMPLE:*** $X = {}^*\mathbf{R}$, Λ the single function $|x - y|$, $x \overset{\Lambda}{\approx} y$ iff $x \approx y$. The finite points are measured from zero, so $\text{fin}_\Lambda({}^*\mathbf{R}) = \mathcal{O}$ and ${}^*\hat{\mathbf{R}}$ the hull equals $\mathbf{R}$.

(8.4.7) ***EXAMPLE:*** $X = {}^*\mathbf{R}$, where Λ is the single function $|\arctan x - \arctan y|$, all points are finite, and the hull is isomorphic to $[-\pi, \pi]$ or $[-\infty, \infty]$.

(8.4.8) ***EXAMPLE:*** $X = {}^*U = {}^*\{z \in \mathbf{C} : |z| < 1\}$ the unit disk. Λ is the function $|z - w|$ so $z \overset{\Lambda}{\approx} w$ iff $z \approx w$. All points are finite and the hull is the closed unit disk.

(8.4.9) ***EXAMPLE:*** ${}^*U$, Λ the single function (see Section 6.6),

$$\eta(z, w) = \frac{1}{2} \log \frac{|1 - \bar{z}w| + |z - w|}{|1 - \bar{z}w| - |z - w|}$$

the hyperbolic metric. Points z so that $|z| \not\approx 1$ are finite, and the hull is the open unit disk. (Here we write $x \overset{\eta}{\approx} y$ for $x \overset{\{\eta\}}{\approx} y$.)

(8.4.10) ***EXAMPLE:*** $X = {}^*U$, Λ the single function,

$$p(z, w) = |z - w|/|1 - \bar{z}w|$$

the pseudo-hyperbolic metric. Now $z \overset{n}{\approx} w$ iff $z \overset{\mathcal{L}}{\approx} w$, but all points in this example are finite unlike (8.4.9). The hull is quite complicated now and, in particular, contains the maximal ideal space of $H^\infty(U)$ in the Gleason parts metric (see Chapter 9).

(8.4.11) ***EXAMPLE:*** The abstract form of these first few examples is where (X, d) is a metric space with d standard. Then $x \overset{d}{\approx} y$ iff $d(x, y) \approx 0$ and the finite points are those x with $d(a, x)$ finite for some preferred point $a \in X$. Notice that (8.4.9) and (8.4.10) are infinitesimally equivalent and nonstandardly inequivalent; **finiteness** is **not** a uniform invariant. $\mathrm{Gal}_d(a)$, *the galaxy in d around a,* is the set of points a finite distance in d from a.

When $(X, d) = ({}^*Y, d)$ is a standard metric space *the infinitesimals around a standard point equal the open monad* of that point (in the topology induced by the metric). If $x \in {}^\sigma X$, then

$$o_d[x] = \mu_{\tau(d)}(x).$$

If $x \notin {}^\sigma X$, equality is usually not the case; in fact, the infinitesimal neighborhood may not even be a monad (of any standard family).

(8.4.12) ***EXAMPLE:*** Let $X = \mathscr{C}^k(\mathbf{R})$ the space of k times continuously differentiable internal functions, Λ the space of standard seminorms $\gamma_n{}^m(f, g) = \sup[|f^{(m)}(x) - g^{(m)}(x)| : x, y \in [-n, n]]$, and $0 \le m \le k$, where n, $k \in {}^\sigma\mathbf{N}$. The finite points are those that are a finite distance from the zero function in all the seminorms $\gamma_n{}^m$, $n \in {}^\sigma\mathbf{N}$ and $m \le k$. A simple description for infinitesimals is

$$f \overset{k}{\approx} g \quad \text{iff} \quad f^{(m)}(x) \approx g^{(m)}(x) \quad \text{for } x \text{ finite} \quad \text{and} \quad m \le k,$$

that is, f and g are close in the first k derivatives at finite points.

(8.4.13) ***REMARK:*** The generalization of the last example is the case where (X, Λ) is a standard uniform space. Then $x \overset{\Lambda}{\approx} y$ in ${}^*X$ if $\lambda(x, y) \approx 0$ for all *standard* pseudometrics $\lambda \in {}^\sigma\Lambda$. *The galaxy around* $a \in {}^*X$, $\mathrm{Gal}_\Lambda(a) = \{x \in {}^*X : \lambda(a, x)$ is finite for all $\lambda \in {}^\sigma\Lambda\}$. Here as in (8.4.11) a galaxy is *not* a uniformly invariant notion. *One must select the right notion for finite to suit the problem independently of the notion of infinitesimal*; when Λ and Γ are equivalent gauges, we may have $\hat{X}_\Lambda \ne \hat{X}_\Gamma$.

The topological and uniform agree at standard points:

$$x \in {}^\sigma X \quad \text{implies} \quad o_\Lambda[x] = \mu_{\tau(\Lambda)}(x) = o_\Gamma[x].$$

The next two examples involve spaces of infinitely differentiable functions that are related to the theory of distributions.

(8.4.14) *EXAMPLE:* $*X = \mathscr{C}^\infty(*\mathbf{R}) = *[\mathscr{C}^\infty(\mathbf{R})]$ the space of infinitely differentiable functions. The infinitesimal relation is

$$f \overset{\mathscr{E}}{\approx} g \quad \text{iff} \quad f^{(k)}(t) \approx g^{(k)}(t) \quad \text{for } t \in \mathscr{O} \quad \text{and} \quad k \in {}^\sigma\mathbf{N},$$

where $f^{(k)}$ denotes the kth derivative.

For example, the function $f(t) \equiv 0$ and the function

$$g(t) = \begin{cases} 0, & t \le \omega \text{ an infinite positive number} \\ \exp[1/(\omega - t)], & t > \omega \end{cases}$$

satisfy $f \overset{\mathscr{E}}{\approx} g$. Also, the function $k(t) = (t/\omega)^\omega$ for ω an infinite natural number is infinitesimally close to zero.

The finite points $\mathscr{E} = \text{fin}(X)$ are those functions that satisfy $f^{(k)}(t) \in \mathscr{O}$ for each finite t and k, that is, those that take on finite values for finite t in all their finite derivatives.

(8.4.15) *EXAMPLE:* $X = \mathscr{C}^\infty(\mathbf{R})$. The infinitesimal relation now weighs behavior at infinite points:

$$f \overset{\mathscr{S}}{\approx} g \quad \text{iff} \quad (1 + |t|^n) f^{(k)}(t) \approx (1 + |t|^n) g^{(k)}(t)$$

for $t \in *\mathbf{R}$ and n, $k \in {}^\sigma\mathbf{N}$. An example of an $\mathscr{S}$-infinitesimal is $\varepsilon \exp(-t^2)$ where $\varepsilon \approx 0$.

The finite points $\mathscr{S} = \text{fin}(X)$ are those functions for which $(1 + |t|^n) f^{(k)}(t)$ is finite for all $t \in *\mathbf{R}$ and for all finite n and $k \in {}^\sigma\mathbf{N}$.

A standard function is finite in $\mathscr{S}$ provided $\lim_{t \to \infty} |t|^n f^{(k)}(t) = 0$, since for infinite t, $[(|t|^n f^{(k)}(t))]$ is finite making the second term infinitesimal.

(8.4.16) *EXAMPLE:* Let Ω be an arbitrary region in the complex plane. Let $X = H(\Omega)$ be the family of holomorphic functions defined on Ω. Then

$$h \overset{\kappa}{\approx} k \quad \text{iff} \quad h(z) \approx k(z) \quad \text{for all } z \in \text{ns}(\Omega).$$

The finite elements are those functions that are finite on $\text{ns}(\Omega)$.

(8.4.17) *EXAMPLE:* In this example we use $H^\infty(U)$ to define an infinitesimal relation on $*U$ as follows:

$$z \overset{H}{\approx} w \quad \text{iff} \quad h(z) \approx h(w) \quad \text{for each } h \in {}^\sigma H^\infty(U).$$

All points of $*U$ are finite.

Behrens [1972] has demonstrated some interesting relations between these and the finer hyperbolic infinitesimals on $*U$ (see Chapter 9).

(8.4.18) ***EXAMPLE:*** Let $X = \mathbf{R}$ and let CP(**R**) be the set of continuous 2π-periodic functions on **R**. Then

$$x \overset{p}{\approx} y \quad \text{iff} \quad f(x) \approx f(y) \quad \text{for } f \in {}^{\sigma}\text{CP}(\mathbf{R}).$$

All points are finite and the hull is the circle.

(8.4.19) ***EXAMPLE:*** See the work of Kugler [1969] for more non-standard theory of uniformly almost periodic functions. Let $X = \mathbf{R}$ and let UAP(**R**) be the set of uniformly almost periodic functions. (A standard function is in UAP(**R**) if all of its translates, finite and infinite, are finite uniformly near-standard S-continuous functions.) The infinitesimal relation is

$$x \overset{B}{\approx} y \quad \text{iff} \quad f(x) \approx f(y) \quad \text{for } f \in {}^{\sigma}\text{UAP}(\mathbf{R}).$$

All points are finite. This gives rise to the Bohr group.

(8.4.20) ***EXAMPLE:*** Let X be a Tychonoff space. BC(X) is the space of bounded continuous functions defined on X. The Stone–Čech infinitesimals are given by

$$x \overset{\beta}{\approx} y \quad \text{iff} \quad f(x) \approx f(y) \quad \text{for each } f \in {}^{\sigma}\text{BC}(X).$$

(8.4.21) ***EXAMPLE:*** Let $X = B$ the unit ball in $\mathbf{R}^3$. Let $h^{\infty}(B)$ be the space of bounded harmonic functions defined on B. The harmonic infinitesimals are

$$x \overset{h}{\approx} y \quad \text{iff} \quad f(x) \approx f(y) \quad \text{for } f \in {}^{\sigma}h^{\infty}.$$

We will have more to say about Examples (8.4.17–8.4.21) below—they give rise to compactifications.

We begin our study of generalized uniform infinitesimals with a closer look at S-continuity, which stands for **standard**-continuity for the following reason.

(8.4.22) THEOREM *Let f be an internal function defined on a set containing the infinitesimals around $a \in {}^*X$, $o_{\Lambda}[a]$. Then the following are equivalent:*

(IC) $\qquad x \approx a \quad$ implies $\quad f(x) \approx f(a)$.

(SC) *For each $\varepsilon \in {}^{\sigma}\mathbf{R}^+$, there exist $\lambda_1, \ldots, \lambda_n \in {}^{\sigma}\Lambda$ and $\delta \in {}^{\sigma}\mathbf{R}^+$ such that $\lambda_j(x, a) < \delta$ for $j = 1, \ldots, n$ implies $|f(x) - f(a)| < \varepsilon$. Consequently, the hull of a function which is S-continuous at all finite points is a continuous function on the hull.*

REMARK: Notice that $f(x) = x^2$ is S-continuous on the finite points of ${}^*\mathbf{R}$, but not on any internal set containing all the finite points.

PROOF: (IC)$\Rightarrow$(SC): Fix a standard error $\varepsilon \in {}^\sigma\mathbf{R}^+$. Embed ${}^\sigma\Lambda$ in an arbitrary $*$-finite set of internal pseudometrics G and ${}^\sigma\mathbf{R}^+$ in any $*$-finite set of positive tolerances T. In this case $P(\varepsilon)$ holds where

$$P(\varepsilon) = \text{“}\sup[\lambda(x, a) : \lambda \in G] < \inf[\delta : \delta \in T] \quad \text{implies} \quad |f(x) - f(a)| < \varepsilon.\text{”}$$

The $*$-finite families lemma (7.4.4(a)) then says there is a standard finite G and T for which $P(\varepsilon)$ holds, thus (IC)$\Rightarrow$(SC). The converse is nearly the same as the “real-line” version in Section 5.1.

Notice that $P(\varepsilon)$ is internal by the internal definition principle.

(8.4.23) THEOREM *Let f be an internal function. Then the following are equivalent:*

(a) *f is S-continuous on the internal set Y.*

(b) *For every $\varepsilon \in {}^\sigma\mathbf{R}^+$, there exist $\lambda_1, \ldots, \lambda_n \in {}^\sigma\Lambda$ and $\delta \in {}^\sigma\mathbf{R}^+$ such that for every $x, y \in Y$ if $\lambda_j(x, y) < \delta$ for $j = 1, \ldots, n$, then $|f(x) - f(y)| < \varepsilon$. In other words, an internal function that is infinitesimally continuous on an internal set is uniformly continuous and the hull $\hat{f}$ is uniformly continuous on $\hat{Y}$.*

PROOF: In the last proof change $P(\varepsilon)$ to

“for every $x, y \in Y$, $\sup[\lambda(x, y) : \lambda \in G] < \inf[\delta : \delta \in T]$ implies $|f(x) - f(y)| < \varepsilon$.”

Also see Sections 5.2 and 6.6.

(8.4.24) THEOREM *The uniform infinitesimal hull $\hat{X}$ is a complete uniform space whose uniformity is generated by the pseudometrics of ${}^\sigma\Lambda$, that is, internal pseudometrics S-continuous on internal sets containing the finite points are uniformly continuous with respect to ${}^\sigma\Lambda$.*

PROOF: The fact that $\{\hat{\lambda} : \lambda \in {}^\sigma\Lambda\}$ generates the uniformity is similar to the proof of the last result, replace P(ε) with

“for every $x, y, z, w \in Y$, $\sup[\lambda(x, y), \lambda(z, w) : \lambda \in G] < \inf[\delta : \delta \in T]$ implies $|\rho(x, z) - \rho(y, w)| < \varepsilon$.”

Since the cardinality of Λ is small, we need only consider nets in $\hat{Z}$ over standard index sets $\hat{D}$ and $(z_d : d \in \hat{D})$ can be extended to an internal net $(z_d : d \in D)$.

Now for each $\varepsilon \in {}^\sigma\mathbf{R}^+$ and λ there is a standard $d(\varepsilon)$ and an infinite $\delta(\varepsilon)$ so that $\lambda(z_\alpha, z_\beta) < \varepsilon$ whenever $d(\varepsilon) < \alpha, \beta < \delta(\varepsilon)$. Since the intervals $(d(\varepsilon), \delta(\varepsilon))$ have the finite intersection property and small cardinality, there is a ξ in them all and $\hat{z}_\xi$ is the limit of the net.

EXERCISE [A. Nyberg]: State and prove an analog of Robinson's lemma for nets.

As noted above, finiteness is not a uniform invariant and by considering pseudometrics of the form

$$\lambda(x, y)/(1 + \lambda(x, y))$$

we could assume all points are finite. In any event, the finite points will be measured from a standard $x \in X$ for the rest of the section.

(8.4.25) ***OBSERVATION:*** The infinitesimal relation "$x \overset{\Lambda}{\approx} y$ if and only if $(x, y) \approx 0$ for all $\lambda \in {}^\sigma\Lambda$" is a *monadic equivalence relation.* Precisely, $\mu(\Lambda) = \bigcap [{}^*\{(x, y) : \lambda(x, y) < \varepsilon\} : \lambda \in {}^\sigma\Lambda,\ \varepsilon \in {}^\sigma\mathbf{R}^+]$ is the graph of $\overset{\Lambda}{\approx}$, $x \overset{\Lambda}{\approx} y$ if and only if $(x, y) \in \mu(\Lambda)$. The filter associated with this monad $u = \{U : {}^*U \supseteq \mu(\Lambda)\}$ is called the **uniformity** and the elements $U \in u$ are called **entourages** of the uniformity. The sets $\{(x, y) : \lambda(x, y) < \varepsilon\}$ form a basis for the entourages. Any monadic equivalence relation gives rise to a uniformity, though we shall omit the proof of this.

When Λ is clear we will write $x \approx y$ for $x \overset{\Lambda}{\approx} y$.

The morphisms of the uniform category are the uniformly continuous maps. The main ideas are to generalize uniform continuity, Cauchy sequence, completeness, and total boundedness from metric spaces. Most of the "topological spaces" of analysis are in fact uniform spaces. For example, in a group the notion of closeness is made uniform by translation to the identity. In standard chauvinist terms, the idea of infinitesimals is a uniform notion rather than a topological one (see the exercises at the end of the last section).

Uniform spaces can be made into topological spaces by assigning neighborhood monads $\mu_{\tau(\Lambda)}\{x\} = o_\Lambda[x]$ for $x \in {}^\sigma Y$ (a functor). We caution the reader that $\mu_\tau\{x\}$ may not even be a subset of $o[x]$ for nonstandard points. (Neighborhoods of standard points are enough to define τ.)

REMARK: The reader should notice that in general the set of points an infinitesimal from some set $A \subseteq {}^*X$, $o[A]$, is not the intersection of the standard ε-neighborhoods of the set. If the set A is internal, these sets are related:

$$\begin{aligned} \mathrm{st}(A) &= {}^\circ\{x : x \approx a \text{ for some } a \in A\} = {}^\circ o[A] \\ &= {}^\circ(\bigcap [\{y : \lambda(y, a) < \varepsilon \text{ for some } a \in A\} : \lambda \in {}^\sigma\Lambda,\ \varepsilon \in {}^\sigma\mathbf{R}^+]). \end{aligned}$$

The external set $\{(1/n) : n \in {}^\sigma\mathbf{N}\}$ does not have this property in the space of real numbers. Since this standard part operation is the topological one associated with the uniformity, the reader can now apply the results of the preceding section on standard parts. For example, the closure of $B \subseteq X$ is $\mathrm{st}({}^*B)$.

The generalizations of a Cauchy sequence in a metric space are Cauchy filters and Cauchy nets in a uniform space. The idea is that a Cauchy sequence is one that is "eventually close together," so a Cauchy filter or net satisfies the appropriate statement, where "closeness" is measured by the uniformity. "Eventually" for filters is in the sense of subset and for nets is in the sense of sufficiently large indices. Recall, a net is a mapping $x_\alpha: A \to X$, where A is a directed set, i.e., there is a transitive relation $\prec$ so that if a, $b \in A$, there exists $c \in A$ with $a \prec c$ and $b \prec c$. *The infinite elements* of $*A$ are those $\omega \in *A$ that satisfy $a \prec \omega$ for each $a \in {}^\sigma A$. They exist since we can embed ${}^\sigma A$ in a $*$-finite set.

(8.4.26) THEOREM

(F) *Let $\mathscr{F}$ be a standard filter of subsets of X. The following are equivalent:*

(a) *$\mathscr{F}$ is a Λ-Cauchy filter.*
(b) *$\mu(\mathscr{F}) \subseteq o_\Lambda[a]$ for one $a \in \mu(\mathscr{F})$ and hence all $a \in \mu(\mathscr{F})$, or, all the elements of the monad of $\mathscr{F}$ are within an infinitesimal.*

(N) *Let $(x_\alpha : \alpha \in A)$ be a standard net in X. The following are equivalent:*

(a) *$(x_\alpha : \alpha \in A)$ is a Λ-Cauchy net.*
(b) *$x_\alpha \overset{\Lambda}{\approx} x_\beta$ for all infinite indices α and β.*

PROOF that (b) implies (a):

(F): $\mu(\mathscr{F}) \times \mu(\mathscr{F}) \subseteq \mu(\Lambda)$, since all the elements of $\mu(\mathscr{F})$ are within an infinitesimal. Now if $\lambda \in \Lambda$ and $\varepsilon \in \mathbf{R}^+$ are given

$$*\{(x, y) : \lambda(x, y) < \varepsilon\} \supseteq \mu(\Lambda) \supseteq \mu(\mathscr{F}) \times \mu(\mathscr{F}),$$

so by Cauchy's principle there is an $F \in \mathscr{F}$ with $*F \times *F$ between.

(N): Let $\lambda \in \Lambda$ and $\varepsilon \in \mathbf{R}^+$ be given. The set

$$\{a \in *A : \alpha, \beta > a \text{ implies } \lambda(x_\alpha, x_\beta) < \varepsilon\}$$

contains all the infinite elements, so it is nonempty. By the standard definition principle x_α is Cauchy.

That (b) implies (a) we leave to the reader.

If $\mathscr{F}$ is a Cauchy filter, we can fatten it up as follows without destroying the fact that it is a Cauchy filter. For the fat Cauchy filter, $\Lambda(\mathscr{F})$ is generated by the sets

$$\{y \in X : \lambda(x, y) < \varepsilon \text{ for some } x \in F\} = \lambda_\varepsilon(F),$$

where $\varepsilon \in \mathbf{R}^+$, $\lambda \in \Lambda$, $F \in \mathscr{F}$. Thus we see that if $o[x]$ contains the monad of some filter, it is itself a *fat Cauchy monad.*

(8.4.27) DEFINITION *The pre-near-standard points of* $*X$, $\mathrm{pns}_\Lambda(*X)$, *are exactly those* $x \in *X$ *whose infinitesimal neighborhood is a monad, and it suffices for the infinitesimals to contain a monad.*

The set $\mathrm{pns}(*X)$ is a τ-chromatic, $\overset{\Lambda}{\approx}$-saturated set, where τ is the topology induced by the uniformity. In fact, $x \in \mathrm{pns}(*X)$ if and only if $\mu_\tau\{x\} \subseteq o[x]$.

We remark that for nonstandard pre-near-standard points, the neighborhood monad may be a proper subset of the infinitesimals. This is easy to demonstrate near a nonisolated standard point.

On the other hand if $o[x]$ is a monad, there will be some y so that $\mu_\tau\{y\} = o[x] = o[y]$. (This must be a prime τ-monad.)

A complete uniform space is one in which every Cauchy filter converges. This is the basis of our terminology "pre-."

(8.4.28) THEOREM *The nonstandard hull of the pre-near-standard points is the completion of* X. *Consequently, a space is complete if* $\mathrm{pns}(*X) \subseteq \mathrm{ns}(*X)$.

PROOF: By (8.4.24) we need only show that the closure of ${}^\sigma X$ in $\hat{X}$ (taking all points finite, say) is the hull of $\mathrm{pns}(*X)$. For this we shall use the following characterization of $\mathrm{pns}(*X)$.

(8.4.29) THEOREM *A monad* μ_0 *is Cauchy if and only if for each entourage* $U \in u$ *there is a standard* $y \in {}^\sigma X$ *such that*

$$\mu_0 \subseteq *U[y] = \{x : (x, y) \in *U\}.$$

PROOF: If $\mathrm{Fil}(\mu_0)$ is Cauchy, then for each $U \in u$ there exists $*F \supseteq \mu_0$ so that $F \times F \subseteq U$ and $F \subseteq U[x]$ for $x \in F$. Conversely, if $\mu_0 \subseteq *U[x]$, the family $\{E \in *\mathrm{Fil}(\mu_0) : E \subseteq *U[x]\}$ is nonempty. By the standard definition principle it contains a standard $*F$ and then $F \times F \subseteq U \circ U \in u$.

END OF PROOF OF **(8.4.28)**: The standard map y_U given by $U \in u \to y$ where $*U[y] \supseteq o[x]$ for $x \in \mathrm{pns}(*X)$ is a Cauchy net (ordering u by inclusion).

(8.4.30) THEOREM *A standard function* f: $X \to \mathbf{R}$ *has a continuous extension to the completion of* X *if and only if* $*f$ *is S-continuous on* $\mathrm{pns}(*X)$. *The extension is* $\hat{f}$ *on the hull of* $\mathrm{pns}(*X)$. *In particular, if* $*f$ *is S-continuous at a point of* $\mathrm{pns}(*X)$, *it is finite there.*

PROOF: See the work of Fenstad and Nyberg [1969] and (8.4.23). The fact that $*f$ is necessarily finite follows from the fact that $f(y_U)$ is Cauchy in $\mathbf{R}$, with y_U as in the proof of (8.4.29).

Let $\mathrm{UC}_\Lambda(X)$ be the space of all Λ-uniformly continuous real-valued functions on Y. Suppose we consider the uniformity generated by $x \overset{U}{\approx} y$ if and only if $f(x) \approx f(y)$ for all $f \in {}^\sigma\mathrm{UC}_\Lambda(X)$. Since the functions are uniformly continuous $x \overset{\Lambda}{\approx} y$ implies $x \overset{U}{\approx} y$.

(8.4.31) THEOREM *The set of uniformly finite points*

$$\{x \in {}^*X : f(x) \sim 0 \text{ for all } f \in {}^\sigma\mathrm{UC}(X)\}$$

equals $\mathrm{pns}_U({}^*X)$.

PROOF: If $f(x)$ is finite, then the set $\{y : |f(y) - f(x)| < \varepsilon\}$ contains $\mu_{\mathscr{D}}\{x\}$, since $y \in \mu_{\mathscr{D}}\{x\}$ always implies $f(y) \in \mu_{\mathscr{D}}\{f(x)\}$ and $\mu_{\mathscr{D}}\{f(x)\} \subseteq \mu_\tau\{f(x)\} \subseteq o[f(x)]$ for finite $f(x)$. By Cauchy's principle the set contains a standard set, in particular there is a $y \in {}^\sigma Y$ with $|f(x) - f(y)| < \varepsilon$.

REMARK: We see now that $\mathrm{pns}_\Lambda({}^*X)$ consists of uniformly finite points, but it is not necessarily true that they are the same. For this and the following uniform notion of finite see the work of Henson [1972]. A point y is chain finite if for each $U \in u$ there exists a finite sequence $x_0, \ldots, x_n$ with $x_0 \in {}^\sigma X$, $(x_j, x_{j+1}) \in {}^*U$, and $x_n = y$.

We can turn the situation around and still maintain (8.4.31). Suppose Y is a topological space (Tychonoff as in Section 8.3). We can declare all continuous functions uniformly continuous by inflicting Y with the infinitesimals: "$x \overset{C}{\approx} y$ if and only if $f(x) \approx f(y)$ for all $f \in {}^\sigma\mathrm{C}(Y)$." We could also only declare bounded continuous functions uniformly continuous by taking: "$x \overset{\beta}{\approx} y$ if and only if $f(x) \approx f(y)$ for all $f \in {}^\sigma\mathrm{BC}(Y)$." The topology of these uniformities is the original one again (see the beginning of Section 8.3.)

(8.4.32) THEOREM *The* C-*pre-near-standard points of* ${}^*Y$ *are the points where all the functions of* ${}^\sigma\mathrm{C}(Y)$ *are finite. All points of* ${}^*Y$ *are* β-*pre-near-standard.*

PROOF: Let $f \in {}^\sigma\mathrm{C}(y)$ (respectively, ${}^\sigma\mathrm{BC}(Y)$). A point $x \in \mathrm{pns}({}^*Y)$ if for each f and $\varepsilon \in {}^\sigma\mathbf{R}^+$, there is a $y \in {}^\sigma Y$ with $|f(x) - f(y)| < \varepsilon$. Apply the appropriate part of the proof of (8.4.31) for this.

(8.4.33) EXERCISE: *Let* $\mathscr{F}$ *be a family of bounded functions on* Y *and define a uniformity by the semimetrics* $|f(x) - f(y)|$. *Show that* $\mathrm{pns}({}^*Y) = {}^*Y$.

The next topic in uniform spaces is total boundedness. A set A is totally bounded if for arbitrarily short leashes we can guard the set by tying a finite number of dogs at appropriate points, or more precisely, for each $U \in u$, there exist $x_1, \ldots, x_n \in A$ so that

$$A \subseteq \bigcup\,[U[x_i] : i = 1, \ldots, n].$$

Totally bounded sets are also called *precompact* since they are exactly the sets whose completion is compact.

(8.4.34) THEOREM *Let B be an internal subset of $*X$. If $B \subseteq \text{pns}(*X)$, then B is S-totally bounded and $\hat{B}$ is compact. In particular, a standard set A is totally bounded if and only if $*A \subseteq \text{pns}(*X)$. The whole space X is totally bounded if and only if* pns$(*X)$ *is an internal set and thus equal to $*X$, being chromatic, also if and only if $\mu(\Lambda)$ is macled.*

PROOF: $B \subseteq \bigcup [*U[y] : y \in {}^{\sigma}X]$, being a set of pre-near-standard points. Apply Cauchy's principle to obtain a finite subcover. Precompactness follows from this and the next result, since $\hat{B}$ is now totally bounded and complete.

If $\mu(\Lambda)$ is macled, then given any entourage $U \in u$, there exist A_i, B_i, $i = 1, \ldots, n$ so that $A_i \times B_i \subseteq$ U. Let $C_i = A_i \cap B_i$, $i = 1, \ldots, n$. C_i is a finite cover of X and $C_i \times C_i \subseteq U$, so X is totally bounded. We leave the proof that if X is totally bounded, then $\mu(\Lambda)$ is macled to the reader (see Haddad [1972]).

EXERCISE: Show that A is totally bounded if and only if every ultrafilter on A is Cauchy.

(8.4.35) THEOREM *X is compact if and only if X is complete and totally bounded. Totally bounded spaces are precompact.*

PROOF: $\text{ns}(*X) = \text{pns}(*X) = *X$. The first equality is completeness, the second is total boundedness, and everything near-standard implies compactness.

(8.4.36) DEFINITION The set of *precompact points* of $*X$ is the discrete union monad of pns$(*X)$, $\text{pcpt}(*X) = \nu_{\mathscr{D}}\, \text{pns}(*X)) = \bigcup [*C : *C \subseteq \text{pns}(*X)] = \bigcup [*C : C \text{ is precompact in } X]$.

(8.4.37) THEOREM *X is locally precompact if and only if $\text{pcpt}(*X) \supseteq \text{ns}(*X)$. The completion of X is locally compact if and only if $\text{pcpt}(*X) = \text{pns}(*X)$.*

PROOF: Left for the reader as an exercise.

Let X be a standard set and Σ a collection of subsets of X. Let $\mathscr{F}(X)$ denote the real- (or complex-) valued functions defined on X. We introduce an infinitesimal relation on $*(\mathscr{F}(X))$ as follows: $f \overset{\Sigma}{\approx} g$ if and only if $f(x) \approx g(x)$ for $x \in \nu(\Sigma)$.

(8.4.38) DEFINITION *We call the uniformity of $\overset{\Sigma}{\approx}$ the Σ-convergence uniformity.*

Examples of this are

(1) *pointwise convergence*, Σ is the set of finite subsets of X and $\nu(\Sigma) = {}^{\sigma}X$;

(2) *uniform convergence*, $\Sigma = \{X\}$ and $\nu(\Sigma) = {}^*X$;

(3) *compact convergence*, Σ is the set of compact subsets of the topological space X and $\nu(\Sigma) = \mathrm{cpt}({}^*X)$;

(4) *precompact convergence*, $\nu(\Sigma) = \mathrm{pcpt}({}^*X)$ where X is a uniform space.

We will use the notation $f \overset{u}{\approx} g$ for uniform infinitesimals and $f \overset{\kappa}{\approx} g$ for compact infinitesimals.

(8.4.39) THEOREM *$\mathscr{F}(X)$ is complete in the Σ-convergence uniformity for any Σ.*

PROOF: Let $f \in \mathrm{pns}({}^*\mathscr{F})$ and define $g(x) = \mathrm{st}(f(x))$, $x \in {}^{\circ}(\nu(\Sigma))$ ($g(x) = 0$ off $\nu(\Sigma)$). Given $S \in \Sigma$ and standard ε, there is a standard h so that

$$|f(x) - h(x)| < \varepsilon/2 \quad \text{on } {}^*S.$$

Also, $f(x) \approx g(x)$ on ${}^{\circ}(\nu(\Sigma))$ so that $|h(x) - g(x)| < \varepsilon$ on S and the Cauchy net of h's converges to g.

(8.4.40) THEOREM *Let X be a uniform space. The uniformly continuous functions* $\mathrm{UC}(X)$ *are complete with respect to uniform convergence.*

PROOF: If $f \in {}^*\mathrm{UC}$ is pre-near-standard, it is near standard since the argument used in (8.4.38) shows that for any ε there is a function $h \in \mathrm{UC}(X)$ such that $|h(x) - g(x)| < \varepsilon/2$ on ${}^*Y$. Now if $g \notin \mathrm{UC}(X)$, then there are $x \overset{\Lambda}{\approx} y$ such that $g(x) \not\approx g(y)$. Since when $h \in \mathrm{UC}(X)$, $h(x) \approx h(y)$, we cannot have $|h(x) - g(x)| < \varepsilon/2$ and $|h(y) - g(y)| < \varepsilon/2$ when $\varepsilon \leq |g(x) - g(y)|$, a contradiction. Thus, if f is pre-near-standard, it is uniformly near its standard part $f(x) \approx g(x)$ for $x \in {}^*X$ where $g(x) = \mathrm{st}(f(x))$ for $x \in {}^{\sigma}X$. This makes f S-continuous everywhere and $g = \hat{f}\,|X$.

(8.4.41) THEOREM *Let X be a locally compact Tychonoff space,* $\mathrm{ns}_{\tau}({}^*X) = \mathrm{cpt}_{\tau}({}^*X)$. *The continuous functions* $\mathrm{C}(X)$ *are complete in the compact convergence uniformity. The compact near-standard functions are exactly those that are finite and S-continuous on* $\mathrm{ns}_{\tau}({}^*X)$.

PROOF: Notice that S-continuity on the near-standard points makes sense topologically: $\mathrm{st}_{\tau}(x) = \mathrm{st}_{\tau}(y)$ implies $f(x) \approx f(y)$. (In other words, $x \approx y$ if and only if $\mathrm{st}_{\tau}(x) = \mathrm{st}_{\tau}(y)$ is the same for compatible uniformities.) The completeness follows the lines of the proof of (8.4.40).

If f is near standard, then $f(x) \approx g(x)$ for $x \in \mathrm{ns}({}^*X)$, with standard g. Also, if y is near a standard point x, then $f(x) \approx g(x) \approx g(y) \approx f(y)$, so f is

S-continuous and finite. Conversely, if f is finite and S-continuous, then at a standard $y \in {}^{\sigma}X$, $*\{x : |\mathrm{st}(f(x)) - \mathrm{st}(f(y))| < \varepsilon\} \supseteq \{x : |f(x) - f(y)| < \varepsilon/2\} \supseteq \mu_\tau\{y\}$ and $g(x) = \mathrm{st}(f(x))$ is standard and continuous. Again, $f(x) \approx g(x)$ for $x \in \mathrm{ns}(*X)$.

A family of functions $\mathscr{G} \subseteq \mathrm{C}(X)$ is said to be *equicontinuous* provided that for each $y \in X$ and for each $\varepsilon \in \mathbf{R}^+$, there exists a neighborhood N of y so that if $x \in N$, then $|f(x) - f(y)| < \varepsilon$ for each $f \in \mathscr{G}$.

Uniform equicontinuity for $\mathscr{G} \subseteq \mathrm{UC}(X)$ means that for each $\varepsilon \in \mathbf{R}^+$, there is an entourage V so that if $(x, y) \in V$, then $|f(x) - f(y)| < \varepsilon$ for all $f \in \mathscr{G}$.

(8.4.42) LEMMA *A standard family $\mathscr{G} \subseteq \mathrm{C}(X)$ is equicontinuous if and only if each $f \in *\mathscr{G}$ is S-continuous on* $\mathrm{ns}(*X)$. *A standard family $\mathscr{H} \subseteq \mathrm{UC}(X)$ is uniformly equicontinuous if and only if each $f \in *\mathscr{H}$ is S-continuous on $*X$.*

PROOF: We show the nonstandard half of the first assertion, the rest is left to the reader. Let $y \in {}^{\sigma}X$ and $\varepsilon \in {}^{\sigma}\mathbf{R}$ be given. Take $N \in *\mathrm{Fil}_\tau(y)$ with $N \subseteq \mu_\tau\{y\}$. Then $(\exists N \in \mathrm{Fil}_\tau(y))[(\forall f \in \mathscr{G})(\forall x \in N)[|f(x) - f(y)| < \varepsilon]]$ holds with $*$'s and by Leibniz' principle therefore holds without $*$'s, and this is equicontinuity.

(8.4.43) THEOREM (Ascoli) *Let X be a locally compact Tychonoff space and $\mathscr{G} \subseteq \mathrm{C}(X)$. The following are equivalent:*

(1) *$\mathscr{G}$ is relatively compact in the compact convergence topology, that is, its closure is compact ($\mathscr{G}$ is κ-precompact).*

(2) $*\mathscr{G} \subseteq \mathrm{ns}_\kappa(*\mathrm{C}(*X))$, *that is, all the functions $f \in *\mathscr{G}$ are finite and S-continuous on* $\mathrm{ns}(*X)$.

(3) *$\mathscr{G}$ is equicontinuous and, for $y \in Y$, $\{f(y) : f \in \mathscr{G}\}$ is bounded (precompact in* $\mathbf{R}$).

PROOF: (1)$\Leftrightarrow$(2) is proved by using (8.3.11).

(2)$\Leftrightarrow$(3) is proved by using the lemma and the previous result. Note that $\{f(y) : f \in *\mathscr{G}\}$ is an internal set of finite points, hence bounded.

EXERCISE: State and prove analogs of (8.4.41–8.4.43) for functions taking values in a complete uniform space and for precompact convergence with functions uniformly continuous on precompact sets.

(8.4.44) ANOTHER NONCOMPACT INFINITESIMAL This can be constructed by taking the function

$$f(x) = (1/\omega)\sin(2\pi\Omega x) \qquad \text{for} \quad \omega \ll \Omega$$

as in (7.6.3) and x on the unit interval. (Compare 8.3.13 and the following example.)

Any standard set K such that $f \in {}^*K$ contains infinitely many functions $(1/m)\sin(2\pi nx)$ for some fixed standard m and cannot be equicontinuous. Let $P(m, n)$ be the property "$(1/m)\sin(2\pi nx) \in K$," and generalize the remarks after (8.3.13) to show

LEMMA *Suppose that $P(m, n)$ is a standard bounded formal property of two natural numbers and that ${}^*P(\omega, \Omega)$ holds for $\omega \ll \Omega$. Then there exists $m \in {}^\sigma\mathbf{N}$ and infinitely many n's $\in {}^\sigma\mathbf{N}$ such that $P(m, n)$ holds.*

No infinite family of the form $\{(1/m)\sin(2\pi nx) : n \in M\}$ can be equicontinuous since its extension contains a function $(1/m)\sin(2\pi\lambda x)$ for an infinite λ.

Continuous functions are by no means complete with respect to pointwise convergence. Consider the following example. Let $\omega = \prod_{k=1}^{\lambda}(2k+1)$ for an infinite λ. Let $f(x) = \exp(i2\pi\omega x)$ and define $g(x) = \mathrm{st}(f(x))$ for $x \in {}^\sigma\mathbf{R}$.

(8.4.45) *EXAMPLE:* The function $g\colon \mathbf{R} \to \mathbf{C}$ is not Lebesgue measurable: First, g is discontinuous since if $a_k = 1/[2(2k+1)]$, then

$$f(a_k) = \exp\left(i\left(\frac{3\cdot 5\cdot 7\cdots\lambda}{2(2k-1)}\right)2\pi\right)$$

$$= \exp\left(i\left(v + \frac{1}{2}\right)2\pi\right)$$

$$= \exp(i\pi) = -1,$$

and

$$g(a_k) = -1 \qquad \text{for} \quad \text{standard } k.$$

Also $a_k \to 0$ and $g(0) = 1$. Finally, g is a group character so it is non-measurable. Suppose it is measurable; then

$$2\pi g(x) = \int_{-\pi}^{\pi} g(y)g(x-y)\,dy$$

and this must be uniformly continuous since it is a convolution of bounded functions (see W. Rudin [1974]).

(8.4.46) EXERCISE: Characterize the near-standard points of $\mathscr{E}$ and $\mathscr{S}$ from Examples (8.4.14) and (8.4.15).

(8.4.47) *REMARK:* In our present terminology, (6.6.12) says that the harmonic functions on the unit ball are complete in the compact convergence uniformity. Finite S-continuous $*$-harmonic functions are $C(B)$-compact near standard and (6.6.12) says the standard part (in $C(B)$) is harmonic.

Notice that the same ideas would apply to any subspace of continuous functions with a local integral representation, in particular to holomorphic and meromorphic functions.

In his retiring presidential address to the Association for Symbolic Logic, Abraham Robinson gave twelve problems in metamathematics [1973a]. Problem 8 was to prove what is sometimes referred to as "Bloch's principle," namely, *Those properties which reduce an entire function to a constant, force a family of holomorphic functions to be normal.* Naturally some restrictions must be placed on the properties one considers and Robinson gave four conditions in his formulation of the problem. Two of Robinson's conditions are sufficient to prove the converse that normal properties force entire functions to be constant. Actually the converse is implicit in many classical normal family proofs; we simply point out that it is indeed a metatheorem. Zalcman [1975] recently solved this problem.

Let $P(D, f)$ be a property of a domain D and a holomorphic function f. We shall assume that

(1) *if* $P(D, f)$ *holds and* $D \supset G$, *then* $P(G, f_{|G})$ *also holds and that*
(2) P *is linearly invariant,*

that is, if $P(D, f)$ holds and g: $G \to D$ is a map $g(z) = az + b$, then $P(G, f \circ g)$ also holds. Let $D \subset G$ mean "D is a subdomain of G" (use $A \subseteq B$ for subset) and consider the formal expressions:

$$E(g) : (\forall D \subset C)[P(D, g_{|D})],$$

where g is entire, and

$$N(\mathscr{F}) : (\forall f \in \mathscr{F})(\forall D \subset U)[P(D, f_{|D})],$$

when $\mathscr{F}$ is of a family of meromorphic functions on U.

(8.4.48) A CONVERSE OF BLOCH'S PRINCIPLE *Let P be a property satisfying* (1) *and* (2) *above. If $N(\mathscr{F})$ implies $\mathscr{F}$ is normal, then $E(g)$ implies g is constant.*

PROOF: Let g be an entire function that satisfies $E(g)$. Define a family on the unit disk by $\mathscr{F} = \{f : f(z) = g(rz), r \in \mathbf{R}^+, |z| < 1\}$. Then $N(\mathscr{F})$ holds by (1) and (2), so $\mathscr{F}$ is a normal family. This simply means $\mathscr{F}$ is compact and every function in the nonstandard extension of $\mathscr{F}$, $f \in {}^*\mathscr{F}$, must be S-continuous. Therefore, even for infinite r, $f'(z) = rg'(rz)$ must be finite and thus $g'(\zeta) = (1/r)f'(\zeta/r)$ is infinitesimal for finite ζ, and so g is constant. (Similarly in the case of poles.) Continuity of the property is needed to prove Zalcman's form of Bloch's principle:

(3) *if* $P(D_n, f_n)$ *holds where* $D_1 \subset D_2 \subset \cdots$, $D = \bigcup D_n$, *and* $f_n \xrightarrow{K} f$ *on the sphere, then* $P(D, f)$ *also holds.*

(8.4.49) BLOCH'S PRINCIPLE *Let P be a property satisfying* (1), (2), *and* (3). *If $E(g)$ implies g is constant, then $N(\mathscr{F})$ implies $\mathscr{F}$ is normal.*

PROOF: If $\mathscr{F}$ is abnormal, there is an $f \in {}^*\mathscr{F}$ that is S-discontinuous at a standard point $z \in {}^\sigma D$. By (3) of Theorem 3.1 in Stroyan [1972a] this means that the spherical derivative is infinite at a point $\zeta \approx z$. Let $\Delta = \{w : |w - z| \le r\}$ be a standard disk inside D and suppose ξ is a point of ${}^*\Delta$ satisfying

$$\text{(M)} \qquad [r^2 - |\xi - z|^2] \frac{|f'(\xi)|}{[1 + |f(\xi)|^2]} \ge [r^2 - |w - z|^2] \frac{|f'(w)|}{[1 + |f(w)|^2]}$$

for all $w \in {}^*\Delta$. Since $\zeta \approx z$ and the spherical derivative is infinite at ζ, we know that the maximum is infinite. Let $\rho = |f'(\xi)|/[1 + |f(\xi)|^2]$ and define an internal function $h(x) = f(\xi + x/\rho)$, for x so that $\xi + (x/\rho)$ is in Δ. Even if $|\xi - z| \approx r$, we know $(r^2 - |\xi - z|^2)\rho$ is infinite, so that h is defined for all finite x. The spherical derivative of h at zero is

$$\frac{h'(0)}{[1 + |h(0)|^2]} = \frac{1}{\rho} \frac{|f'(\xi)|}{[1 + |f(\xi)|^2]} = 1$$

while, for finite values of x,

$$\begin{aligned} \frac{h'(x)}{[1 + |h(x)|^2]} &\le \frac{|f'(\xi + (x/\rho))|[r^2 - |\xi + (x/\rho) - z|^2]}{\rho[1 + |f(\xi + (x/\rho))|^2][r^2 - |\xi + (x/\rho) - z|^2]} \\ &\le \frac{[r^2 - |\xi - z|^2]}{[r^2 - |\xi + (x/\rho) - z|^2]}, \quad \text{by maximality (M)} \\ &\le \frac{(r + |\xi - z|)}{(r + |\xi - z| + (|x|/\rho))} \cdot \frac{(r - |\xi - z|)}{(r - |\xi - z| - (|x|/\rho))} \\ &\lessapprox 1 \cdot \frac{1}{1 - |x|/(\rho[r - |\xi - z|])} \lessapprox 1. \end{aligned}$$

Thus $g(x) = \text{st}(h(x))$ defines a nonconstant entire function, since h is S-continuous by Theorem 3.1 of Stroyan [1972a] and g has a nonzero spherical derivative at zero. By virtue of (1), (2), and the infinitesimal form of (3) we see that $E(g)$ also holds.

(8.4.50) EXERCISE: Give the infinitesimal form of the continuity condition (3).

(8.4.51) EXERCISES ON COMPACT MAPS

(a) Let $f\colon X \to Y$ be a standard map between metric spaces. The following are equivalent:

(1) f takes finite points of $^*X$ to near-standard points of $^*Y$.
(2) $\mathrm{cl}(f(B))$ is compact in Y whenever B is bounded in X. (Compare Robinson [1966, 4.64 and 4.6.5].)

(b) Let $f_n\colon X \to Y$ be a standard sequence of compact maps (as in (1) and (2) above). If $f_n \to f$ uniformly, show that f is also compact. We will take this up in a Hilbert space setting below.)

We conclude the section with the following remark: *the infinitesimal hull generally depends on the particular model.* Of course, all the hulls (using saturated models) of a particular uniform space contain the completion, and if $\mathrm{fin}(^*Y) = \mathrm{pns}(^*Y)$, the hull is uniquely the completion. An interesting question that this suggests is "What does it mean when a space only has one hull?" We refer the reader to Henson and Moore [1972] and continue to consider only the fixed model of Section 7.5.

8.5 TOPOLOGICAL INFINITESIMALS AT REMOTE POINTS

This section sketches part of the work of Wattenberg [1971, 1973] and Henson [1973] on the question of what "infinitesimally close" might mean in a topological space away from the near-standard points. At the near-standard points it is natural to say points are infinitesimally close if they belong to the neighborhood monad of the same standard point $\mu_\tau\{x\}$ and to ignore the finer μ_τ-structure inside. This agrees with the uniform notion of "infinitesimal" for compatible uniformities at the near-standard points. The β-uniform infinitesimals of (8.4.32) provide one answer for the restricted collection of bounded continuous functions, one that gives rise to the Čech–Stone compactification. Wattenberg has provided a promising general topological answer and tied it to various specialized topological constructions. Henson has related it to the C-uniform infinitesimals.

(8.5.1) DEFINITION *Let $\mathscr{T}$ be a classical category of topological spaces as in Section* 3.11. *An infinitesimal relation for $\mathscr{T}$ is an equivalence relation $\underset{t}{\approx}$ that satisfies*

(a) *if x is near-standard in $^*X$, then $x \underset{t}{\approx} y$ if and only if* $\mathrm{st}(x) = \mathrm{st}(y)$; *and*

(b) *each continuous standard $\mathscr{T}$-map $f\colon X \to Y$ satisfies t-continuity*

$$x \underset{t}{\approx} y \quad \text{in} \quad {}^*X \qquad \text{implies} \qquad f(x) \underset{t}{\approx} f(y) \quad \text{in} \quad {}^*Y.$$

As far as we are concerned all spaces are still weakly Tychonoff, and the infinitesimals may distinguish normality, paracompactness, or other interesting analytical properties short of compactness.

(8.5.2) REMOTELY DISCRETE INFINITESIMALS $\mathscr{T}$ equals the (weak) Hausdorff spaces.

FINE: $x \underset{d}{\approx} y$ if and only if $\mathrm{st}_\tau(x) = \mathrm{st}_\tau(y)$ if x is near standard, and $x = y$ if x is remote.

COARSE: $x \underset{D}{\approx} y$ if and only if $\mathrm{st}(x) = \mathrm{st}(y)$ if $x \in \mathrm{ns}_\tau(^*X)$, and $\mu_{\mathscr{D}}\{x\} = \mu_{\mathscr{D}}\{y\}$ if $x \in \mathrm{rmt}_\tau(^*X)$.

(8.5.3) METRIZABLE INFINITESIMALS $\mathscr{T}$ equals the metrizable spaces.

$x \underset{m}{\approx} y$ if and only if D is a metric for X and $d(x, y) < f(x)$ for each standard continuous f: $X \to (0, 1)$.

Notice that this is finer than $x \overset{d}{\approx} y$. Consider $\mathbf{R}$ for example and take $f(x) = e^{x^2}$.

For the purpose of describing Wattenberg's topological infinitesimals, let Tyc be the classical category of Tychonoff (completely regular) spaces in a standard model of analysis, say $\mathscr{N}$. Let $\kappa = 2^{\mathrm{card(Tyc)}}$ and take the superstructure based on $X = \mathbf{N} \cup \kappa$. (We could work with $\mathbf{R}^\lambda$ for $\lambda > \mathrm{card}(T)$ for each specific $T \in$ Tyc, but we just fix one huge κ instead.)

Let A be the space of maps from κ into $(0, \infty)$ with the sup metric

$$A = \{f : \kappa \to (0, \infty)\}, \qquad u(f, g) = \sup[\,|f_\alpha - g_\alpha| : \alpha \in \kappa].$$

In $^*A$ we write $f \overset{u}{\approx} g$ if and only if $f_\alpha \approx g_\alpha$ for all $\alpha \in {}^*\kappa$.

(8.5.4) DEFINITION *The W-infinitesimals on a classical Tychonoff space T are defined by*

$x \underset{W}{\approx} y$ if and only if $f(x) \overset{u}{\approx} f(y)$ for every standard and continuous f: $T \to A$.

EXERCISE: Verify (8.5.1) for $\underset{W}{\approx}$ and show that it agrees with (8.5.3) on metrizable spaces.

Henson observed that Wattenberg's Tychonoff infinitesimals are monadic and therefore the uniform infinitesimals of some uniformity. He showed

(8.5.5) THEOREM *If $X \in$ Tyc, then*

$x \underset{W}{\approx} y$ if and only if $d(x, y) \approx 0$ for every standard continuous pseudometric d on X.

Moreover, $\{x : f(x)$ *is finite for every* $f \in {}^\sigma C(X)\}$ *is the pre-near-standard set of the finest uniformity compatible with the topology and for such points* $x \underset{W}{\approx} y$ *if and only if* $x \overset{C}{\approx} y$.

PROOF: Since $\underset{W}{\approx}$ is a monadic equivalence relation, it is the monad of a uniformity by Luxemburg [1969, (3.9.1)]. It is compatible by (8.5.1(a)). (Pseudometrics generate every uniformity.)

The finest compatible uniformity on X is also the coarsest one that makes every continuous map into a metric space uniformly continuous, and since all the constant functions are included in the definition of $\underset{W}{\approx}$, we know $\underset{W}{\approx}$ is coarser than the finest compatible one and still compatible.

The second part of the theorem requires first that we show that if $x \overset{C}{\approx} y$ and x is pre-near-standard for the uniformity of $\underset{W}{\approx}$, that then $x \underset{W}{\approx} y$. Notice that $x \underset{W}{\approx} y$ implies $x \overset{C}{\approx} y$ by (8.5.1(b)). We assume x is pre-near-standard and $y \underset{W}{\not\approx} x$, then there is a continuous pseudometric d so that $d(x, y) > \delta \in {}^\sigma\mathbf{R}^+$. Since x is pre-near-standard, there is a $z \in {}^\sigma X$ such that $d(z, x) < \delta/3$. The standard function $f(x) = d(z, x)$ then shows $x \overset{C}{\not\approx} y$.

It remains to show that the C-pre-near-standard points and W-pre-near-standard points are the same. This relies on the fact that the cardinality of X is not measurable (has no countably additive measure taking only the values 0 and 1). We refer the reader to the work of Henson [1972a, 1973]; subsets of $\mathscr{X}$ do not have measurable cardinality. Henson also showed

(8.5.6) COROLLARY *If* X *is homeomorphic to a subspace of* $\mathbf{R}^n$, *then* $x \underset{W}{\approx} y$ *if and only if* $x \overset{C}{\approx} y$.

PROOF: Since $x \underset{W}{\approx} y$ if and only if $x \underset{m}{\approx} y$, it suffices to show that $\max[|x_i - y_i| : 1 \le i \le n] < f(x)$ when $x \overset{C}{\approx} y$.

Let f: $X \to (0, 1)$ be continuous. Also, $g = 1/f$ is continuous and since $x \overset{C}{\approx} y$, $f(x) \approx f(y)$ and $g(x) \approx g(y)$. Either $f(x)$ or $g(x)$ is not infinitesimal and then $f(y)/f(x) \approx 1$ or $f(x)/f(y) \approx 1$ by the above.

The functions $x_i/f(x)$ are continuous, so $x_i/f(x) \approx y_i/f(y)$ and therefore $|x_i - y_i| < f(x)$.

(8.5.7) DEFINITION *Let* X *be a standard topological space. Two points* x *and* y *are said to have the same order of diffusion on* $^*X$ *if either* $x = y$ *or there is a standard locally finite family* $\mathscr{K}$ *of compact subsets of* X *such that* $x, y \in K \in {}^*\mathscr{K}$, *and then we write*

$$\delta(x) = \delta(y).$$

All standard points have the same order of diffusion since $\{x, y\}$ is a compact set and a singleton family. Compact points (8.3.12) have a standard order of diffusion $x, y \in {}^*K$, but noncompact points do not—they lie in $*$-compact nonstandard sets. In particular, non-locally-compact spaces have near-standard points with nonstandard order (see (8.3.13) and (8.4.44)).

(8.5.8) PROPOSITION

(a) *$\delta(x) = \delta(y)$ is an equivalence relation.*

(b) *If $f: X \to Y$ is a standard homeomorphism onto its image, then $\delta(x) = \delta(y)$ implies $\delta(f(x)) = \delta(f(y))$. If $f(X)$ is closed besides, then $\delta(f(x)) = \delta(f(y))$ implies $\delta(x) = \delta(y)$.*

PROOF: Left as an exercise; see the work of Wattenberg [1973] for help.

(8.5.9) DEFINITION *Let X be a standard space. Two points x and y are said to have the same order of magnitude in ${}^*X$ provided that for every standard continuous $f: X \to (0, \infty)$ there exists a standard continuous $g: X \to (0, \infty)$ such that $g(y) \leq f(x)$* and *$g(x) \leq f(y)$, and then we write*

$$M(x) = M(y).$$

Compare this with (7.1.5).

EXERCISE: Let Ω be infinite and ε be infinitesimal in ${}^*\mathbf{R}$. Does $M(\Omega) = M(\Omega + \varepsilon)$? Does $M(\Omega) = M(\Omega^{\Omega})$?

A connection between diffusion and magnitude of Wattenberg's is:

(8.5.10) THEOREM *Suppose that X is a metric space. Then $\delta(x) = \delta(y)$ implies $M(x) = M(y)$.*

PROOF: Let $\mathscr{K} = \{K_i : i \in I\}$ be a locally finite compact family with $x, y \in K_j$, $j \in {}^*I$. Each $x \in X$ has a neighborhood N_x that meets only finitely many K_j (perhaps none). Each K_i has a finite cover by N_i's, $M_i = N_1 \cup \cdots \cup N_n$. The family $\{M_i : i \in I\} \bigcup \{X \backslash \bigcup [K_i : i \in I]\}$ has a locally finite refinement $\{V_j : j \in J\}$ by paracompactness of X. Each K_i has a finite cover by V_j's. Let $U_i = V_1 \cup \cdots \cup V_n$ so that $\{U_i : i \in I\}$ is a locally finite open cover of $\bigcup [K_i : i \in I]$ with $K_i \subseteq U_i$. Let $\sum h_i + \sum h_j$ be a partition of unity with h_i's subordinate to U_i and $\sum h_j$ subordinate to the part of $\{V_j : j \in J\}$ not used in the cover of $\bigcup K_i$.

Well order I. We define

(1) $g_1: K_0 \to (0, \infty)$ as $\min[f(z) : z \in K_0]$.

(2) Given $g_\beta: \bigcup [K_\alpha : \alpha < \beta] \to (0, \infty)$, define

$$g_{\beta+1}: \bigcup [K_\alpha : \alpha < \beta + 1] \to (0, \infty)$$

as follows: Let $m = \min(\{f(z) : z \in K_\beta\} \cup \{g_\beta(z) : z \in \bigcup [K_\alpha \cap K_\beta : \alpha < \beta]\})$ and for each $z \in \bigcup [K_\alpha : \alpha < \beta + 1]$

$$a(z) = d(z, \bigcup [K_\alpha \backslash U_\beta : \alpha < \beta]), \qquad b(z) = d(z, K_\beta),$$

$$g_{\beta+1}(z) = (b(z)g_\beta(z) + a(z)m)/(a(z) + b(z)).$$

Notice that if $z \notin U_\beta$ that $a(z) = 0$ and $g_{\beta+1}(z) = g_\beta(z)$.

(3) At limit ordinals let

$$g_\lambda(z) = \min[g_\beta(z) : \beta < \lambda].$$

Finally, take $g_I(z) = g_{\sup \beta}(z)$ and let

$$g(z) = \sum g_I(z)h_i(z)$$

for the partition of unity above. We have then that for each i and $z \in K_i$, $g(z) \leq \min f(K_i)$, so that $g(x) \leq f(y)$ and $g(y) \leq f(x)$ when x, $y \in K_j$, $j \in {}^*I$, by Leibniz' principle.

REMARK: Notice that if x, $y \in K_j$, where $\mathscr{K}$ is a *disjoint* standard family, then the result applies in paracompact spaces by a simple partition of unity lemma.

(8.5.11) THEOREM *Suppose that X is a metric space. Then $x \underset{m}{\approx} y$ implies $M(x) = M(y)$.*

PROOF: Left as an exercise.

Let $m(x) = \{y : x \underset{m}{\approx} y\}$ the metrizable infinitesimals (or monad) around x. Then:

(8.5.12) THEOREM *Suppose x, $y = x + t \in {}^*\mathbf{R}$ and $|x| < |y|$. The following are equivalent:*

(a) $M(x) = M(y)$.

(b) $m(x) + t = m(y)$.

(c) *There is a standard continuous f: $[0, \infty) \to [0, \infty)$ such that $|x| < |y| < {}^*f(|x|)$.*

(d) $\delta(x) = \delta(y)$.

PROOF: (a)$\Rightarrow$(b): Suppose $M(x) = M(y)$ and $z \underset{m}{\approx} x$. Let f: $\mathbf{R} \to (0, \infty)$ be continuous. We know there is a g: $\mathbf{R} \to (0, \infty)$ with $f(y) \geq g(x) > |z - x| = |z + t - y|$; hence $(z + t) \underset{m}{\approx} y$.

(b)$\Rightarrow$(c): We know $y + 1/y \underset{m}{\not\approx} y$ (since $1/y^2$ is continuous and standard away from zero). By (b), there is some standard h: $\mathbf{R} \to (0, \infty)$ such that $h(x) \leq 1/y$. Let $g(t) = \frac{1}{2}\min[h(s) : 0 \leq s \leq t]$, so g is standard, continuous, and $g(x) < 1/y$. Let $f(x) = 1/g(x)$. This proves (c).

(c) ⇒ (d): Let f be as in (c) and take $K_n = [n, M_n]$ with $M_n = \max[f(x) : 0 \leq x \leq n]$. We may as well assume y is infinite so that $M_n \to \infty$ monotonically and K_n is locally finite. Now $x, y \in K_\Omega$ with Ω equal to the greatest integer in x.

(d) ⇒ (a): By (8.5.8).

Now we turn to new applications of Wattenberg's infinitesimals.

(8.5.13) THEOREM *Let X be a classical Tychonoff space. Then X is paracompact if and only if $^*X$ is "internally normal" in the sense that whenever K is an internal set of remote points, $K \subseteq \mathrm{rmt}(^*X)$, there exists a W-continuous internal k: $^*X \to {}^*[0, 1]$ such that $k(K) = \{1\}$ and $k(\mathrm{ns}(^*X)) = \{0\}$.*

PROOF: Assume X is paracompact, so by (8.3.15) each internal $K \subseteq \mathrm{rmt}(^*X)$ is separated from $\mathrm{ns}(^*X)$ by a standard locally finite open cover. Let $\sum_{i \in I} h_i$ be a subordinate partition of unity, so that the carriers (supports) $\mathrm{carr}(h_i) \perp K$ for $i \in {}^\sigma I$. The standard function

$$H(x, y) = \sum_I |h_i(x) - h_i(y)|$$

is a continuous standard semimetric on X and if $J \subseteq I$,

$$\left| \sum_J h_j(x) - \sum_J h_j(y) \right| \leq \sum_I \left| h_i(x) - h_i(y) \right|.$$

Take J to be the internal set $\{i \in {}^*I : \mathrm{carr}(h_i) \cap K \neq \varnothing\}$, which is internal by the internal definition principle

$$k(x) = \sum_J h_j(x).$$

Conversely, by (8.3.15) it suffices to show that whenever K is an internal set of remote points, then there is a locally finite standard cover disjoint from K. Let $k(x)$ be the required separating function and let

$$h(x, y) = |k(x) - k(y)|.$$

The internal semimetric h is W-continuous and therefore

$$0 \leq h(x, y) \lessapprox H(x, y), \qquad x, y \in {}^*X,$$

for some bounded continuous standard semimetric by the infinitesimal hull theorem. Also, H is not identically zero since $h(x, y) = 1$ if $x \in {}^\sigma X$ and $y \in K$.

By A. H. Stone's theorem (X, H) is paracompact, and since H separates K from the standard points, there is a locally finite H-open covering separating K and $\mathrm{ns}_H(X^*)$ by (8.3.15). This cover is topologically open since H is continuous and this proves the theorem.

(8.5.14) THEOREM *Let $X \in \text{Tyc}$ (so $\text{card}(X)$ is not measurable). The following are equivalent:*

(a) *X is topologically complete.*
(b) *The W-infinitesimals around remote points are never $\bigcap$-monads.*
(c) *For each remote point $x \in \text{rmt}(^*X)$, there exists a standard continuous real-valued function whose value at x is infinite.*
(d) *X is real compact.*
(e) *For each remote $x \in \text{rmt}(^*X)$ there exists an internal W-continuous $f\colon {}^*X \to {}^*[0, 1]$ such that $f(\text{ns}\ {}^*X) = \{1\}$ and $f(x) = 0$.*
(f) *For each remote $x \in \text{rmt}(^*X)$ there is a standard partition of unity $\sum_I h_i$ such that $h_i(x) = 0$ for $i \in {}^\sigma I$.*

PROOF: Topologically complete means there is some compatible complete uniformity. Then by (8.4.27) and (8.4.28)

$$\text{ns}(^*X) \subseteq \text{pns}_W(^*X) \subseteq \text{pns}_u(^*X) \subseteq \text{ns}(^*X).$$

Real compact means C-complete and by (8.4.32) and (8.5.5), $\text{pns}_W(^*X) = \text{pns}_C(^*X)$, as in (c), so (a–d) are equivalent.

By (8.4.29) when X is complete, each $x \in \text{rmt}(^*X)$ has $\mu_{\mathscr{D}}\{x\} = \mu_x$ **non-**Cauchy so there is an entourage $^*U \in u$ such that $\mu_x {}^*U[y] \neq \varnothing$ for every $y \in {}^\sigma X$. Since all points of μ_x have the same standard properties by (8.1.6), we actually have $x \notin {}^*U[y]$. Let $d(x, y)$ be a standard u-pseudometric that is one outside U and let $f(z) = d(x, z)$ for $z \in {}^*X$. This shows that (a) implies (e).

Now we show that (e) implies (f). Take $x \in \text{rmt}(^*X)$ with internal separating function f. Let $F(x, y) = |f(x) - f(y)|$. By the infinitesimal hull theorem (8.4.24) $F \lessapprox d$, a standard continuous pseudometric. The open cover by d-balls of radius $\frac{1}{2}$ has a subordinate standard partition of unity by Stone's theorem, and this proves (f).

Condition (f) implies (a) by refining the semimetrics $F(x, y) = |f(x) - f(y)|$, $f \in C(X)$ for compatibility with the following. For each remote $z \in \text{rmt}(^*X)$ take the standard semimetric $H(x, y) = \sum_I |h_i(x) - h_i(y)|$ for the partition of unity associated with z. When x is standard only a standard finite set of h_i's are nonzero at x and for those $h_i(z) = 0$, consequently, $H(x, z) \geq 1$ for $x \in {}^\sigma X$. By (8.4.29), z is not pre-near-standard. Therefore, the F's and H's together define a complete compatible uniformity.

There is a good deal more one can do with these ideas in set theoretic topology and we refer the reader to M. E. Rudin [1975] for problems to work on.

CHAPTER 9 COMPACTIFICATIONS

The purpose of this chapter is to give infinitesimal analysis constructions of various compactifications. This shows how such abstract constructions are subsumed by polysaturated models and that in turn has two sides. On one side, it shows what the models "look like" to someone who knows (in some sense) what each compactification "looks like" and conversely gives a partial picture of the compactifications. On the other side, it indicates how infinitesimals might be used in problems like those where abstract compactifications have been successful—it tends to pull nonstandard methods up by the bootstraps. We feel that one promising aspect of the model-theoretic approach is that it carries along abundant structure associated with a compactification in a natural way, whereas some "standard" methods tend to present difficulties in that respect.

9.1 DISCRETE ČECH–STONE COMPACTIFICATION OF N

Take *$\mathbf{N}$ with $\mu_{\mathscr{D}}$ as a closure operator. This is a quasi-compact space by (8.1.7).

(9.1.1) ***REMARK:*** *Closures of points are either identical or disjoint*, since when n is standard $\mu_{\mathscr{D}}\{n\} = \{n\}$ and when ω is infinite and $\omega \in \mu_{\mathscr{D}}\{\Omega\}$, then $\mu_{\mathscr{D}}\{\omega\} = \mu_{\mathscr{D}}\{\Omega\}$. If $\Omega \notin \mu_{\mathscr{D}}\{\omega\}$, ω and Ω can be distinguished by a standard property: $\Omega \notin {}^*D$ and $\omega \in {}^*D$. The complement $E = \mathbf{N}\backslash D$ satisfies $\Omega \in {}^*E$ and thus $\omega \notin \mu_{\mathscr{D}}\{\Omega\}$. We therefore define an *infinitesimal relation* by

$$\omega \overset{\pounds}{\approx} \Omega \qquad \text{if and only if} \qquad \mu_{\mathscr{D}}\{\omega\} = \mu_{\mathscr{D}}\{\Omega\}.$$

(9.1.2) ***NOTATION:*** We denote the quotient space mod this relation by

$$\beta\mathbf{N} = \langle {}^*\mathbf{N}, \mu_{\mathscr{D}}\rangle / \overset{\pounds}{\approx}\,.$$

(9.1.3) ***REMARK:*** *The space $\beta\mathbf{N}$ is a compact Hausdorff space* since two points not in the same discrete monad are separated by standard sets, for example, ${}^*D$ and ${}^*E = {}^*\mathbf{N}\backslash{}^*D$. All standard sets are closed and open (clopen) in $\langle {}^*\mathbf{N}, \mu_{\mathscr{D}}\rangle$, so $\beta\mathbf{N}$ is Hausdorff.

(9.1.4) ***REMARK:*** *Addition and multiplication do not extend continuously to $\beta\mathbf{N}$* since the open sets $\{0\}$ and $\{1\}$ go, respectively, to $\{\omega\}$ under $0 + \omega$

and $1 \cdot \omega$. An infinite singleton cannot be $^*D$ for any $D \subseteq \mathbf{N}$, and thus $\{\omega\}$ is not open. (We mention group compactifications below; also see Robinson [1969].)

We let $\ell^\infty(\mathbf{N})$ denote the algebra of bounded sequences. Since **N** carries the discrete topology every $f \in \ell^\infty$ is a continuous function, or $\ell^\infty(\mathbf{N}) = \mathrm{BC}(\mathbf{N})$, the algebra of bounded continuous functions.

(9.1.5) ***OBSERVATION:*** *The nonstandard extension $^*f$ for $f \in {}^\sigma l^\infty(\mathbf{N})$ is β-S-continuous,* $\omega \overset{\beta}{\approx} \Omega$ in $^*\mathbf{N}$ implies $f(\omega) \approx f(\Omega)$. If $f(\omega) \not\approx f(\Omega)$, then

$$\omega \in f^{-1}[a, b] \qquad \text{and} \qquad \Omega \notin f^{-1}[a, b]$$

where

$$a = \mathrm{st}(f(\omega)) - \tfrac{1}{2}\,\mathrm{st}|f(\omega) - f(\Omega)|, \qquad b = \mathrm{st}(f(\omega)) + \tfrac{1}{2}\,\mathrm{st}|f(\omega) - f(\Omega)|.$$

Consequently we have the following

(9.1.6) ***PROPOSITION:*** *The standard part of each standard $f \in {}^\sigma\ell^\infty$,*

$$\hat{f}(\mu_{\mathscr{D}}\{\omega\}) = \mathrm{st}\, f(\omega)$$

is a well-defined continuous extension of f to $\beta\mathbf{N}$.

EXERCISE: Show that if g is finite, internal, and β-S-continuous, $\hat{g}$ is continuous on $\beta\mathbf{N}$.

(9.1.7) ***REMARK:*** *The compactification $\beta\mathbf{N}$ can also be viewed as a uniform completion* as in (8.4.32) and (8.4.34) since $\omega \overset{\beta}{\approx} \Omega$ if and only if $f(\omega) \approx f(\Omega)$ for every standard bounded (discretely continuous) function f on **N**.

PROOF: We already saw that $\omega \overset{\beta}{\approx} \Omega \Rightarrow f(\omega) \approx f(\Omega)$, so we assume $\omega \overset{\beta}{\not\approx} \Omega$. Then $\omega \in {}^*D$ and $\Omega \in {}^*E = {}^*\mathbf{N} \setminus {}^*D$. Let f be the indicator function of D; $f(n) = 0$, if $n \notin D$ and $f(n) = 1$, if $n \in D$. We have $f(\omega) = 1$ and $f(\Omega) = 0$.

(9.1.8) ***REMARK:*** Still another interpretation of $\beta\mathbf{N}$ is as the *maximal ideal space* of $\ell^\infty(\mathbf{N})$. (To simplify the algebra take sequences with complex values as ℓ^∞.) On a commutative Banach algebra with unit, maximal ideals and *multiplicative linear functionals* correspond, so we can make the maximal ideal interpretation in $^*\mathbf{N}$ in two steps: (1) standard parts of point evaluations (at standard or nonstandard points) are multiplicative linear functionals (compare Section 6.4); (2) every standard algebra functional has such a representation. Finally, the factorization $^*\mathbf{N}/\overset{\beta}{\approx}$ only identifies points associated with the same functionals so $\beta\mathbf{N}$ "is" the maximal ideal space.

To each $\omega \in {}^*\mathbf{N}$ associate the homomorphism

$$\varphi_\omega : {}^\sigma\ell^\infty(\mathbf{N}) \to \mathbf{C}$$

given by

$$\varphi_\omega(f) = \mathrm{st}(f(\omega)).$$

EXERCISE: Each φ_ω is an algebra homomorphism, $\varphi(af + bg) = a\varphi(f) + b\varphi(g)$ and $\varphi(fg) = \varphi(f) \cdot \varphi(g)$. If $\omega \overset{\beta}{\approx} \Omega$, then $\varphi_\omega = \varphi_\Omega$ (notice that both domains are ${}^\sigma\ell^\infty$).

It remains only to show that each nontrivial standard algebra functional $\psi : \ell^\infty(\mathbf{N}) \to \mathbf{C}$ has a representation as one of the φ_ω's. Let $f,\ g \in \ker(\psi)$, since a kernal is a proper ideal, $|f(n)|$ and $|g(n)|$ cannot be bounded below. (If they were, $F(n) = 1/f(n)$, for example, would be an inverse in the algebra and the ideal would be the whole space.) Let ε and δ be positive standard reals, $\varepsilon_f = |f^{-1}|([0,\ \varepsilon])$ and $\delta_g = |g^{-1}|([0,\ \delta])$. Then $\varepsilon_f \cap \delta_g \neq \varnothing$, for suppose otherwise and let $\chi(n) = 0$ if $|f(n)| \leq \varepsilon$ or if $|g(n)| \leq \delta$, but $\chi(n) = 1$ for different n. We contend that $\psi(\chi) = 0$ and also that $\psi((1 - \chi)(f + g) + \chi) = 0$, a contradiction, since the argument function is bounded below. Proof that χ is in the ideal follows from the observation that if it is not in the ideal, it is invertible mod the ideal, so that $(\chi + I)(\xi + I) = (1 + I)$ for some ξ. However, $\chi(n)\xi(n) = 0$ on $\varepsilon_f \cup \delta_g$ so that $1 - \chi$ must belong to the ideal in order that $(\chi\xi + I) = (1 + I)$. Therefore $f + g + (1 - \chi)$ belongs to the ideal and is never small, since $(1 - \chi)$ is big where f and g are small, hence χ is in the ideal instead. The contradiction shows that the sets ε_f have the finite intersection property. Since ${}^*\mathbf{N}$ is polysaturated,

$$\bigcap[{}^*\varepsilon_f : \varepsilon \in {}^\sigma\mathbf{R}^+, f \in {}^\sigma\ker(\psi)] \neq \varnothing.$$

For any ω in the intersection $\psi = \varphi_\omega$ and this concludes the proof that $\beta\mathbf{N}$ is a space of maximal ideals. (Infinitesimal relations on spaces of functionals will be treated in the next chapter, in case the reader is interested in having the functionals carry the topology more directly.)

By now we hope the reader already has an intuitive picture of ${}^*\mathbf{N}$, something like the following heuristic discussion. First, we know ${}^\sigma\mathbf{N}$ looks like $\mathbf{N}$. Beginning at an infinite ω we could add or subtract any finite integer, so the picture of ${}^*\mathbf{N}$, so far, looks like a copy of $\mathbf{N}$ and infinitely far to the right around ω, a copy of $\mathbf{Z}$. The greatest integer in ω/n for any finite n is also an *infinite* integer, so between $\mathbf{N}$ and the copy of $\mathbf{Z}$ around ω there is a dense sequence of copies of $\mathbf{Z}$. (There is even more and Robinson [1966] proves that the order type is that of $\mathbf{N}$ plus a dense ordinal times that of $\mathbf{Z}$.) There are also copies of $\mathbf{Z}$ around multiples of ω, powers of ω, and so on.

To obtain a picture of $\beta\mathbf{N}$ we have to make identifications $\omega \overset{\mathcal{L}}{\approx} \Omega$, and we saw in (8.1.5) that these identifications are made infinitely far apart.

STANDARD PROBLEM: Draw a picture of $\beta\mathbf{N}$. Unfortunately we cannot draw the standard compactification even to the extent to which we "drew" $^*\mathbf{N}$ above.

9.2 MEASURABLE INFINITESIMALS

It is no surprise that $\beta\mathbf{N}$ is extremely disconnected—we began with a base of clopen (closed and open) sets defining the topology. Moreover, this phenomenon is typical of L^∞-spaces (though not characteristic; see (9.2.6)). Extremely disconnected spaces are sometimes called *Stonian spaces* after M. H. Stone. The purpose of measure theory is to extend the class of functions beyond only continuous ones while maintaining the nice features of Riemann integrals of continuous limits and functions. On the other hand, Littlewood's first two principles of integration maintain that measurable sets are almost finite unions of intervals and measurable functions are almost continuous. Using an idea of Loeb [1972] we shall construct the maximal ideal space of $L^\infty(d\theta)$ simply by refining the infinitesimals on the $*$-circle in such a way as to make standard measurable functions S-continuous. Each time we allow a discontinuity we produce a disconnection in the carrier space. We choose the unit circle and $d\theta$-measure because we want to use it in the third example of this section. Loeb [1972, 1974, 1975] and Bernstein and Loeb [1974] carry out this construction in greater generality.

(9.2.1) ***REMARK:*** Littlewood's third principle (Egoroff's theorem) to date has no intrinsic infinitesimal description, see Robinson [1966, p. 132]. In fact, polysaturated models may be the wrong framework for a model theoretical form of this principle of analysis. Loeb [1974] offers an alternative.

We begin with the sigma-algebra $\mathscr{A}$ of arclength Lebesgue-measurable sets on C, the unit circle. An $\mathscr{A}$-partition of C is a finite set of disjoint measurable sets $\{B_i \in \mathscr{A} : 1 \le i \le n\}$ such that $B_i \perp B_j$ $(B_i \cap B_j = \varnothing)$ and $\bigcup_{i=1}^n B_i = C$. The common refinement of a finite set of measurable partitions is formed by taking intersections of the sets in the various partitions. The refinement relation is concurrent in the sense of Chapter 7.

(9.2.2) DEFINITION By (7.2.6(b)) we can embed the set of *all standard* $\mathscr{A}$-partitions in a $*$-finite set and then form the $*$-refinement, or by (7.2.6(a)) we can directly find a $*$-partition finer than all standard partitions. This $*$-finite partition is certainly not unique, but *each one*

$$\{\Delta_i : 1 \le i \le \omega\}$$

satisfies

$$\Delta_i \perp \Delta_j \qquad and \qquad {}^*C = \bigcup_{i=1}^{\omega} \Delta_i$$

and whenever $\{B_i : 1 \le i \le n\}$ *is a standard partition either*

$$\Delta_i \perp {}^*B_j \qquad or \qquad \Delta_i \subseteq {}^*B_j.$$

We shall refer to such partitions as hyperfine partitions of $\mathscr{A}$.

The Δ_i's play the same role in integration of bounded measurable functions that infinitesimal intervals (which they refine) play for Riemann integrals of continuous functions, namely,

(9.2.3) MEASURABLE INFINITESIMALS THEOREM *Let* $f: C \to \mathbf{R}$ *be a standard bounded function. Then* f *is* $\mathscr{A}$*-measurable if and only if for each hyperfine partition* $\{\Delta_i : 1 \le i \le \omega\}$ *and each* i, *whenever* $x, y \in \Delta_i$, *then* $f(x) \approx f(y)$. *In other words, if we define:*

$x \overset{M}{\approx} y$ *if and only if* $x, y \in \Delta_i$ *of some hyperfine partition, then the standard bounded measurable functions are characterized by* M*-continuity*

$$x \overset{M}{\approx} y \qquad implies \qquad f(x) \approx f(y).$$

Moreover, if $f(x) \approx f(y)$ *for all bounded measurable functions, then* $x \overset{M}{\approx} y$.

PROOF: We require the standard lemma that a bounded function is measurable if and only if for every $\varepsilon > 0$ there is a partition of C such that $(\sup[f(x) : x \in B_i] - \inf[f(x) : x \in B_i]) < \varepsilon$ for each B_i in the partition. We shall not prove the lemma (the reader is referred to a text on measure theory).

Notice that since f is standardly bounded and the Δ_i's are internal, the numbers $\varepsilon_i = (\sup[f(x) : x \in \Delta_i] - \inf[f(x) : x \in \Delta_i])$ exist internally. Moreover, the $*$-finite maximum ε of the ε_i, $1 \le i \le \omega$, is attained and clearly $\varepsilon \approx 0$ for each hyperfine partition if and only if f is M-continuous.

The argument of (5.1.1) or (5.2.1) shows how to use Leibniz' principle to go back and forth between the standard lemma and the infinitesimal condition.

NOTE: $x \overset{M}{\approx} y$ is still **external** since it refers to all the hyperfine refinements, while the variation number ε is an internal function of the hyperfine partition and f.

Now we verify the final claim of the theorem. Embed the standard bounded measurable functions in a $*$-finite set of $*$-bounded measurable functions $\mathscr{F}$. Let

$$\varepsilon = \min[|f(x) - f(y)| : f(x) \neq f(y) \;\&\; f \in \mathscr{F}].$$

By $*$-transform of the variation lemma above, to each f in $\mathscr{F}$ there corresponds a $*$-partition with f-variation less than ε. The common refinement of these as f runs over $\mathscr{F}$ will be denoted by P. In case x and y lie in distinct sets of P, take a new partition $\{\Delta_i : 1 \le i \le \omega\}$ with $\Delta_i = B_i$ for $x, y \notin B_i \in P$, $\Delta_i = B_x \cup B_y$ if $x \in B_x$ and $y \in B_y$ in P. We claim $\{\Delta_i\}$ is hyperfine and therefore $x \overset{M}{\approx} y$. The hyperfineness of $\{\Delta_i\}$ follows by considering the bounded measurable indicator functions, $\chi_A(x) = 0$, $x \notin A$, and $\chi_A(x) = 1$, $x \in A$. If Δ_i meets both $*A$ and $*(C \backslash A)$, the variation on Δ_i is one, whereas our construction assures infinitesimal variation.

EXERCISE: Show that if $x \overset{M}{\approx} y$, then $x \approx y$ metrically. (Consider, for example, the function $f(\theta) = \theta^2$ on C with radian measure $-\pi < \theta \le \pi$. Alternately, use Riemannian partitions.)

EXERCISE: Generalize the theorem to an abstract measurable space $(X, \mathscr{A})$.

(9.2.4) ***NOTATION:*** At this stage we introduce the arc-length-Lebesgue measure $d\theta$ on C. Each standard set N of measure zero is in the standard partition $\{N, (C \backslash N)\}$, hence $\bigcup[\Delta_i : d\theta(\Delta_i) = 0] \supseteq *N$. We denote the circle less the null sets of $\{\Delta_i\}$ by

$$\mathscr{S} = \mathscr{S}\{\Delta_i\} = \bigcup[\Delta_i : d\theta(\Delta_i) \neq 0],$$

where $\mathscr{S}$, for Stone or Shilov, will be a Shilov boundary in another maximal ideal space below; it is the premaximal ideal space of $L^\infty(d\theta)$.

A first approximation of what $\mathscr{S}$ looks like can be obtained by embedding ${}^\sigma C$ in a $*$-finite set and taking the complementary $*$-finite union of open infinitesimal intervals. Naturally, more of $*C$ is removed to actually obtain $\mathscr{S}$, but the first step is rather badly disconnected, having all the standard points removed!

(9.2.5) ***REMARK:*** *The infinitesimal hull* $\hat{\mathscr{S}} = (\mathscr{S}/\overset{M}{\approx})$ *is a compact Hausdorff space* by (8.4.32) and (8.4.34). (Notice that $\mathscr{S}$ is an *internal* pre-near-standard set.) *The hull* $\hat{f}$ *of every* $f \in {}^\sigma L^\infty(d\theta)$ *is continuous on* $\hat{\mathscr{S}}$ by the same general theory and the last part of (8.4.23). *Standard maximal ideals of* ${}^\sigma L^\infty(d\theta)$ correspond precisely to standard parts of point evaluations in $\mathscr{S}$ with M-infinitesimally nearby points producing the same algebra functional. (By removing the null sets of $\{\Delta_i\}$ from $*C$ to obtain $\mathscr{S}$, point evaluation makes sense for equivalent ${}^\sigma L^\infty$-functions.)

EXERCISE: Show that for every standard measurable set, $*A \cap \mathscr{S}$ is clopen. Prove that such sets form a base for the nonstandard hull topology.

EXERCISE: Show that every standard algebra homomorphism of $L^\infty(d\theta)$ has a point evaluation representation in $\mathscr{S}$. (*Hint:* See the proof for $\beta\mathbf{N}$ and use saturation.)

(9.2.6) A SPACE WITHOUT 'NORMAL MEASURES' Take $*C$ with Robinson's Q-topology generated by all internal intervals $(\varphi, \psi) = \{e^{i\theta} : \varphi < \theta < \psi\}$. The $*$-topology of $*C$ consists of the internal unions of such intervals, whereas the Q-topology admits all unions, external or not. Any countable intersection of decreasing Q-open sets will contain a Q-open set by $\aleph_1$-saturation, hence there cannot be a measure assigning open sets positive measures and meager sets zero measure (since that would violate countable continuity of the measure).

Dixmier [1951] shows that the maximal ideal space of the algebra of bounded measurable functions on a space without normal measures is extremely disconnected, but **not** a maximal ideal space of an L^∞-space.

(9.2.7) *REMARK:* *Integration of ${}^\sigma L^\infty$-functions* or even internal M-continuous functions on $\mathscr{S}$ can be done in the following simple manner. Let $x_i \in \Delta_i$ be an internal selection of points from the hyperfine partition and let $\delta_i = d\theta(\Delta_i)$. Then

$$\int_C f(\theta)\, d\theta = \mathrm{st}\left(\sum_{i=1}^{\omega} f(x_i)\delta_i\right).$$

All the $*$-finite sums are infinitesimally close together for different choices $x_i \in \Delta_i$ as well as for different hyperfine partitions. In fact the arguments of Chapter 5 for Riemann integrals of continuous functions extend to this case by simply replacing classical infinitesimals with measurable ones and applying the theorem above.

The maximal ideal spaces of L^∞-spaces tell you that measurable functions can be made continuous if you mess up the carrier space very badly. Our description simply says this in terms of infinitesimals while carrying along the integral in a form like the infinitesimal Riemann integral. We think it helps provide a partial picture of these spaces as well.

Compactifications can also be used to describe boundary behavior, as was alluded to at the end of Section 6.6. Our next few examples are directed more toward that kind of question and carry correspondingly more structure.

9.3 THE SAMUEL COMPACTIFICATION OF THE HYPERBOLIC PLANE

We begin with the unit disk

$$U = \{z \in \mathbf{C} : |z| < 1\}$$

and the pseudo-hyperbolic metric

$$p(z, w) = |(z - w)/(1 - \bar{z}w)|,$$

a bounded version of the hyperbolic metric, discussed at length in Section 6.6 for three dimensions; see Examples (8.4.9) and (8.4.10). Theorems (6.6.8–6.6.14) also apply in two dimensions and give us boundary value information for compactifications built from $*U$—the details of this transfer are left as an exercise. The discussion of (6.6.2) specializes to give us a "picture" of the p-infinitesimal hull of $*U$ as a standard disk $\mathrm{Gal}(0)/\overset{p}{\approx}$ with infinitely many isometrically equivalent copies $\mathrm{Gal}(\alpha)/\overset{p}{\approx}$ under the maps $((z - \alpha)/(1 - \alpha z))$. We want to see how much this structure is preserved in the following compactifications, $\mathscr{P}$, $\mathscr{N}$, $\mathscr{F}$, $\mathscr{M}$.

We denote the p-infinitesimal hull of $*U$ by $\hat{U} = \hat{U}_p$. This space is noncompact as can be computed, for example, with the $*$-sequence $z_n = (1 - (1/n))$ for $n \in *\mathbf{N}$,

$$p(z_m, z_n) = \frac{1 - (m/n)}{1 + ((m - 1)/n)},$$

and is nearly 1 for $m = \omega$ and $n = \omega^\omega$. The cardinality of $\hat{U}$ is also outrageously large for the purpose of describing boundary behavior of standard families of functions.

(9.3.1) THE SAMUEL COMPACTIFICATION OF ($*U$, p) $\mathscr{P}$ is essentially the finest compactification compatible with p. The monadic *closure* is

$$\mu^p(A) = \mu_{\mathscr{D}}(o_p(\mu_{\mathscr{D}}(A)))$$

and $*U$ with μ^p as closure and μ^p-equivalent points identified as denoted by $\mathscr{P}$. Theorem (8.1.7) assures us that $\mathscr{P}$ is compact once we show that

$$x \overset{\mathscr{P}}{\approx} y \qquad \text{if and only if} \quad \mu^p\{x\} = \mu^p\{y\}$$

is an equivalence relation, which we leave as an exercise (see Stroyan [1972b, p. 259]).

Observe that for any standard sequence $*(z_n : n \in \mathbf{N})$, $z_\omega \overset{\beta}{\approx} z_\Omega$ when $\omega \overset{\beta}{\approx} \Omega$ in $*\mathbf{N}$. Simply consider the set $D = \{n : z_n \in E\}$ whenever $z_\omega \in *E$ to see $\Omega \in *D$ and $z_\Omega \in *E$ also. This gives us a connection between $\mathscr{P}$ and $\beta\mathbf{N}$. The terms of the sequence $z_n = (1 - (1/n)^n)$ grow increasingly farther apart hyperbolically, and infinite terms are all in different hyperbolic galaxies since $p(z_\omega, z_{\omega+1}) \approx 1$. In this case, $z_\omega \overset{\mathscr{P}}{\approx} z_\Omega$ only if $\omega \overset{\beta}{\not\approx} \Omega$ as well, since if $\omega \in {}^*D$, $\Omega \in {}^*E$, and $D \perp E$, then $\{x : p(x, z_D) < \frac{1}{2}\} \perp \{y : p(y, z_E) < \frac{1}{2}\}$.

EXERCISE: Does $z_\omega \overset{\mathscr{P}}{\approx} z_\Omega$ for $z_n = (1 - (1/n))$ imply $\Omega \overset{\beta}{\not\approx} \omega$?

Compactifying $\hat{U}$ into $\mathscr{P}$ has identified points in different hyperbolic galaxies and p-infinitesimals were already considered equivalent. Now we show that finitely separated points in the same galaxy are **not** identified.

We begin with a tesselation—a triangular type is shown in Fig. 9.3.1. Tesselations are tied strongly to the geometry, the group of motions, and the analytic structure, so while we could do without them for $\mathscr{P}$, we will not. They become more essential in $\mathscr{N}$ below. We number the tiles in a manner like discretized polar coordinates—say in the (2, 3, 7) example. Beginning at the zero vertex number each light–dark pair (0, 1), (0, 2), ..., (0, 7).

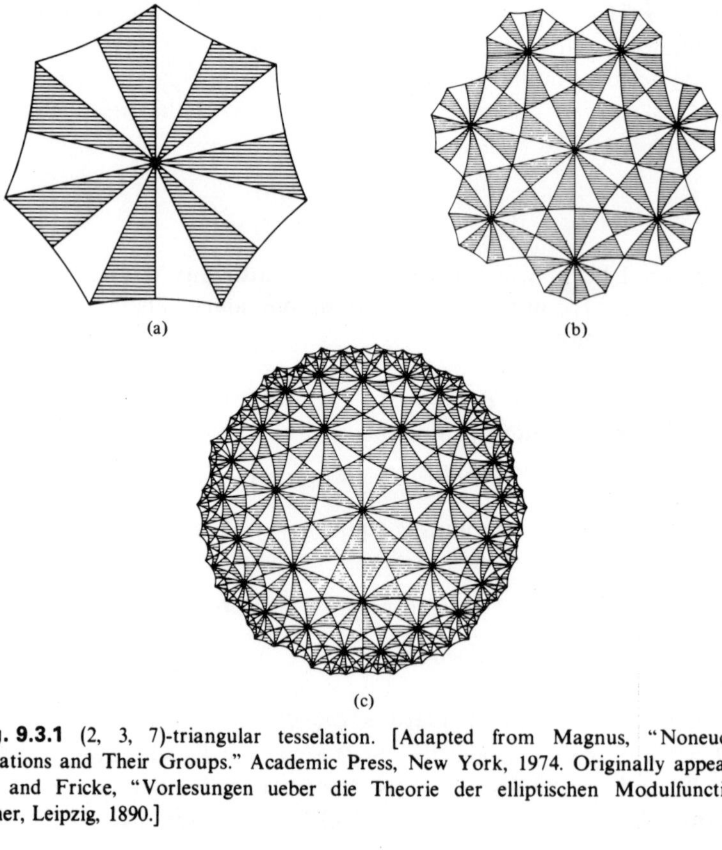

Fig. 9.3.1 (2, 3, 7)-triangular tesselation. [Adapted from Magnus, "Noneuclidean Tesselations and Their Groups." Academic Press, New York, 1974. Originally appeared in Klein and Fricke, "Vorlesungen ueber die Theorie der elliptischen Modulfunctionen." Teubner, Leipzig, 1890.]

Remove the numbered triangles in the central heptagon and number the next layer (1, 1, 1), ..., (1, 1, 7) around the first 7-vertex above the positive x-axis, then (1, 2, 1), ..., (1, 2, 7) around the next heptagon, and so on counterclockwise around to (1, 7, 7). Remove this layer of triangulated heptagons and number the next layer (2, 1, 1), ..., (2, 1, 7)–(2, 21, 7). Each triangle thus gets a numbering (ρ, θ, v) with ρ roughly proportional to the hyperbolic distance to zero and θ roughly proportional to the hyperbolic arc length from the plus-x-axis on a circle centered at zero and $1 \le v \le 7$. Now consider the standard map that sends a point to its tesselation number:

$$x \mapsto (\rho(x), \theta(x), v(x)).$$

First, if the hyperbolic distance $\eta(z, w)$ is finite and not infinitesimal, then $\rho(z)$ and $\rho(w)$ are finitely far apart. If $\rho(z) \neq \rho(w)$, then we know by (8.1.5) that $\rho(z) \overset{\mathscr{B}}{\not\approx} \rho(w)$ and therefore $z \notin \mu_{\mathscr{D}}\{w\}$ by considering $\rho^{-1}({}^*D)$ where $\rho(z) \in {}^*D$ and $\rho(w) \notin {}^*D$. On the other hand, if $\rho(z) = \rho(w)$, then $\theta(z)$ is finitely far from $\theta(w)$ so again, either $\theta(z) = \theta(w)$ or $z \notin \mu_{\mathscr{D}}\{w\}$. If $\rho(z) = \rho(w)$, $\theta(z) = \theta(w)$, and $v(z) = v(w)$, we must subdivide our tiles each into a standard finite number of pieces small enough to distinguish points whose hyperbolic distance is $\frac{1}{2}\,\mathrm{st}\,\eta(z, w)$ to see finally that $z \notin \mu_{\mathscr{D}}\{w\}$ whenever $\eta(z, w)$ is finitely nonzero.

The map $f(z) = (z - \alpha)/(1 - \bar{\alpha}z)$ restricted to Gal(α) and factored mod p-infinitesimals is a coordinate chart in $\hat{U}$ for the disk Gal$(\alpha)/\overset{\mathscr{P}}{\approx}$. Moreover, if $\beta \approx \alpha$, $(z - \beta)/(1 - \bar{\beta}z)$ factors to the same hull mapping and thus gives the same coordinate chart on Gal(α). Finally, suppose $\alpha \in \mu_{\mathscr{D}}\{\beta\}$ so that $\alpha \overset{\mathscr{P}}{\approx} \beta$. For each standard $w \in U$ the standard property "there exists a unique z with $p(z, \beta) < 1$ and $w = (z - \beta)/(1 - \bar{\beta}z)$" holds for β whence for α by (8.1.4), of course, $z = (w + \beta)/(1 + \bar{\beta}w)$ and $y = (w + \alpha)/(1 + \bar{\alpha}w)$ and $y \in \mu_{\mathscr{D}}\{z\}$ by inversion inside the corresponding properties. Therefore we see that $(z - \alpha)/(1 - \bar{\alpha}z)$ restricted to Gal(α) can be factored mod $\overset{\mathscr{P}}{\approx}$ giving a well-defined chart for an analytic disk in $\mathscr{P}$. We have demonstrated the first statement of:

(9.3.2) THEOREM *$\mathscr{P}$ is a compact Hausdorff space containing $2^{2^{\aleph_0}}$ analytic disks. Every standard function f taking values in a compact space which is hyperbolically uniformly continuous extends to $\mathscr{P}$ by $\hat{f} = {}^*f/\overset{\mathscr{P}}{\approx}$.*

It remains only to show that the factorization of $^*f$ is well defined. If $z \in \mu_{\mathscr{D}}\{w\}$, then the standard properties describing estimates of the location of $f(z)$ and $f(w)$ are the same. Compactness assures us that such properties exist, that is, $f(z)$ is near standard for each z.

9.4 NORMAL MEROMORPHIC FUNCTIONS AND ANALYTIC DISKS

A meromorphic function $f\colon U \to \mathbf{S} = \mathbf{C} \cup \{\infty\}$ (see Section 4.7) is said to be *normal* if the family of hyperbolic translates of f

$$\left\{f\left(e^{i\theta} \cdot \frac{z-\alpha}{1-\bar{\alpha}z}\right) : |\alpha| < 1, \theta \in \mathbf{R}\right\}$$

is a normal family, that is, relatively compact in the compact convergence topology. By (8.4.42) this is equivalent to each $*$-translate being S-continuous on Gal(0) and since the motions are rigid and homogeneous over U it is equivalent to f being p-S-continuous everywhere on $^*U$ with the spherical metric on the Riemann sphere $^*\mathbf{S}$. This in turn is equivalent to uniform continuity by (8.4.23). Therefore, by (9.3.2), normal meromorphic functions have natural continuous extensions to $\mathscr{P}$ and the two-dimensional versions of (6.6.10–6.6.14) apply.

Normal functions have been studied infinitesimally by Behrens [1972] and Stroyan [1972a]; they have many interesting properties that are easily accessible by infinitesimal methods. Also see Robinson's [1966] chapter on analytic functions. We shall give one classical-type property essentially due to Behrens and then study the "abstract boundary" which normal functions impose on U. Further applications are in the references above and an extensive standard literature on the subject, which we must leave to the reader. Classical boundary behavior and limits on the "abstract boundary" are clearly tied together by the infinitesimal approach. We begin with some reformulations of classical ways of measuring boundary behavior.

Let $A\colon [0, 1) \to U$ be a *boundary arc* in U, that is, $\lim_{t\to 1} |A(t)| = 1$, or equivalently $t \approx 1$ implies that $A(t)$ is hyperbolically infinite or $|A(t)| \approx 1$. This could be a ray terminating on a point of the unit circle, or a spiral tending to the whole unit circle, or an arc oscillating to a segment of the unit circle, and so on.

(9.4.1) DEFINITION *A function $f\colon U \to \mathbf{S}$ is said to have a limit along $A(t)$ if $\lim_{t\to 1} f(A(t))$ exists and since all points of $^*\mathbf{S}$ are nearstandard, it suffices that f have infinitesimal spherical variation on $\{A(t) : t \approx 1\}$.* We let $L(f; A)$ equal the set of limits, which could be empty.

(9.4.2) *REMARK:* The cluster set of $f : U \to \mathbf{S}$ through D is given infinitesimally by

$$\text{Clust}(f; D) = \text{st}\{f(z) : z \in {}^*D\backslash\text{Gal}(0)\},$$

that is, the standard parts (on $\mathbf{S}$) of the values of f for points of $^*D$ infinitesimally close to the unit circle.

EXERCISE: Describe Clust(f; D) as the set of sequential limits $f(d_n) \to k$ for $\{d_n\} \subseteq D$ with $\lim_n |d_n| = 1$ as well as by $\bigcap_{r<1} \text{cl}[f(D \cap \{z : |z| > r\})]$, that is, prove that the three descriptions are equivalent.

(9.4.3) ***REMARK:*** The *cluster set of f through D at c,*

$$\text{Clust}(f; D, c) = \text{st}\{f(d) : d \in {}^*D \text{ and } |d - c| \approx 0\},$$

the standard parts of values of f on points of $^*D$ infinitesimally near c.

EXERCISE: Reformulate this in terms of sequences and epsilons and deltas.

The reader should notice some connections between our various limit sets. If $f \to k$ along A, then Clust(f; $\{A(t) : t \in [0, 1)\}$) $= \{k\}$ and further if $A(t) \to c$, Clust(f; $A[0, 1)$, c) $= \{k\}$.

(9.4.4) THEOREM *If f is analytic (harmonic, meromorphic, holomorphic) and uniformly p-continuous and if $f \to k$ along a boundary arc A, then $\hat{f}$ is constant on the analytic disks*

$$\text{Gal}(A(t))/ \overset{\mathscr{P}}{\approx}$$

for $t \approx 1$.

PROOF: Since analytic functions have integral representations (Cauchy or Poisson) a variation on (6.6.12) proves standard parts of S-continuous ones remain analytic. The $*$-analytic function $f((z + \alpha)/(1 + \bar{\alpha}z))$, $\alpha = A(t)$ with $t \approx 1$, is S-continuous on Gal(0) and has infinitesimal variation on $A[0, 1) \cup \text{Gal}(\alpha)$, hence its analytic standard part is constant on a set with an adherent point inside U. This forces f to be nearly constant on the whole disk $\text{Gal}(\alpha)/ \overset{\mathscr{P}}{\approx}$, since $\text{st}(f((z + \alpha)/(1 + \bar{\alpha}z))$ is constant on U.

EXERCISE: Prove a standard analog of this theorem. (*Hint:* See (6.6.14), but be careful.)

As a result of this theorem we make the following

(9.4.5) DEFINITION *The hyperbolic cluster set of f through D:*

$$\text{HC}(f; D) = \text{st}\{f(z) : z \in \text{Gal}(d), d \in {}^*D \backslash \text{Gal}(0)\},$$

the standard parts of values taken on by f finitely far from the infinite portion of $^*D$.

If $f \to k$ along $A(t)$, then HC(f; $A[0, 1)$) $= \{k\}$ by the last theorem.

EXERCISE: Describe HC(f; D) or HC(f; $A[0, 1)$) in terms of neighborhoods.

(9.4.6) DEFINITION *The hyperbolic range of f on D is the set of actual standard values taken on infinitely often in finite neighborhoods of D, in terms of the discrete standard part,*

$$\mathrm{HR}(f; D) = {}^{\circ}\{f(z) : z \in \mathrm{Gal}(d), d \in {}^{*}D \backslash \mathrm{Gal}(0)\}.$$

Intuitively, the cluster set is big, while the set of limits on arcs and the range are small, perhaps even empty.

(9.4.7) DEFINITION We weaken the limits along arcs to read: *the hyperbolic limits of f along $A(t)$ are those $w \in \mathbf{S}$ such that, "for every small $\varepsilon \in \mathbf{R}^{+}$, and big $r \in \mathbf{R}^{+}$, there exists a $t \in [0, 1)$ such that $\eta(z, A(t)) < r$ implies* $s(f(z), w) < \varepsilon$," where $s(\cdot, \cdot)$ denotes spherical arc length. Notice in particular when $A(t)$ is rectifiable this means f gets closer and closer to w over longer and longer segments of $A(t)$. We denote the set of hyperbolic limits along A by

$$\mathrm{HL}(f; A).$$

This is contained in the set of principal values.

(9.4.8) THEOREM *Let f be a normal meromorphic function and let A: $[0, 1) \to U$ be a boundary arc, then*

$$\mathrm{HC}(f; A[0, 1)) = \mathrm{HL}(f; A) \cup \mathrm{HR}(f; A).$$

PROOF: Case 1: Suppose f varies only infinitesimally on $\mathrm{Gal}(\alpha)$, where $c = \mathrm{st}(f(\alpha))$ and $\eta(\alpha, A(t_0))$ is finite. By (9.4.4), f is nearly c on $\mathrm{Gal}(\alpha)$ so that for any standard ε and r, there does exist a t, namely t_0, satisfying the definition of hyperbolic limit above. By Leibniz' principle

$$c \in \mathrm{HL}(f; A).$$

Case 2: Suppose f varies finitely on $\mathrm{Gal}(\alpha)$, $\mathrm{st}(f(\alpha)) = c$, and $\eta(\alpha, A(t_0))$ is finite. The following useful infinitesimal form of the open mapping theorem concludes the proof.

(9.4.9) INFINITESIMAL OPEN MAPPING THEOREM *Suppose f is $*$-meromorphic and S-continuous on a finite neighborhood of $a \in {}^{\sigma}U$. If $f(z)$ varies finitely in the neighborhood, that is, if* $\mathrm{st}(f(z))$ *is nonconstant near a, then* $\mathrm{st}(f(a))$ *is attained by f on the monad of a, there exists $z \approx a$ such that* $f(z) = \mathrm{st}(f(a))$.

PROOF: Without loss of generality by S-continuity we may assume $f(z)$ is finite in $^{*}\mathbf{C}$ for z finitely near a. Let $g(z) = \mathrm{st}(f(z))$. On a subneighborhood perhaps, g is holomorphic and assumes the value $b = g(a)$ *only* at a. The winding number or index of g applied to a small finite circle around a

therefore is one with respect to the value $g(a) = b$. Now since both $f(z)$ and $f'(z)$ have integral representations in terms of $f(z)$, $f(z) \approx g(z)$ and $g'(z) \approx f'(z)$ along the circle, so we know the index of the small circle under f is also one with respect to $g(a)$,

$$\frac{1}{2\pi i}\oint \frac{g'(z)\,dz}{g(z)-b} = \frac{1}{2\pi i}\oint \frac{f'(z)\,dz}{f'(z)-b} = 1$$

(with equalities since winding numbers are integers).

(9.4.10) EXERCISE (Behrens): Suppose f is $*$-meromorphic and S-continuous on Gal(0). If st($f(z)$) is nonconstant and $f(z)$ injective on $^*A$, then st($f(z)$) is injective on the interior of A.

Behrens [1972, 1974b] gives several applications of infinitesimal open mapping. He also proves many beautiful cluster set theorems, which we regret lacking space for.

(9.4.11) NORMAL COMPACTIFICATION $\mathscr{N}$ OF THE UNIT DISK This can be described by applying (8.4.33) and (8.4.34) to the infinitesimal relation:

$$z \overset{\mathscr{N}}{\approx} w \qquad \text{if and only if} \qquad f(z) \approx f(w) \quad \text{on } ^*\mathbf{S}$$

for every standard normal meromorphic f.

(The slight generalization in (8.4.33) and (8.4.34) going from bounded to compact **S** is left to the reader.) We know by (9.3.2) that $\mathscr{P} \supseteq \mathscr{N}$ and $z \overset{\mathscr{P}}{\approx} w$ implies $z \overset{\mathscr{N}}{\approx} w$ since normal functions are uniformly continuous.

The proof of (9.3.2) used the tesselation of Figure 9.3.1 to show that two points z and w in different tiles of the tesselation which are only finitely separated are not discretely identified: if $\eta(z, w)$ is finite and their tesselation numbers are different, then $z \overset{\mathscr{P}}{\not\approx} w$. Inside the tiles we subdivided to see this, but we can also use the analytic structure to show this with a normal function constructed as follows.

The group H_1 of rigid motions that preserve the whole tesselation makes identifications as shown in Fig. 9.4.1 on the edges of each tile. The reader can easily see that gluing the edges as shown produces a sphere. Since rational functions separate the points on a sphere we can pull back a function to U by requiring f to be *automorphic* with respect to H_1 and take the values of the rational function on the sphere $U/H_1 \cong R_1$/edge identifications. That is, on U we have invariance under H_1:

$$f(hz) = f(z); \qquad h \in H_1, \quad z \in U.$$

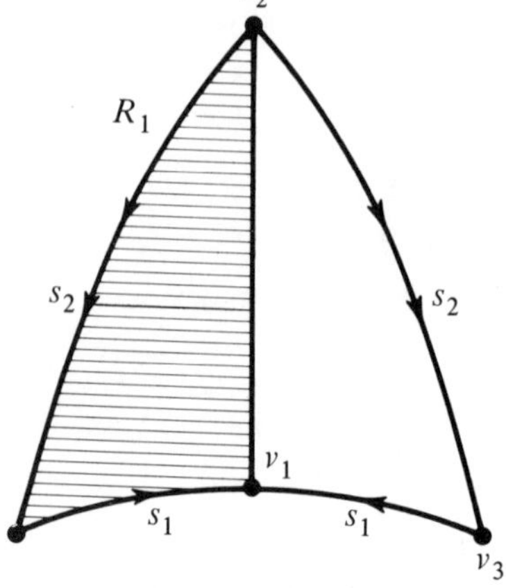

Fig. 9.4.1

Now by extension to $^*U$, we can separate any finitely separated pair of points in the same tile, simply by translation under H_1 to a standard tile. Notice that f is normal since continuous on a compact set R_1 implies uniformly continuous.

To prove that any hyperbolically finitely separated points of $^*U$ can be separated we shall rely on the theory of automorphic functions to provide the details of a generalization of the argument we just gave. We select finitely generated groups $H_1 \supset H_2 \supset \cdots$ with increasing compact fundamental regions $R_1 \subset R_2 \subset R_3 \subset \cdots \bigcup R_n = U$ so that the regions exhaust the unit disk. In each case the automorphic functions separate the points of R_n. Given a finite distance d, one of the groups will have a fundamental region of greater diameter and not identify z with anything close to w if $\eta(z, w) \leq d$. A standard automorphic function associated with H_n separates the points. The details would take us too far afield although we encourage our reader to investigate this beautiful branch of mathematics at least to verify these details. This proves that $\mathcal{N}$ consists entirely of analytic disks.

9.5 THE FATOU–LINDELÖF BOUNDARY

The Fatou–Lindelöf boundary is obtained by further requiring our normal meromorphic functions to have *bounded Nevanlinna characteristic* (or type). This class $\mathcal{F}$ includes the spherical area finite functions and especially the bounded holomorphic functions which we discuss in subsequent sections. Functions of bounded characteristic have radial limits almost everywhere $[d\theta]$, normal ones therefore have uniform nontangential limits almost everywhere $[d\theta]$ by (6.6.10) or (9.4.4). (This is essentially the classical theorems of Fatou and Lindelöf combined.) Along a radius where a limit exists we know a function is nearly constant on the whole radial galaxy by (9.4.4). Far enough out and for directions pointing to $\mathscr{S}$ of Section 9.2 we shall see that this collapsing takes place for *all* ${}^{\sigma}\mathscr{F}$.

One way to say radial limits for f exist except for a null set N of C is: For $\varepsilon > 0$, there exists a map $\lambda \mapsto r_\lambda(f, \varepsilon)$ such that, if $t > r_\lambda$, then $s(f(t \cdot \lambda), f(r_\lambda \cdot \lambda)) < \varepsilon$ $(\lambda = e^{i\theta})$. This is a concurrent relation in the following sense: Given $f_1, \ldots, f_n \in {}^{\sigma}\mathscr{F}$ and $\varepsilon \in {}^{\sigma}\mathbf{R}^+$, there exists an internal $\lambda \mapsto r_\lambda$ defined everywhere on $\mathscr{S}$ (of Section 9.2) satisfying spherical variation less than epsilon as above. The concurrence follows by taking the maximum of the $r_\lambda(f_j, \varepsilon)$ and restricting to $\mathscr{S}$ whereon all are defined since ${}^*N \perp \mathscr{S}$.

(9.5.1) DEFINITION An internal radial function $\lambda \mapsto r_\lambda$ on $\mathscr{S}$ such that for every $f \in {}^{\sigma}\mathscr{F}$ and every $\varepsilon \in {}^{\sigma}\mathbf{R}$, $s(f(t \cdot \lambda), f(r_\lambda)) < \varepsilon$ defines a *ray of very remote galaxies* at $\lambda \in \mathscr{S}$, those containing points a finite distance of some $t \cdot \lambda$, $t > r_\lambda$, $\lambda = e^{i\theta}$.

In summary we have:

(9.5.2) THEOREM *Each ray of very remote galaxies pointing toward $\mathscr{S}$ collapses to a point under identification by the infinitesimal relation*

$$z \overset{\mathscr{F}}{\approx} w \quad \text{if and only if} \quad f(z) \overset{S}{\approx} f(w) \quad \text{for every } f \in {}^{\sigma}\mathscr{F}.$$

(9.5.3) THEOREM *A meromorphic $f\colon U \to \mathbf{S}$ is a normal function of bounded Nevanlinna characteristic if and only if there exist holomorphic functions $g\colon U \to U$, $h\colon U \to U$ such that $f = g/h$ and the hyperbolic distance between the zeros of h and g, $\eta(h^{-1}(0), g^{-1}(0)) > 0$. Consequently, $z \overset{\mathscr{F}}{\approx} w$ if and only if $z \overset{H}{\approx} w$.*

PROOF: If f has bounded type, the classical representation theorem says there exist holomorphic g and h bounded by one such that $f = g/h$. If poles and zeros lie infinitesimally close f cannot be S-continuous, therefore, the distance condition is met also when f is normal.

Conversely, if $f = g/h$, since the characteristic satisfies:

$$T(r, g/h) \leq T(r, g) + T(r, 1/h) \leq T(r, g) + T(r, h) + \mathcal{O} \cdot 1,$$

we see that f has bounded type. Bounded functions are S-continuous by Pick's theorem (see the beginning of the next section), so when the poles and zeros are separated f is S-continuous and normal.

The last statement follows easily from this representation and is left as an exercise in the definition of $(\overset{H}{\approx})$ given in the next section. Observe then also that $f \in {}^\sigma\mathscr{F}$ implies f has a continuous extension to $\mathscr{M}$ of the next section. Normal functions generally only extend to the portion of $\mathscr{M}$ that does not collapse under the additional requirement of bounded characteristic, however, a meromorphic function is normal if and only if it extends continuously to that part, called $\mathscr{G}$ below.

(9.5.4) SUMMARY *$\mathscr{P} = \mathscr{N}$ and this projects onto $\mathscr{M}$.*

We know that $\mathscr{P}$ "looks like" titles of a tesselation identified in the same manner as $*\mathbf{N}$ is identified to obtain $\beta\mathbf{N}$ (taking the sum of the tesselation numbers, for example). Now suppose z is not $\mathscr{P}$-infinitesimally close to w. If $\eta(z, w)$ is finite, we already know z is not $\mathscr{N}$-close to w so we may assume $\eta(z, w)$ is infinite. Discreteness of the group of the tesselation means that the rigid motion sending one tile to another keeps inequivalent points hyperbolically separated, so there are two standard sequences of disks, hyperbolically separated, one containing z, the other w. $\mathscr{N}$ distinguishes these with an automorphic function.

9.6 BOUNDED HOLOMORPHIC FUNCTIONS AND GLEASON PARTS

Now we specialize our class of functions further to bounded holomorphic functions. Pick's invariant form of Schwarz' lemma says that *if $f\colon U \to U$ is holomorphic, either $\eta(f(z), f(w)) < \eta(z, w)$ for all $z \neq w$ or f is a rigid motion, $f(z) = e^{i\theta} \cdot (z - \alpha)/(1 - \bar{\alpha}z)$.* Consequently, by scaling down, if $|f(z)| < K$ for all $z \in U$, then f is uniformly continuous. Of course, a constantly bounded function also has bounded characteristic, so Sections 9.4 and 9.5 apply; in particular, the infinitesimal relation on $*U$

$$z \overset{H}{\approx} w \qquad \text{if and only if} \qquad h(z) \approx h(z) \qquad \text{for all} \quad h \in {}^\sigma H^\infty(U),$$

where $H^\infty(U)$ denotes the space of bounded holomorphic functions, is coarser than $\overset{D}{\approx}$, $\overset{\mathscr{P}}{\approx}$, $\overset{\mathscr{N}}{\approx}$, and $\overset{\mathscr{F}}{\approx}$. Moreover, (8.4.33) and (8.4.34) imply that $\mathscr{M} = *U/\overset{H}{\approx}$ is a compactification of U ($f(z) = z$ separates ${}^\sigma U$).

One of the reasons H^∞ is additionally interesting is the fact that it is a Banach algebra under pointwise operations and the sup norm ($f \overset{u}{\approx} g$ in $*H^\infty(U)$ if and only if $f(z) \approx g(z)$ for all $z \in *U$). Point evaluations

$$\varphi_z(f) = \mathrm{st}(f(z)), \qquad \varphi_z\colon {}^\sigma H^\infty(U) \to \mathbf{C}$$

are algebra homomorphisms corresponding to standard maximal ideals of $H^\infty(U)$. Thus $\mathscr{M}$ is a compact space of maximal ideals. We are naturally led to ask: Does every standard maximal ideal arise in this manner? This question is equivalent to the corona problem that was raised some time ago by function algebraists. The *corona problem* simply asks whether the closure of U in the Gelfand space of all maximal ideals is the whole space or whether there is a "corona" left over. Carleson [1962] proved that the unit disk has no corona, that is, our $\mathcal{M}$ equals the Gelfand space. We think it is fair to say that function–algebra theory has contributed nothing to the solution of that problem, although it brought attention to the question. Infinitesimal analysis has not contributed to this question in the disk to date, but Abraham Robinson felt that a nonstandard proof might be possible and, in light of the difficulty of Carleson's proof, that would be quite interesting (as would any conceptual proof.)

Behrens [1971] made important contributions toward solution of the planar corona problem by constructing the first corona-free infinitely connected regions and by proving a reduction for the planar problem. (As far as we know the question is open for the bidisk and 2-ball, for example.) Behrens' work was inspired by his reformulation of Hoffman's [1967] characterization of the analytic structure of $\mathcal{M}$. Unfortunately, only a little of Behrens' [1974a] infinitesimal methods have appeared. We find that work fascinating, but will confine ourselves to the earliest portion (Behrens [1969]), which might be called "only" a reformulation of Hoffman's [1967] work. We think the reformulation adds important clarity and simplicity. In the next section we take up later developments of Behrens [1972] for the disk. We highly recommend Hoffman's paper and that of Kerr-Lawson [1965] to our reader as well.

Theorem (9.5.2) says many whole rays of galaxies *collapse* to points under $z \overset{H}{\approx} w$.

The Gleason parts metric (arising from the norm in the dual space of H^∞) can be described by:

$$\hat{g}(z, w) = \sup[|h(z)| : h \in {}^\sigma H^\infty(U), \|h\| \leq 1 \;\&\; h(w) = 0]$$

(using the corona theorem for U). In the principal galaxy, Gal(0), we know that $g(z, w) \approx p(z, w)$, since $h(z) = (z - \alpha)/(1 - \bar{\alpha}z)$ satisfies the constraints in the definition of $g(\cdot, \cdot)$ with $\alpha = \mathrm{st}(w)$. Naturally, Pick's theorem says $g(\cdot, \cdot) \leq p(\cdot, \cdot)$, but it may happen that the parts metric is *deficient*, as in the case of collapsing galaxies. A Gleason part is a set of the form $\{\hat{z} : \hat{g}(\hat{z}, \hat{\alpha}) < 1\}$ **in** $\mathcal{M}$ (again using no corona).

When standard bounded holomorphic functions have the maximal variation on a galaxy,

$$\sup[\mathrm{st}|h(z) - h(w)| : h \in {}^\sigma H^\infty(U), \|h\| \leq 1] = p(z, w) \qquad \text{for} \quad z, w \in \mathrm{Gal}(\alpha),$$

we say that H^∞ is *full* for Gal(α). It is clear by what we have done in this chapter that a full galaxy is an isometric copy of U, hence an analytic disk in $\mathcal{M}$. Consequently, a full galaxy gives rise to an analytic disk

Gleason part in $\mathcal{M}$. A collapsing galaxy gives a point part in $\mathcal{M}$. There are no partially deficient galaxies in the unit disk.

(9.6.1) THEOREM *Every hyperbolic galaxy of* $^*U$ *is either full and gives rise to a disk part in* $\mathcal{M}$ *or collapses to a point.* Gal(x) *is full if and only if there is a standard interpolating sequence* $(z_n : n \in \mathbf{N})$ *such that* $z_\omega \in$ Gal(x) *for* $\omega \in {}^*\mathbf{N}$ *if and only if there is a standard interpolating sequence* $(w_n : n \in \mathbf{N})$ *such that* $w_\Omega \overset{\rho}{\approx} x$ *for some* $\Omega \in {}^*\mathbf{N}$.

A sequence $(\alpha_n : n \in \mathbf{N}) \subseteq U$ is an *interpolating sequence* provided that the Blaschke product

(1) $B(z) = \prod_{n=0}^{\infty} \alpha_n/|\alpha_n| \cdot (z - \alpha_n)/1 - \bar{\alpha}_n z)$ converges or, equivalently,

$$\textstyle\sum_{n=0}^{\infty} (1 - |\alpha_n|) < \infty,$$

and

(2) $\inf[(1 - |\alpha_n|^2)\, dB/dz(\alpha_n) : n \in \mathbf{N}] \equiv \delta(B) > 0$ or equivalently,

$$\prod_{m \neq n} |(\alpha_m - \alpha_n)/(1 - \bar{\alpha}_n \alpha_m)| \geq \delta.$$

This condition simply says $B(z)$ has a finite rate of growth at its infinite zeros with respect to the hyperbolic metric. In this case we also call $B(z)$ an *interpolating Blaschke product.*

PROOF: Suppose Gal(x) does not collapse, so there is a standard function f such that

$$h(w) = \mathrm{st}\left(f\left(\frac{x - w}{1 - \bar{x}w}\right)\right)$$

is nonconstant on U. Let $z \in U$ be such that $h'(z) \neq 0$ and let $y = (x - z)/(1 - \bar{x}z)$ so $y \in$ Gal(x) with positive growth at y. Then

$$\begin{aligned}
\frac{d}{dw}\left(f\left(\frac{y - w}{1 - \bar{y}w}\right)\right)\bigg|_{w=0} &= f'(y)(\bar{y}y - 1) \\
&= \left[f'\left(\frac{x - z}{1 - \bar{x}z}\right)\frac{1 - \bar{x}x}{(1 - \bar{x}z)^2}\right] \cdot \left[(1 - \bar{x}z)^2 \frac{(1 - \bar{y}y)}{(1 - \bar{x}x)}\right] \\
&= \frac{d}{dw}\left(f\left(\frac{x - w}{1 - \bar{x}w}\right)\right)\bigg|_{w=z} \cdot \left[(1 - \bar{x}z)^2 \frac{(1 - |y|^2)}{(1 - |x|^2)}\right] \\
&\approx h'(z) \cdot \left[\frac{1 - |y|^2}{1 - |x|^2}\right][1 - \bar{x}z]^2,
\end{aligned}$$

a finite nonzero number by the choice of z above. Notice that since x and y lie in the same galaxy, $[(1-|y|^2)/(1-|x|^2)]$ is finite by a simple calculation like the one preceding (6.6.1). By the infinitesimal open mapping theorem (9.4.9), there is a $\xi \approx 0$ such that $f((y-\xi)/(1-\bar{y}\xi)) = \text{st}(f(y)) = b$. Now let $g(w) = f(w) - b$ so that if $\zeta = (y-\xi)/(1-\bar{y}\xi)$, $g(\zeta) = 0$ and the hyperbolic growth $g'(\zeta)[1-|\zeta|^2]$ is not infinitesimal. Let $\delta = \frac{1}{2}\,\text{st}(g'(\zeta)[1-|\zeta|^2])$.

The standard sequence (α_n) given by

$$\{z : g(z) = 0 \quad \text{and} \quad |g'(z)[1-|z|^2]| \geq \delta\}$$

is an interpolating sequence that contains ζ and approximates y. Let B be the Blaschke product of (α_n) and $h = g/B$, so $\|h\| = \|g\|$ and

$$\begin{aligned}
\delta &\leq |g'(\alpha_n)[1-|\alpha_n|^2]| \\
&\leq \left| \frac{d}{dw}\left[g\left(\frac{\alpha_n - w}{1-\bar{\alpha}_n w}\right)\right]_{w=0} \right| \\
&\leq \left| \frac{d}{dw}\left[B \cdot h\left(\frac{\alpha_n - w}{1-\bar{\alpha}_n w}\right)\right]_{w=0} \right| \\
&\leq |B(\alpha_n)| \cdot \left| \frac{d}{dw}\left[h\left(\frac{\alpha_n - w}{1-\bar{\alpha}_n w}\right)\right]_{w=0} \right| + |h(\alpha_n)| \left| \frac{d}{dw}\left[B\left(\frac{\alpha_n - w}{1-\bar{\alpha}_n w}\right)\right]_{w=0} \right| \\
&\leq \|h\| \cdot |B'(\alpha_n)|[1-|\alpha_n|^2],
\end{aligned}$$

so that $\delta(B) \geq (\delta/\|h\|) = \varepsilon$. This justifies the statement that (α_n) is interpolating with points near y, in particular, in $\text{Gal}(x)$.

It remains to show that $\text{Gal}(x)$ is full and that every point can be approximated by a standard interpolating sequence. Behrens' proof requires the lemma from Hoffman [1967] that says that if $B(z)$ interpolates, there exist Blaschke products $B_1(z)$ and $B_2(z)$ such that $B = B_1 B_2$ and $\delta(B_j) \geq \delta(B)^{1/2}$. We omit the proof of this lemma.

Factor B and suppose $B_1(\zeta) = 0$, since either one factor must satisfy this. Also $\delta(B_1) \geq \varepsilon^{1/2}$. Now factor B_1 with B_2 being the term satisfying $B_2(\zeta) = 0$ and $\delta(B_2) \geq \varepsilon^{1/4}$. Continue by induction obtaining Blaschke factors B_n with $B_n(\zeta) = 0$ and $\delta(B_n) \geq \varepsilon^{1/2^n}$. An infinite term in this sequence satisfies $B_\omega(\zeta) = 0$ and

$$\frac{d}{dw}\left[B_\omega\left(\frac{\zeta - w}{1-\bar{\zeta}w}\right)\right]_{w=0} \approx 1.$$

By Schwarz' lemma, $\text{st}(B_\omega((\zeta - w)/(1-\bar{\zeta}w))) = w$ on U and therefore B_n converges completely to the coordinate function on $\text{Gal}(x)$ making it full and finishing the proof since the coordinate function has unit derivative.

A more complete picture of the analytic disks in $\mathscr{M}$ is obtained by reformulating another lemma of Hoffman [1967] as follows ($\stackrel{\beta}{\approx}$ is from Section 9.1.)

(9.6.2) THEOREM *The nonstandard extension of an interpolating sequence* $(\alpha_n : n \in \mathbf{N}) \subseteq U$ *induces a homeomorphism onto its image:*

$$^*\alpha_n : {}^*\mathbf{N}/\stackrel{\beta}{\approx} \to {}^*U/\stackrel{H}{\approx} \qquad \textit{equals} \qquad \hat{\alpha}_n : \beta\mathbf{BN} \to \mathscr{M}$$

and specifically, $\alpha_\omega \stackrel{H}{\approx} \alpha_\Omega$ *if and only if* $\omega \stackrel{\beta}{\approx} \Omega$.

PROOF: Hoffman [1967] shows that for every δ, $0 \le \delta \le 1$, there exist ε and r with $\varepsilon \to 1$ as $\delta \to 1$ such that if B is a Blaschke product with zeros (α_n) and $\delta(B) = \delta$, then $B^{-1}\{z : |z| < \varepsilon\}$ is contained in the disjoint union of hyperbolic disks of radius r and center α_n whereon B is one-to-one.

Let B have zeros at the (α_n) of the theorem. If $z \stackrel{H}{\approx} \alpha_\Omega$, then certainly $|B(z) - 0| < \varepsilon$ of the lemma and therefore z lies in a disk of radius r about α_ω. If $\omega \in {}^*D \perp {}^*E \ni \Omega$, then the subsequences $(\alpha_n : n \in D)$ and $(\alpha_n : n \in E)$ interpolate and separate z from α_Ω, a contradiction. Hence $z \stackrel{H}{\approx} \alpha_\Omega$ implies $z \stackrel{\beta}{\approx} \alpha_\omega$ with $\omega \stackrel{\beta}{\approx} \Omega$.

If $\omega \stackrel{\beta}{\approx} \Omega$, clearly $f(\alpha_\omega) \approx f(\alpha_\Omega)$ for bounded holomorphic functions, since this is true for arbitrary bounded functions.

(9.6.3) THEOREM *For each* $x \in {}^*U$, $[\mathrm{Gal}(x)/\stackrel{H}{\approx}]$ *is a Gleason part. Moreover, it is a point part if and only if whenever* V *is a subset of* U *such that some hyperbolic* ε*-neighborhood of* V *covers* U, *then* $x \stackrel{H}{\approx} v \in {}^*V$.

PROOF: By the above results, a full galaxy gives rise to a disk part, the difficult case is showing that a collapsing galaxy gives a point part. Suppose $y \stackrel{H}{\approx} x$ and $\mathrm{Gal}(x)$ is deficient. We know there is an $f \in {}^\sigma H^\infty$ with $\|f\| \le 1$ and $f(y) \not\approx f(x) \approx 0$. Let $f = h \cdot B$ where h has no zeros and B is a Blaschke product. If $h(x) \approx 0$, since $h^{1/n} \in H^\infty$ and $h^{1/n}(x) \approx 0$ while $h^{1/n}(y) \to 1$, we know that the parts metric satisfies

$$\hat{g}(\hat{x}, \hat{y}) \ge 1.$$

This proves that x and y lie in different Gleason parts in the case where $h(x) \approx 0$.

Suppose $h(x) \not\approx 0$, so that $B(x) \approx 0$ and $B(y) \not\approx 0$. By the results above we know that $\mathrm{Gal}(x)$ contains no zeros of interpolating subsequences of $B^{-1}(0)$. Two further cases now arise: $\mathrm{Gal}(x) \cap B^{-1}(0)$ is empty or not and quite different techniques yield in either case that x is an infinitesimal of infinite order:

$$B(z) = \prod_{k=1}^{\infty} A_k(z),$$

where the A_k are standard Blaschke products and $A_k(x) \approx 0$ for $k \in {}^\sigma\mathbf{N}$. (Hoffman [1967, Theorems 3.2 and 3.3 in the nonempty case and Theorems 5.2 and 5.3 in the empty case]; details omitted.) Since $B(y) \not\approx 0$ we have

$$\left| \prod_{k \geq n} A_k(y) \right| \to 1$$

as a standard limit in n therefore

$$\hat{g}(\hat{x}, \hat{y}) \geq 1$$

so x and y lie in different parts.

The last part of the theorem can be seen as follows. If Gal(x) contains points of an interpolating sequence approximating x, then $B^{-1}(|z| > \varepsilon)$ finitely nearly covers U by the lemma of Hoffman mentioned above. Conversely, if V nearly covers U and Gal(x) collapses, then ${}^*V \cap \text{Gal}(x) \neq \varnothing$ and $v \overset{H}{\approx} x$ for any $v \in \text{Gal}(x)$. •

(9.6.4) EXERCISE (Behrens): *If A is a boundary arc such that the maximal ideal m lies in the closure of* $A[0, 1)$ *in* $\mathcal{M}$, *show that* f*(Gleason part (m))* $\subseteq$ HR(f; A), *the hyperbolic range defined above. Show that* $\lambda \in$ HL(f, A) *if and only if* $\hat{f} \equiv \lambda$ *on a Gleason part in the* $\mathcal{M}$*-closure of* A. Compare this with Theorem (9.4.8).

Now we consider the map φ

$$[z/\overset{\mathcal{N}}{\approx}] \to [z/\overset{H}{\approx}]$$

or

evaluation at $n \in \mathcal{N} \to H^\infty$-restricted evaluation in $\mathcal{M}$.

We shall denote the set of all disk Gleason parts of $\mathcal{M}$ by $\mathcal{G}$.

(9.6.5) THEOREM *The map* φ: $\mathcal{N} \to \mathcal{M}$ *is closed and continuous. The map* $\varphi^{-1}|_{\mathcal{G}}$ *is an embedding, that is, the interpolating galaxies naturally look alike in both* $\mathcal{M}$ *and* $\mathcal{N}$.

PROOF: We know $z \overset{\mathcal{N}}{\approx} w$ implies $z \overset{H}{\approx} w$, since bounded functions are normal, therefore φ is continuous since it is the hull of an S-continuous internal map.

Also φ is injective on interpolating galaxies since our results above show that $\mathcal{G}$ is isomorphic to $\mathcal{P}$ on those galaxies—we only identify mod $\beta\mathbf{N}$ along interpolating sequences. $\mathcal{G}$ is exhausted by sets $B^{-1}\{|z| < \varepsilon\}^\wedge$ for interpolating Blaschke products and those sets are compact. Since a continuous injection of compact sets is a homeomorphism this shows $\mathcal{G}$ is the same in $\mathcal{N}$ and $\mathcal{M}$.

The map φ is closed since monads generate the topology of $\mathcal{N}$ and standard parts of monads (taken from $^*U \to \mathcal{M}$) are always closed.

The Stone space $\mathcal{S}$ is defined in Section 9.2. A very remote r_λ, for $\lambda \in \mathcal{S}$ is defined in Section 9.5.

(9.6.6) THEOREM *The map $R(\lambda) = r_\lambda e^{i\theta}$, $\lambda = e^{i\theta}$, defines an embedding of $\mathcal{S}$ into $^*U$ such that*

$$\mathcal{S}\hbar \equiv R(\mathcal{S})/\overset{H}{\approx} \quad \text{is homeomorphic to} \quad \mathcal{S}/\overset{M}{\approx} = \hat{\mathcal{S}}.$$

Moreover, $\mathcal{S}\hbar$ is the Shilov boundary of $\mathcal{M}$, the smallest closed subset on which each $f \in {}^\sigma H^\infty(U)$ attains its maximum.

PROOF: Bounded homomorphic functions have bounded measurable radial limits almost everywhere $d\theta$, so when $\psi \in \mathcal{S}$, $f(se^{i\psi}) \approx \rho f(\psi)$, the radial limit at ψ for all standard bounded functions provided r is close enough to one and $r < s$. Therefore $\psi \overset{M}{\approx} \varphi$ implies $re^{i\psi} \overset{H}{\approx} re^{i\varphi}$ and $\hat{\boldsymbol{R}}$ is continuous. $\hat{R}$ is injective as well since boundary values of holomorphic functions separate points of $\hat{\mathcal{S}}$ (or equivalently, are distinct when they differ on sets of positive measure). The easiest way to show this is through boundary values of Poisson integrals

$$f(z) = \frac{1}{2\pi}\int_{-\pi}^{\pi} \frac{e^{i\theta} + z}{e^{i\theta} - z}\, u(\theta)\, d\theta$$

and $h(z) = e^{f(z)}$ is a bounded holomorphic function.

This technique also shows that $\mathcal{S}\hbar$ is a minimal boundary by taking indicator functions of standard sets of positive $d\theta$ measure, for $u(\theta)$, $|\rho h(\theta)| \le 1$ a.e. off the carrier of u, while $|\rho h(\theta)| = e$ a.e. on carr(u).

$\mathcal{S}\hbar$ is a maximum-value "boundary" since the sets

$$\{\theta : |\rho f(\theta)| > \|f\| - \varepsilon\}$$

have positive measure and therefore meet $\mathcal{S}$. If $\psi_j \in \Delta_j$ of the hyperfine partition defining $\mathcal{S}$, $\|f\| = \mathrm{st}[\max(|\rho f(\psi_j)| : j \in I)]$.

(9.6.7) DEFINITION *Let θ be a standard angle in ${}^\sigma(-\pi, \pi]$. The fiber of $\mathcal{M}$ at $\lambda = e^{i\theta}$ is given by*

$$\mathcal{M}_\lambda = \mu(\theta)/\overset{H}{\approx}, \qquad \textit{where} \quad \mu(\lambda) = \{z \in {}^*U : |z - e^{i\theta}| \approx 0\}$$

the Euclidean monad of $e^{i\theta}$ inside $^*U$. Notice that these are precisely the points $m \in \mathcal{M}$ where the hull of the identity function satisfies $\widehat{\mathrm{id}}(m) = e^{i\theta}$.

(9.6.8) DEFINITION *The lens points of $\mu(\lambda)$ are those $z \approx e^{i\theta} = \lambda$ within a finite hyperbolic distance of some $\rho e^{i\theta}$*, that is, the points of the *radial galaxies* pointing to $e^{i\theta}$. By basic hyperbolic geometry they lie inside the standard lens-shaped equidistant regions from the radius or inside standard cones or "Stoltz angles" pointing to $e^{i\theta}$. A point z is *tangential* to λ if the vector from zero to λ and the vector from z to λ are nearly perpendicular; see (6.6.11). The hull of the λ-lens points will be denoted by $\mathscr{R}_\lambda$, the set in $*U$ by R_λ, $R_\lambda/\overset{H}{\approx} = \mathscr{R}_\lambda$.

(9.6.9) DEFINITION *A point z in $\mu(\lambda)$ is barely tangential to $\lambda = e^{i\theta}$ if z is in the intersection monad of standard interior convex subregions tangent to $e^{i\theta}$ and z is not a lens point. More specifically, at $\theta = 0$, a standard arc $\gamma(\varphi) = 1 + ir(\varphi)e^{i\varphi}$, $0 \le \varphi < \pi/2$, is convex and tangent to 1 when $r(\varphi)$ is increasing, continuous, and $r(0) = 0$.* If $z = 1 + i\rho e^{i\psi}$ with $\psi \approx 0$ and $\rho < r(\psi)$ for every such standard r, then z is barely tangential to one. The hull of the barely tangential points at λ is denoted $\mathscr{B}_\lambda$, the set in $*U$ by B_λ.

(9.6.10) DEFINITION *The oricyclic points of $\mu(\lambda)$ are those $z \approx e^{i\theta} = \lambda$ that lie between standard circles tangent to the unit circle at $e^{i\theta}$.* The hull of the λ-oricyclic points will be denoted by $\mathscr{C}_\lambda$, the set by C_λ in $*U$, $\mathscr{C}_\lambda = C_\lambda/\overset{H}{\approx}$.

(9.6.11) DEFINITION *The points of $\mu(\lambda)$ that are extremely tangential to $\lambda = e^{i\theta}$ are those points lying outside every standard convex curve tangent to $e^{i\theta}$.* At $\theta = 0$, in the manner of (9.6.9), $z = 1 + \rho e^{i\psi}$ is extremely tangential if $\rho > r(\psi)$ for every standard r, $\psi \approx 0$. The extremely tangential points in $\mu(\lambda)$ are denoted E_λ, the hull $\mathscr{E}_\lambda = E_\lambda/\overset{H}{\approx}$.

(9.6.12) THEOREM *For standard λ, λ-radial and λ-oricyclic galaxies are full. Moreover, $\mathscr{R}_\lambda$ and $\mathscr{C}_\lambda$ are open subsets of $\mathscr{M}_\lambda$. There is a collapsing barely tangential galaxy.*

PROOF: Left as a problem for the reader; see Kerr-Lawson [1965] and Hoffman [1967]. Notice that a barely tangential galaxy is **not** in $\mathscr{S}\mathscr{h}$; its radius is *not far enough out.*

(9.6.13) THEOREM *For standard λ, $\mathscr{B}_\lambda$ is the topological boundary of $\mathscr{R}_\lambda$ in $\mathscr{M}_\lambda$ and $\mathscr{E}_\lambda$ is the boundary of $\mathscr{M}_\lambda$ in $\mathscr{M}\backslash\hat{U}$. Also, $\mathscr{E}_\lambda \supseteq \mathscr{S}\mathscr{h} \cap \mathscr{M}_\lambda$, but there are both full and collapsing extremely tangential galaxies.*

PROOF: Left as a problem for the reader; note that hulls of monads are always closed.

Theorem (9.6.2) can be extended to the following more geometrical result: Let $L_\alpha(z) = (z + \alpha)/(1 + \bar{\alpha}z)$.

(9.6.14) THEOREM *Locally $\mathscr{G}$ looks like a disk times $\beta\mathbf{N}$, that is, each interpolating point α_Ω has a neighborhood $\{|z - \alpha_\Omega| < \varepsilon\}$ that factors along $\{\alpha_n \mathbf{N}\}$ like $^*\mathbf{N}$ factors into $\beta\mathbf{N}$. When $\alpha_n \to \infty$ very rapidly, for example, if $\prod_{k \neq n} p(\alpha_k, \alpha_n) \to 1$, the hull of L_α is a homeomorphism of U onto the part at $\hat{\alpha} = \hat{\alpha}_\Omega$.*

PROOF: Hoffman [1967, Section 6].

(9.6.15) THEOREM *There is a full galaxy* Gal(α) *such that $\hat{L}_\alpha$ is not a homeomorphism on U* (in the $\overset{H}{\approx}$-topology).

PROOF: Hoffman [1967, p. 109]. Essentially what goes wrong is that regions from different identified galaxies around a point accumulate back on themselves.

9.7 FIXED POINTS OF ANALYTIC MAPS ON $\mathscr{M}$

(9.7.1) ***REMARK:*** Pick's invariant form of Schwarz' lemma says that a holomorphic map on the disk is either a strict hyperbolic contraction or a rigid motion; h: $U \to U$ implies either $\eta(h(z), h(w)) < \eta(z, w)$, for all z, w in U, or $h(z) = e^{i\theta} \cdot (z - \alpha)/(1 - \bar{\alpha}z)$. Consequently, *a map that fixes two points is the identity* $h(z) \equiv z$, since a rigid motion that fixes two points is the identity.

(9.7.2) ***REMARK:*** Similarly, in an infinite galaxy, *if* $h(z) \overset{p}{\approx} z$ and $h(w) \overset{p}{\approx} w$ *where* $z \overset{p}{\approx} w$ in Gal(α), *then h is nearly the identity on* Gal(α). The map $L_\alpha^{-1} \circ h \circ L_\alpha$ is near a standard map with two fixed points on U, where $L_\alpha(z) = (z + \alpha)/(1 + \bar{\alpha}z)$.

This section is devoted to Behrens' [1969, 1972, 1974a] work on studying generalizations of this theorem in $\mathscr{M}$. A holomorphic map h: $U \to U$ factors:

$$^*h\colon {}^*U/\overset{H}{\approx} \to {}^*U/\overset{H}{\approx}$$

to the hull taken in $\mathscr{M}$

$$\mathscr{H}\colon \mathscr{M} \to \mathscr{M}$$

by taking $\mathscr{H}([z/\overset{H}{\approx}]) = [h(z)/\overset{H}{\approx}]$. If we denote the hull in $\mathscr{M}$ of L_α by $\mathscr{L}_\alpha$ for α in a full galaxy

$$\mathscr{L}_\alpha(z) = [(z + \alpha)/(1 + \bar{\alpha}z)/\overset{H}{\approx}],$$

then $\mathscr{L}_\alpha$ is an "analytic" injection into the Gleason part determined by Gal(α),

$$\hat{f} \circ \mathscr{L}_\alpha = (f \circ L_\alpha)\hat{}\,,$$

the right-hand term clearly being analytic for $f \in {}^{\sigma}H^{\infty}(U)$. If Gal($\alpha$) collapses, $\mathcal{L}_\alpha$ is constant on U. When the hull of h is taken in $\mathcal{M}$,

$$\mathcal{L}_\beta^{-1} \circ \mathcal{H} \circ \mathcal{L}_\alpha(z)$$

is an "analytic" map on U if it is defined at a single point, since either $\beta = h(\alpha)$ lies in a full or collapsing galaxy and $h(\text{Gal}(\alpha)) \subseteq \text{Gal}(h(\alpha))$. Because U is dense in $\mathcal{M}$, hulls of standard holomorphic maps from U to U constitute the full class of "analytic" maps of $\mathcal{M}$ that preserve U.

(9.7.3) *OBSERVATION:* The simplest fixed point theorem for $\mathcal{M}$ is that *if $\mathcal{H}$ fixes two points in a single Gleason part, then $\mathcal{H}$ is the identity on that part*. If a part has two points it is a disk and

$$\mathcal{L}_\alpha^{-1} \circ \mathcal{H} \circ \mathcal{L}_\alpha$$

is a map on U with two fixed points.

Recall that $\mathcal{G}$ denotes the disk Gleason parts of $\mathcal{M}$. Then $\mathcal{G}$ is an open dense subset, of course, the part U itself is dense and, in fact, the other parts have small closures—they lie in single fibers and do not reach the Shilov boundary. The next result allows us to "lift" fixed points into $^*U$.

(9.7.4) THEOREM *$\mathcal{H}$ fixes a point of $\mathcal{G}$ if and only if* $\inf[\eta(z, h(z)):$ $z \in U] = 0$. *If $m = [z/\overset{H}{\approx}]$ is the fixed point of $\mathcal{H}$, $^*h(z) \overset{\eta}{\approx} z$.*

PROOF: Since $m \in \mathcal{G}$ we know that $z \overset{H}{\approx} w$ implies either $z \overset{\eta}{\approx} w$ in the same galaxy or z and w lie in different galaxies which are identified mod $\overset{H}{\approx}$. Let $z \overset{\eta}{\approx} \alpha_\omega$ where $\{\alpha_n\}$ is an interpolating sequence and let $h(\alpha_\omega) = w$, which, since $\mathcal{H}(m) = m$, is near some α_Ω. By (9.6.2) and (8.1.7), ω and Ω are either infinitely far apart or $\omega = \Omega$. If $\omega = \Omega$, we are done and otherwise we argue as follows supposing $\omega < \Omega$.

Interpolating sequences must be hyperbolically separated, so let δ be such that the disks of hyperbolic radius δ and center α_n are disjoint. Let T be the map $T(\alpha_m) = \alpha_n$ if $m < n$ and $\eta(h(\alpha_m), \alpha_n) < \delta$. Then T is a standard map and $T(\alpha_\omega) = \alpha_\Omega$, so T is defined on an infinite set. Partition $\{\alpha_n\}$ by the equivalence $T^r(\alpha_m) = T^s(\alpha_n)$ for r, s positive integers. When α_m and α_n are equivalent, further distinguish the cases of odd or even equivalence according to whether the smallest $(r + s)$ is even or odd. Now α_ω and α_Ω are odd equivalent since $\eta(h(\alpha_\omega), \alpha_\Omega) < \delta$. On the other hand, the set of points even equivalent to α_ω is standard and therefore contains α_Ω since $\omega \overset{\beta}{\approx} \Omega$. This contradiction (and a similar argument for $\Omega < \omega$) shows that $\omega = \Omega$ and $h(z) \overset{\eta}{\approx} z$. The rest of the theorem is a clear exercise on infinitesimals.

(9.7.5) LEMMA *Let f be an internal hyperbolic contraction on $^*U$. Suppose that $f(z) \overset{\eta}{\approx} z$ and $f(w) \overset{\eta}{\approx} w$. Then for every x between z and w on the hyperbolic line joining z and w, $f(x) \overset{\eta}{\approx} x$.*

PROOF: Let Δ be the closed hyperbolic disk with center z and radius $\eta(z, x) = \rho$ and D the disk with center w and radius $\eta(w, x) = r$. We know that $f(\Delta)$ lies inside the disk of center $f(z)$ and radius ρ while $f(D)$ lies inside the disk of radius r around $f(w)$. The standard parts of $f(\Delta)$ and $f(D)$ lie, respectively, inside the horocycles through x tangent to the unit circle at the opposite ends of the line through zw (see Fig. 9.7.1). The unique overlapping point is x.

(9.7.6) EXERCISE (Behrens): Is this lemma true in Euclidean geometry?

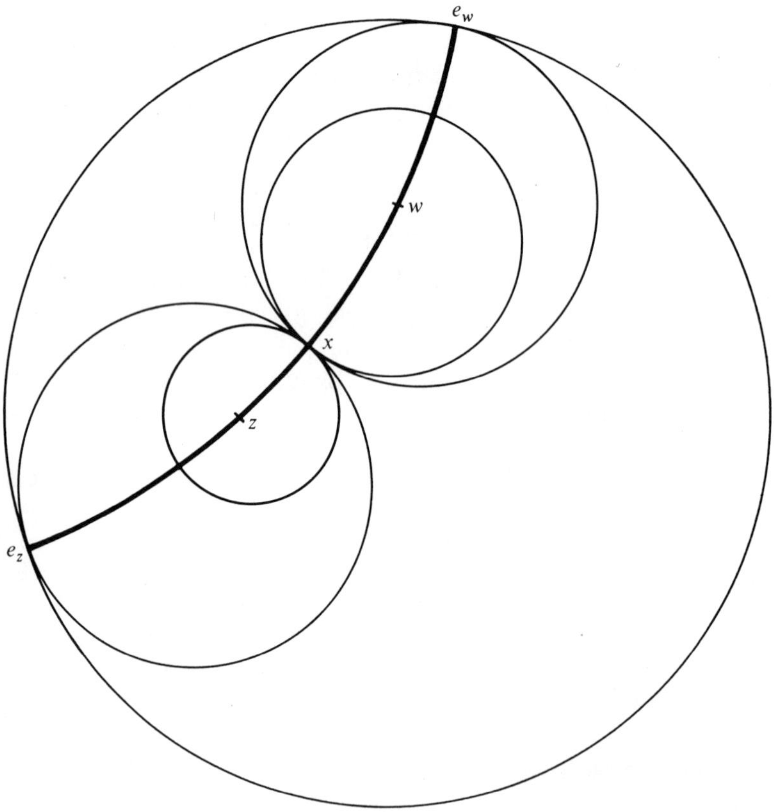

Fig. 9.7.1

(9.7.7) THEOREM *If $\mathcal{H}$ fixes points of $\mathcal{G}$ that lie in distinct fibers of $\mathcal{M}$, then $h \equiv z$.*

PROOF: By (9.7.4), $h(z) \approx z$, $h(w) \approx w$ and the line through z and w intersects U. On that line inside U, h is the identity since a standard map cannot move a standard point only infinitesimally.

(9.7.8) THEOREM *Suppose $\mathcal{H}$ fixes a point of $\mathcal{G} \cap \mathcal{M}_\lambda$ for a standard λ. Then $\mathcal{H}$ maps the $\mathcal{M}$-closure of tangent horodisks at λ into themselves and fixes every lens point in $\mathcal{R}_\lambda$. Moreover $\mathcal{H}$ fixes a point of $\mathcal{G} \cap \mathcal{M}_\lambda$ if and only if the angular derivative of h is one at λ, that is, the limit* $\lim(h(z) - e^{i\theta})/(z - e^{i\theta}) = 1$ *uniformly inside lenses at $e^{i\theta} = \lambda$.*

PROOF: Suppose h nearly fixes the interpolating point $z \overset{e}{\approx} 1$. A standard horocycle at 1 is a circle tangent to the unit circle at 1; call the disk inside a *horodisk* D. Let Δ be the hyperbolic disk of infinite radius R and center z that intersects the real axis at the same standard point as the horocycle. The disk of radius R and center $h(z)$ has the same closed hull in $\mathcal{M}$ as D and Δ contains $h(\Delta)$.

There are many points that represent $[z/\overset{H}{\approx}]$, in particular, there is a $w \overset{e}{\approx} 1$ such that $z \overset{H}{\approx} w$ and $|(1 - w)/(1 - z)| < \varepsilon \approx 0$. The hyperbolic line joining z and w comes infinitesimally near x on the real axis and by the contraction lemma above $h(x) \approx x$. We can work back and forth on an interpolating sequence to show every x on the real axis near 1 is a fixed point and (9.7.3) shows that every lens point is fixed.

The sufficiency of the limit condition follows from the fact (9.6.12) that $\mathcal{R}_\lambda \subseteq \mathcal{G}$; see (6.6.10) for discussion of nontangential limits.

(9.7.9) COROLLARY *If a Blaschke product B with zeros $\{\alpha_n\}$ satisfies*

$$\sum (1 - |\alpha_n|^2)/|e^{i\theta} - \alpha_n|^2 \neq 1,$$

then B fixes no point of $\mathcal{G} \cap \mathcal{M}_\lambda$. The series equals one if and only if B fixes every point of $\mathcal{R}_\lambda$.

PROOF: The angular derivative of a Blaschke product is given by the series. The interested reader should compare Caretheódory [1954, Vol. 2, p. 28].

Next we look for fixed points in a sequence of iterates of our map h. If h has no fixed points in U, then $|h^n(\alpha)| \to 1$ for $\alpha \in U$, $h^n = h \circ h \circ \cdots \circ h$, n times. By Pick's theorem,

$$\eta(h^{n+1}(\alpha), h^n(\alpha)) < \eta(h^n(\alpha), h^{n-1}(\alpha)) < \eta(h(\alpha), \alpha)$$

unless h is a rigid motion.

(9.7.10) THEOREM *Any point in the closure of $\{h^n(\alpha)\}$ that is a point part is fixed for $\mathscr{H}$.*

PROOF: Left as an exercise. (*Hint:* Apply contraction and what it means to be in a collapsing galaxy.)

(9.7.11) THEOREM *If* $\lim \eta(h^{n+1}(\alpha), h^n(\alpha)) = 0$, *then each point of $\mathscr{M}\backslash U$ in the closure of $\{h^n(\alpha)\}$ is fixed for $\mathscr{H}$.*

PROOF: Apply the infinitesimal formulation of limit, and complete the proof as an exericse.

(9.7.12) THEOREM *If $\{h^n(\alpha)\}$ is interpolating, then no point of the closure of $\{h^n(\alpha)\}$ is fixed for $\mathscr{H}$.*

PROOF: The sets $\{h^{2n}(\alpha)\}$ and $\{h^{2n-1}(\alpha)\}$ have disjoint closures when $\{h^n(\alpha)\}$ interpolates.

When the hull of h: $U \to U$ is taken in $\mathscr{M}$, the values are harder to relate to the Shilov boundary $\mathscr{S}\mathscr{h}$. Inner functions still behave nicely.

(9.7.13) LEMMA *If h: $U \to U$ is a holomorphic map and f is bounded and measurable on the unit circle, then the boundary values of $f \circ h$ equal the boundary values of $h \circ f$ a.e., where f is harmonic inside U.*

PROOF: Let f be real and g be harmonic conjugate to f. If A is a set of measure 2π where h and $(f + ig) \circ h$ have radial limits and if $e^{i\theta} \in A$, then $f + ig$ converges as z goes to $h(e^{i\theta})$ along the curve $r \mapsto h(re^{i\theta})$. If $h(e^{i\theta}) \in U$, the limit equals $(f + ig) \circ h(e^{i\theta})$. If $h(e^{i\theta})$ is on the unit circle, by Lindelöf's theorem, $f + ig$ has a radial limit at $h(e^{i\theta})$ and that limit equals the limit of $f + ig$ along $r \mapsto h(re^{i\theta})$.

(9.7.14) COROLLARY *The composition of inner functions is inner. Moreover, h is inner if and only if $\mathscr{H}(\mathscr{S}\mathscr{h}) \subseteq \mathscr{S}\mathscr{h}$.*

PROOF: The first part is apparent from the theorem. We know that $\mathscr{S}\mathscr{h}$ is characterized by the property that every inner function has modulus one on $\mathscr{S}\mathscr{h}$. Let g be another inner function. Then $|g(\mathscr{H}(m))| = |g \circ h(m)| = 1$ and $\mathscr{H}(m) \in \mathscr{S}\mathscr{h}$. If h is outer, there is an $\varepsilon < 1$ such that the set $E = \{\lambda : |h(\lambda)| < \varepsilon\}$ has positive measure and $R(E)$ is mapped into $\{|z| \leq \varepsilon\}$ so $\mathscr{H}(\mathscr{S}\mathscr{h}) \nsubseteq \mathscr{S}\mathscr{h}$.

(9.7.15) THEOREM *If h is inner, fixes a point in U, and $\mathscr{H}$ fixes another point of $\mathscr{S}\mathscr{h}$, then $h \equiv z$.*

LEMMA *If h is inner and $h(0) = 0$, then $d\theta\, h^{-1}(E) = d\theta(E)$.*

PROOF OF THE FIRST LEMMA: $E \to d\theta\, h^{-1}(E)$ is a measure on the circle by the rule $\int g\, d\theta\, h^{-1} = \int g \circ h\, d\theta$, for $g \in L^{\infty}(d\theta)$. If $g = e^{in\theta}$ $\int e^{in\theta}\, d\theta\, h^{-1} = \int h^n\, d\theta = h^n(0) = 0$, so $d\theta\, h^{-1} = d\theta$.

LEMMA *An inner h with $h(0) = 0$ is ergodic; $h(E) \subseteq E$ implies $d\theta(E) = 0$ or 2π.*

PROOF OF THE SECOND LEMMA: It suffices to show that whenever $g \circ h = g$ almost everywhere for $g \in L^{\infty}(d\theta)$ that g is constant almost everywhere. By the above the harmonic extension of g satisfies $g(h(z)) = g(z)$, $z \in U$, so by induction, $g(h^n(z)) = g(z)$. By Schwarz' lemma $h^n(z) \to 0$ inside U and by the continuity of g, $g(h^n(z)) \to 0$, forcing $g(z) \equiv g(0)$.

PROOF OF THE THEOREM: Reduce to the case $h(0) = 0$. Let $I = \partial U \cap \{\mathrm{Re}(z) > 0\}$, $A_1 = I \cap h^{-1}(\partial U \backslash I)$, and $A_n = I \cap h^{-1}(A_{n-1})$. Let $E_1 = I \backslash \bigcup A_j$, $E_2 = A_1 \cup A_3 \cup A_5 \cup \cdots$, and $E_3 = A_2 \cup A_4 \cup A_6 \cup \cdots$. Then the A_j's are disjoint, the E_j's are disjoint, $f(E_3) \subseteq E_2$ and $f(E_2) \subseteq E_3 \subseteq (\partial U \backslash I)$. Let $R(A) = R(A \cap \mathcal{S})$, etc., so the $R(E_j)$ are disjoint clopen sets in $\mathcal{S}\hbar$. Since $\mathcal{H}(R(E_3)) \subseteq R(E_2)$ and $\mathcal{H}(R(E_2)) \subseteq R(E_3 \cup [\partial U \backslash I])$, $\mathcal{H}$ has no fixed point in $R(E_2 \cup E_3)$. We know $h(E_1) \subseteq E_1$, so $d\theta(E) = 0$ and $R(E_1) = \varnothing$, since $E_1 \perp \mathcal{S}$.

(9.7.16) THEOREM *The function z^n, $n > 1$, only fixes zero in $\mathcal{M}$.*

PROOF: We give the proof in case n is even and leave n odd to the reader. By simple geometry fixed points are excluded everywhere except in the Euclidean monad of one, whose hull is the fiber $\mathcal{M}_1$ and which are fixed by a calculation of the angular derivative (9.7.8). We conclude the proof by showing that tangential points cannot be fixed either.

Fix φ, $0 < \varphi < \pi/2$, and let

$$A = \{e^{i\theta} : \varphi/n^{(2m+1)} < |\theta| < \varphi/n^{2m};\ m = 1, 2, 3, \ldots\}.$$

Then $z^n(A)$ is contained in

$$B = \{e^{i\theta} : \varphi/n^{2m} < |\theta| < \varphi/h^{(2m-1)};\ m = 1, 2, 3, \ldots\}.$$

Let Φ be the harmonic extension of the indicator function of A. Since $z^n(A) \subseteq B$ and $A \cup B$ is all but a set of measure zero around one,

$$\Phi(z^n) + \Phi(z) \approx 1 \qquad \text{for} \quad z \approx 1.$$

This means that a nearly fixed z must satisfy

$$2\Phi(z) \approx \Phi(z^n) + \Phi(z) \approx 1$$

and hence be in the set

$$\{z : \tfrac{1}{4} < \Phi(z) < \tfrac{3}{4}\}.$$

Under infinite magnification a segment of this set looks like the circular segments shown in Fig. 9.7.2 (see Carathéodory [1954, Vol. 1, p. 154]). Since the same argument applies for all standard angles φ, we see that a fixed z must be nontangential. It lies in two of the overlapping crescents and the angle from 1 to the midpoint of the central right angle circle is nearly $\arctan((n-1)/2)$.

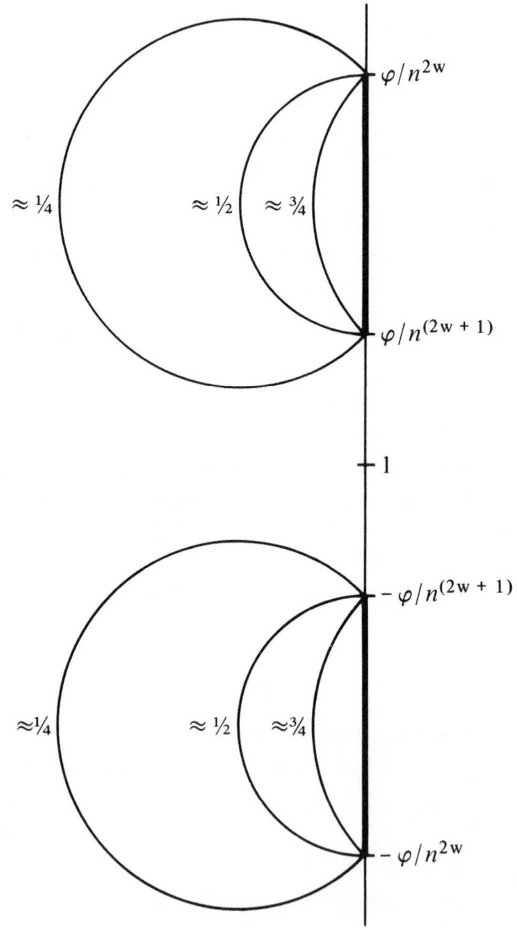

Fig. 9.7.2

9.8 THE BOHR GROUP

We hope the last few sections demonstrate how Robinson's infinitesimals naturally carry analytic, geometric, and classical boundary behavior theorems into compactifications. We close the chapter with an example where algebraic structure is also involved by considering Bohr's elegant theory of almost periodic functions.

Kugler [1969] first gave an account of almost periodic functions using infinitesimals, but we shall take a different approach. Robinson [1969] considers group compactifications more generally.

We ask our reader to sketch the development of Example (8.4.18) along the lines of what we do for almost periodic functions. The reader may wish to consult the work of Luxemburg [1972a] for ideas on infinitesimals in harmonic analysis.

(9.8.1) DEFINITION *An internal function f: $^*\mathbf{R} \to {}^*\mathbf{C}$ is an S-uniformly almost periodic function $f \in \mathrm{SUAP}$ provided f is finite and S-continuous everywhere on $^*\mathbf{R}$ and provided there is an $\bigcap$-monad $\mathscr{P}(f)$ of near-periods of f such that each infinitely long interval $[t, T] \subseteq {}^*\mathbf{R}$ contains a near-period $P \in \mathscr{P}(f)$, $t \leq P \leq T$, satisfying*

$$f(x + P) \approx f(x)$$

for each $x \in {}^*\mathbf{R}$.

(9.8.2) THEOREM *S-uniformly almost periodic functions are uniformly near their standard part, that is, if $f \in \mathrm{SUAP}$ and $g(x) = \mathrm{st}\, f(x)$ for $x \in {}^\sigma\mathbf{R}$, then $g \in \mathrm{SUAP}$ and $f(x) \approx g(x)$ for all $x \in {}^*\mathbf{R}$. Moreover, the near-periods of f and g form the same additive monadic subgroup of $^*\mathbf{R}$, $\mathscr{P}(f) = \mathscr{P}(g)$.*

PROOF: The standard part g is bounded and uniformly continuous since g equals $\hat{f}\,|\mathbf{R}$, the infinitesimal hull restricted to $\mathbf{R}$, by (8.4.23) [compare (5.2.1) and (5.2.2)]. It is easy to see that f is finitely bounded on $^*\mathbf{R}$, since its set of bounds is internal and contains every infinite number [compare (5.1.5)]. The same finite bound works for g.

Let ε be a standard positive number in ${}^\sigma\mathbf{R}^+$. The set of ε-periods of f is internal by the internal definition principle: $\mathscr{P}(f, \varepsilon) = \{p \in {}^*\mathbf{R} : \sup_x |f(x + p) - f(x)| < \varepsilon\}$. Every interval of length L meets $\mathscr{P}(f, \varepsilon)$ whenever L is infinite since the near-periods $\mathscr{P}(f)$ are contained in $\mathscr{P}(f, \varepsilon)$. Since $\mathscr{P}(f, \varepsilon)$ is internal, some finite $L(\varepsilon)$ satisfies this property as well.

A standard number p in $\mathscr{P}(f, \varepsilon)$ satisfies

$$|f(x + p) - g(x)| < 2 \cdot \varepsilon$$

for $x \in \mathbf{R}$, that is, ${}^{\circ}\mathscr{P}(f, \varepsilon) \subseteq {}^{\circ}\mathscr{P}({}^*g, 2 \cdot \varepsilon) = \mathscr{P}(g, 2\varepsilon)$. Every interval of length $L(\varepsilon)$ in $\mathbf{R}$ contains such a point since if $p \approx q$ and $q \in \mathscr{P}(f, \varepsilon)$, then $p \in \mathscr{P}(g, 2\varepsilon)$ by simple estimates left to our reader. This implies that the monad of near-periods of g is nonempty:

$$\mathscr{P}(g) = \bigcap [\mathscr{P}({}^*g, \varepsilon) : \varepsilon \in {}^{\sigma}\mathbf{R}^+].$$

This shows that $g \in SUAP$. Notice that we have not used the fact that $\mathscr{P}(f)$ is monadic yet.

Since $\mathscr{P}(f)$ is an intersection monad,

$$\mathscr{P}(f) \subseteq {}^*E \subseteq \mathscr{P}(f, \varepsilon) \qquad \text{for some} \quad {}^*E$$

by Cauchy's principle and therefore by the above

$${}^{\circ}E \subseteq {}^{\circ}\mathscr{P}({}^*g, 2\varepsilon)$$

so that

$$E \subseteq \mathscr{P}(g, 2\varepsilon) \qquad \text{and} \qquad \mathscr{P}(f) \subseteq \mathscr{P}(g).$$

Now for any standard positive $\varepsilon \in {}^{\sigma}\mathbf{R}^+$, there exists a standard positive $\delta \in {}^{\sigma}\mathbf{R}^+$ such that

$$\mathscr{P}(f, \delta) \subseteq \mathscr{P}(g, \varepsilon)$$

since

$$\mathscr{P}(f) = \bigcap [\mathscr{P}(f, \delta) : \delta \in {}^{\sigma}\mathbf{R}^+].$$

Then within a finite $L(\delta)$ of $z \in {}^*\mathbf{R}$, there is a p in $\mathscr{P}(f, \delta)$ satisfying

$$|f(w + p) - f(w)| < \delta \qquad \text{and} \qquad |g(w + p) - g(w)| < \varepsilon$$

$$\text{for all} \quad w \in {}^*\mathbf{R}.$$

Since $(z - p)$ is finite, $g(z - p) \approx f(z - p) \approx f(z - p + p) \approx f(z)$ and $|g(z - p + p) - g(z - p)| < \varepsilon$ so $|g(z) - f(z)| \lessapprox \varepsilon$. Since $\varepsilon \in {}^{\sigma}\mathbf{R}^+$ and $z \in {}^*\mathbf{R}$ were arbitrary, $f(z) \approx g(z)$ for all $z \in {}^*\mathbf{R}$.

Since f and g are infinitesimally close on all ${}^*\mathbf{R}$ a near-period of one is clearly a near-period of the other so $\mathscr{P}(f) = \mathscr{P}(g)$.

$\mathscr{P}(f)$ is a group since if $P, Q \in \mathscr{P}(f)$, then $f(x) \approx f(x + P) \approx f(x + Q)$ and $f([x - Q] + P) \approx f([x - Q]) \approx f([x - Q] + Q) \approx f(x)$, so $(P - Q) \in \mathscr{P}(f)$.

(9.8.3) PROBLEM: Can you relax the condition that $\mathscr{P}(f)$ is a monad and still have (9.8.2)?

The fundamental concurrence relation of almost periodic functions is:

(9.8.4) LEMMA *Two functions f, $g \in$ SUAP have a common near-period on any infinitely long interval $[t, T]$, that is, one P, $t \leq P \leq T$, such that both*

$$f(x+P) \approx f(x) \qquad \text{and} \qquad g(x+P) \approx g(x)$$

for all $x \in {}^*\mathbf{R}$. In particular, $\mathscr{P}(f) \cap \mathscr{P}(g) \neq \varnothing$.

PROOF: Without restriction by (9.8.2) we may assume f and g are standard, so a conventional method applies. The proof is left to the reader as a problem.

(9.8.5) ***REMARK:*** The space UAP of standard uniformly almost periodic functions is *complete in the uniform norm.* A uniform convergence near-standard function g is bounded and S-continuous directly from the definitions. It is almost periodic as well. When $[t, T]$ is an infinite interval and $f \in {}^{\sigma}$UAP with $\sup|f(x) - g(x)| < \varepsilon/2$, the set $\{p : \sup|f(x+p) - f(x)| < \varepsilon\}$ contains the near-periods of f since they satisfy $|f(x+P) - f(x)| < \varepsilon/2$. By saturation there is a common Q in the standard (f, ε)-approximations to g which is a near-period. (This uses the preceding lemma for concurrence of successive approximations.) In particular, a uniformly convergent standard sequence of almost periodic functions has an almost periodic limit.

(9.8.6) ***REMARK:*** A more fundamental consequence of Lemma (9.8.4) is the fact that *finite sums and products of almost periodic functions are also almost periodic.* This is clear when the near-periods are identical.

(9.8.7) EXERCISE: Prove that

(a) a globally (on ${}^*\mathbf{R}$) S-uniformly differentiable almost periodic function has an almost periodic derivative (see (5.7.5));

(b) the indefinite integral of an almost periodic function is almost periodic provided it is bounded;

(c) if F is uniformly continuous and g is almost periodic, $F(g(x))$ is almost periodic;

(d) if $f(x)$ is almost periodic and bounded below by a finite positive constant, then $1/f(x)$ is almost periodic;

(e) give a near-period for $f(x) = \exp(ix) + \exp(i2\pi x)$;

(f) if f is almost periodic, so are all the functions $f_a(x) = f(x-a)$ for $a \in {}^*\mathbf{R}$, consequently by (9.8.2) $\operatorname{st}\{f_a : a \in {}^*\mathbf{R}\}$ is compact in the uniform norm (see (8.3.11)). Also, $\mathscr{P}(f_a) = \mathscr{P}(f)$ for all $a \in {}^*\mathbf{R}$;

(g) if f has a finite near period, show that $\operatorname{st}(f)$ is periodic.

The concurrence lemma (9.8.4) tells us that the monad of near-periods of all uniformly almost periodic functions

$$\mathscr{P} = \bigcap[\mathscr{P}(f) : f \in {}^\sigma\text{UAP}] = \bigcap[\mathscr{P}(f) : f \in S\text{UAP}]$$

meets every infinitely long interval.

(9.8.8) DEFINITION *The Bohr infinitesimals on* $^*\mathbf{R}$ *are given by* $x \overset{\text{B}}{\approx} y$ *if and only if* $f(x) \approx f(y)$ *for every* $f \in S\text{UAP}$, *or if and only if* $f(x) \approx f(y)$ *for every* $f \in {}^\sigma\text{UAP}$. Hence, if $P \in \mathscr{P}$ and $y = x + P$, $y \overset{\text{B}}{\approx} x$. Conversely, if $x \overset{\text{B}}{\approx} y$, then $f_z(x) \approx f_z(y)$ where $f_z(w) = f(w + z - y)$, $f_z(x) = f(x + z - y) \approx f_z(y) = f(z)$ so that $(x - y) \in \mathscr{P}$.

By (8.1.7), $^*\mathbf{R}$ with $\mu_{\mathscr{D}_\text{B}}$ as a closure is a quasi-compact space. If $x \in \mu_{\mathscr{D}}\{y\}$, then clearly $x \overset{\text{B}}{\approx} y$ since x and y satisfy the standard properties: $|f(x) - b| < \varepsilon$, $b = \text{st}(f(x))$, $f \in {}^\sigma\text{UAP}$, by (8.1.5). Since $\mathscr{P}$ is a monad, it is a closed subset of $(^*\mathbf{R}, \mu_{\mathscr{D}})$ and therefore

(9.8.9) THE BOHR GROUP

$$\mathbf{BR} = (^*\mathbf{R}, \mu_{\mathscr{D}})/\overset{\text{B}}{\approx}$$

is a compact topological group to which each $f \in S\text{UAP}$ has a continuous extension given by the infinitesimal hull

$$\hat{f}(\hat{x}) = \text{st}\, f(x).$$

The Bohr group **BR** is also the infinitesimal hull of $^*\mathbf{R}$ mod the Bohr infinitesimals, thus a compact group by (8.4.32) and (8.4.34).

The intuitive "picture" of **BR** this construction gives is a real line wrapped in an infinite number of loops nearly touching at distances of P apart for each $P \in \mathscr{P}$. While this has some intuitive appeal it is quite complicated.

(9.8.10) *REMARK:* In analogy with the periodic case we can represent **BR** as the points in one period interval, $\mathbf{BR} \cong [0, P)/\overset{\text{B}}{\approx} \cong [P, Q)/\overset{\text{B}}{\approx}$ for $P, Q \in \mathscr{P}$. The main change is that we can no longer pick a minimum P, the infinite interval $[0, \sqrt{P})$ also contains a near-period Q. It is not hard to show that if $f \in {}^\sigma\text{UAP}$ and $[s, S)$ and $[t, T)$ are any infinite intervals,

$$1/(S - s)\int_s^S f(x)\, dx \approx 1/(T - t)\int_t^T f(y)\, dy \approx 1/P\int_0^P f(z)\, dz,$$

so a common standard part $M(f)$ exists and standardly equals $\lim_{L\to\infty} (1/L) \int_a^{a+L} f(x)\, dx$ uniformly in a. (See the work of Kugler [1969] for hints; note that if $Q \in [0, T] \cap \mathscr{P}$ satisfies $(T - Q)/T \approx 0$, say $Q \in [0, \sqrt{T})$, then the T-mean is nearly the Q-mean.)

(9.8.11) THEOREM *If f is a standard almost periodic function, there is at most a countable sequence $\{\lambda_n\}$ of numbers in ${}^\sigma\mathbf{R}$ such that*

$$M(f(x)\exp(-i2\pi\lambda_n x)) \neq 0.$$

Conversely, given a sequence $\{\lambda_n\} \in \mathbf{R}$, there is an $f \in {}^\sigma$UAP with precisely that spectrum $a(\lambda) = 0$ unless $\lambda = \lambda_n$ for some n.

SKETCH OF PROOF: Let $a(\lambda) = M(f(x)e^{-i2\pi\lambda x})$. A calculation using $(1/P)\int_0^P dx$ yields

$$M\left(\left|f(x) - \sum_{m=1}^{n} b_m e^{i2\pi\lambda_m x}\right|^2\right) = M(|f(x)|^2) - \sum_{m=1}^{n} |a(\lambda_m)|^2 + \sum_{m=1}^{n} |b_m - a(\lambda_m)|^2.$$

Hence taking $b_m = a(\lambda_m)$ we see that

$$\sum_{\lambda \in \mathbf{R}} |a(\lambda)|^2 \leq M(|f(x)|^2).$$

Conversely, given $\{\lambda_n\}$, the function

$$f(x) = \sum_{k=1}^{\infty} (1/k^2)e^{i2\pi\lambda_k x}$$

converges uniformly, hence is almost periodic.

Naturally, this suggests relating an almost periodic function to its Fourier series

$$f(x) \sim \sum_{n=1}^{\infty} a(\lambda_n)e^{i2\pi\lambda_n x}, \qquad a(\lambda_n) \neq 0,$$

analogous to the harmonic case. Most of the basic periodic theory carries over simply by replacing the harmonic $(1/2\pi)\int_0^{2\pi} f(x)\,dx$ integrals by means $M(f) \approx (1/P)\int_0^P f(x)\,dx$ and ignoring infinitesimal differences.

The fundamental classical theorems on uniformly almost periodic functions are the equivalent uniqueness theorem and Parseval's equation:

UNIQUENESS THEOREM *If f, $g \in$ UAP and $M(f(x)e^{i2\pi\lambda x}) = M(g(x)e^{i2\pi\lambda x})$ for all $\lambda \in \mathbf{R}$, then $f(x) \equiv g(x)$.*

PARSEVAL'S EQUATION *If $f \in$ UAP, then*

$$M(|f(x)|^2) = \sum |a(\lambda)|^2.$$

A simple proof using infinite means can be found in the work of Kugler [1969]. A key lemma there is the exercise that

$$F(x) = (1/T)\int_0^T f(x+t)\bar{f}(t)\,dt$$

is in SUAP when $f \in$ UAP.

Infinite means are also useful to help obtain a "picture" of the near-periods $\mathscr{P}(f)$ when

$$f(x) \sim \sum_{n=1}^{\infty} A_n e^{i2\pi\lambda_n x}.$$

In that case, if $P \in \mathscr{P}(f)$ and T is infinite,

$$A_n \approx (1/T)\int_0^T f(x)e^{-i2\pi\lambda_n x}\,dx \approx (1/T)\int_P^{T+P} f(x)e^{-i2\pi_n x}\,dx$$

$$= (1/T)\int_0^T f(x+P)e^{-i2\pi\lambda_n(x+P)}\,dx \approx e^{-i2\pi\lambda_n P}(1/T)\int_0^T f(x)e^{-i2\pi\lambda_n x}\,dx.$$

In other words, $A_n \approx e^{-i2\pi P\lambda_n} \cdot A_n$ and $e^{-i2\pi P\lambda_n} \approx 1$. Therefore $P\lambda_n$ is infinitesimally close to some hyperinteger Ω_n in $^*\mathbf{Z}$. We have shown one half of:

(9.8.12) THEOREM *Let $f \in$ UAP with Fourier exponents $\{\lambda_n\}$. Then $P \in \mathscr{P}(f)$ if and only if for each standard n there is a hyperinteger $\Omega_n \in {}^*\mathbf{Z}$ such that $P\lambda_n \approx \Omega_n$.*

Conversely, if $P\lambda_n \approx {}^*\mathbf{Z}$ for $n \in {}^\sigma\mathbf{N}$,

$$(1/T)\int_0^T f(x+P)e^{-i2\pi\lambda_n x}\,dx \approx (e^{i2\pi P\lambda_n}/T)\int_0^T f(x)e^{-i2\pi\lambda_n x}\,dx,$$

so that $f(x)$ and $\operatorname{st} f(x+P)$ have the same Fourier coefficients. They are the same function by the uniqueness theorem stated above and therefore $P \in \mathscr{P}(f)$.

(9.8.13) ***REMARK:*** *The reals are a dense subspace of* **BR** *embedded as the Bohr hull of* $^\sigma\mathbf{R}$. *They are not topologically embedded because infinite intervals always contain near-periods that are infinitesimally near each other.* A standardized version of this says there are points of $^\sigma\mathbf{R}$ far apart in the metric of **R**, but within epsilon for **BR**.

(9.8.14) ***REMARK:*** The additive group **R** acts *uniformly continuously on* **BR** by the formula: $\hat{x} \in \mathbf{BR}$, $t \in \mathbf{R}$

$$\hat{x} + t = \widehat{(x + t)}$$

continuity coming from

$$x \overset{B}{\approx} y \quad \text{and} \quad s \approx t \qquad \text{implies} \qquad \hat{x} + s \overset{B}{\approx} \hat{y} + t.$$

This action tells us that if g is a continuous function on **BR**, g arises as the hull of a uniformly almost periodic function

$$f(t) = g(\hat{0} + t).$$

Clearly, f is bounded, uniformly continuous, and agrees with g on a dense subset of **BR**, the epsilons and deltas of ε-periods for f can be lifted from g.

(9.8.15) ***REMARK:*** The *Haar measure* on **BR** is induced by the infinite means

$$(1/T) \int_0^T f(x)\, dx$$

since this defines it for the continuous functions and the measure can be extended from these. It might be worthwhile to construct the Haar measure more directly on the points of $^*\mathbf{R}$ and to study its properties at that level, for example, the ergodicity of the **R**-action. We leave this as a problem.

(9.8.16) ***REMARK:*** A corollary of (9.8.12) is $P \in \mathscr{P}$ *if and only if* $P \cdot {}^\sigma\mathbf{R} \approx {}^*\mathbf{Z}$, *that is, for each* $\lambda \in {}^\sigma\mathbf{R}$, $P\lambda \approx \Omega \in {}^*\mathbf{Z}$. Also, we know $x \overset{B}{\approx} y$ if and only if $(x - y) \in \mathscr{P}$, so each hull

$$\Lambda(\hat{x}) = (e^{i2\pi\lambda x})^\wedge$$

is a continuous group character on **BR**. If $\lambda \neq \tau$ (both in ${}^\sigma\mathbf{R}$), then $\sup_x |e^{i2\pi\lambda x} - e^{i2\pi\tau x}| = 1$ so **R** is discrete in the group of characters of **BR** under this "hull of the exponential" embedding.

On the other hand, the group of characters of the discrete group **R** contains all the exponentials

$$e^{i2\pi p \cdot x}.$$

In the extension, such characters are infinitesimal precisely when

$$e^{i2\pi p \cdot x} \approx 1 \qquad \text{for all} \quad x \in {}^\sigma\mathbf{R},$$

in other words precisely when $p \in \mathscr{P}$. Every internal exponential is near standard in this topology since $\chi(x) = \mathrm{st}(e^{i2\pi p \cdot x})$ is a standard character (compare (8.4.45)). The standard part of an internal set of near-standard characters is always compact by (8.3.11). Hence the dual of the discrete additive reals **bR** contains **BR**.

There are many ways to see that $\mathbf{bR} \cong \mathbf{BR}$, for example, by using **R**-invariant measures or the algebra of continuous functions, but all the proofs carry the same essential content as Kronecker's approximation theorem in the statement that the standard exponentials are (pointwise) dense in **bR** (we know ${}^\sigma\mathbf{R}$ is dense in **BR**). Density of the exponentials means that if $\chi(x)$ is any character on **R** there exists a $\beta \in {}^*\mathbf{R}$ such that

$$\chi(x) \approx e^{i2\pi\beta x} \qquad \text{for all} \quad x \in {}^\sigma\mathbf{R}.$$

If f is an additive function on **R**,

$$f(x) + f(y) = f(x + y),$$

then

$$\chi(x) = e^{i2\pi f(x)}$$

is a character and there is a β in ${}^*\mathbf{R}$ such that

$$\beta \cdot x \approxeq f(x) \quad (\text{mod } {}^*\mathbf{Z})$$

for all $x \in {}^\sigma\mathbf{R}$.

(9.8.17) THEOREM *Let f: $\mathbf{R} \to \mathbf{R}$ be a standard additive function. If f is bounded on some interval, then $f(x) = x \cdot f(1)$ for all $x \in \mathbf{R}$.*

PROOF: By transfer ${}^*f$ is additive on ${}^*\mathbf{R}$ and $f(q \cdot x) = q \cdot f(x)$ for $x \in {}^*\mathbf{R}$ and $q \in {}^*\mathbf{Q}$. Let a be an interior point of the interval where f is bounded so $x \approx a$ implies $|f(x)| < B \in {}^\sigma\mathbf{R}$, or $\varepsilon \approx 0$ implies $|f(\varepsilon)| \leq A + |f(a)|$ by additivity. If $n \in {}^\sigma\mathbf{N}$ and $\varepsilon \approx 0$, $n\varepsilon \approx 0$, so that $|f(\varepsilon)| \leq (A + f(a))/n$ for all $n \in {}^\sigma\mathbf{N}$ or rather

$$\varepsilon \approx 0 \qquad \text{implies} \qquad f(\varepsilon) \approx f(0).$$

We know ${}^*\mathbf{Q}$ is dense in ${}^*\mathbf{R}$, so for each $x \in {}^\sigma\mathbf{R}$, let $x = q + \varepsilon$ with $q \in {}^*\mathbf{Q}$ and ε infinitesimal. Then

$$f(x) = f(q + \varepsilon) = f(q) + f(\varepsilon) = q \cdot f(1) + \delta$$

with δ infinitesimal. Since q is finite

$$f(x) \approx x \cdot f(1)$$

and both sides are standard so we actually have equality.

(9.8.18) EXERCISE: A discontinuous additive function has a graph that is dense in the plane. (*Hint*: Suppose $bf(a) \neq af(b)$. Solve $ax + by = z$ and $f(a)x + f(b)y = w$. Giggle to ${}^*\mathbf{Q}$.)

(9.8.19) EXERCISE: Let $F(x) = \beta \cdot x$ for some infinite $\beta \in {}^*\mathbf{R}_\infty$. Show that $F(\mathscr{P})$ is Bohr dense in $^*\mathbf{R}$, in other words, F mixes the space up very badly [and in particular $G(\beta)$ below is **not** $^*\mathbf{Q}$].

Let f: $\mathbf{R} \to \mathbf{R}$ be a standard additive discrete automorphism of $\mathbf{R}$ and let $e^{i2\pi\beta x}$ be a Kronecker approximation to $e^{i2\pi f(x)}$. Each internal set

$$A(\varepsilon) = \{x \in {}^*\mathbf{R} : |e^{i2\pi\beta x} - e^{i2\pi f(x)}| < \varepsilon\}$$

contains ${}^\sigma\mathbf{R}$ when ε is standard and positive in ${}^\sigma\mathbf{R}^+$. Hence, the family $\{{}^*\mathbf{Q}, A(\varepsilon), {}^*E \colon {}^*E \supseteq \mathscr{P}\}$ has the finite intersection property. Let

$$G = \bigcap [A(\varepsilon) \cap {}^*\mathbf{Q} : \varepsilon \in {}^\sigma\mathbf{R}^+]$$

be the approximation group of β. The Bohr hull of G is $\mathbf{BR}$ since G contains ${}^\sigma\mathbf{Q}$ and is an intersection of hulls of internal sets which are closed. The map

$$F: G \to {}^*\mathbf{R}$$

given by restricting $\beta \cdot x = F(x)$ to G is Bohr continuous since, if $P \in G \cap \mathscr{P}$,

$$\begin{aligned} e^{i2\pi\beta P} &\approx e^{i2\pi f(P)}, && P \in A(\varepsilon) \quad \text{for all} \quad \varepsilon \in {}^\sigma\mathbf{R}^+ \\ &= e^{i2\pi P f(1)}, && P \in {}^*\mathbf{Q} \\ &= (e^{i2\pi P})^{f(1)}, && \\ &\approx (1)^{f(1)}, && P \in \mathscr{P} \\ &\approx 1, && \end{aligned}$$

so $\beta \cdot P \in \mathscr{P}$. Therefore the Bohr hull of F induces an automorphism

$$\hat{F}: \mathbf{BR} \to \mathbf{BR}.$$

This proves one implication in:

(9.8.20) THEOREM *The automorphisms of the Bohr group* $\mathbf{BR}$ *are in one-to-one correspondence with discrete additive automorphisms of* $\mathbf{R}$. *They can be represented by multiplication by a nonstandard* β *acting on a dense subgroup* $G(\beta)$ *through the Kronecker approximation.*

Proof of the converse portion can be seen by using duality theory to show that an automorphism of $b\mathbf{R}$ corresponds to one on discrete $\mathbf{R}$.

(9.8.21) PROBLEM (Bohr): If f is almost periodic and the Fourier exponents are linearly independent over $\mathbf{Q}$, show that the Fourier series converges uniformly to f and $\sup|f| = \sum |a_n|$. (*Hint:* Use Kronecker approximation.)

CHAPTER 10 **LINEAR INFINITESIMALS AND THE LOCALLY CONVEX HULL**

10.1 BASIC THEORY

In this chapter we discuss the notions of finite and infinitesimal compatible with locally convex linear spaces. The first section is aimed at the most basic results, later ones at more particular results. We want to stress internal spaces, internal norms, and an additional natural generalization of infinitesimals that will permit simple descriptions of distributions and Mackey topologies in particular; otherwise this section specializes Section 8.4.

In this chapter **K** will denote a field taken to be either ***R** or ***C**, $\mathcal{O}$ will denote the finite elements of **K**, and $a \approx b$ means the corresponding infinitesimal relation in **K** (see Chapter 4). We will let **E**, **F**, etc., denote internal **K**-vector spaces, for example, standard ones such as $*L^1[0, 1]$, $*H^\infty(\mathrm{U})$, etc.

(10.1.1) DEFINITION A $*$-*seminorm on* **E** *is an internal map* $\gamma: \mathbf{E} \to *[0, \infty)$ *satisfying:* $x, y \in \mathbf{E}$, $\gamma(x) \geq 0$

$$\text{(H)} \qquad \gamma(ax) = |a|\gamma(x), \qquad a \in \mathbf{K}$$

$$\text{(A)} \qquad \gamma(x + y) \leq \gamma(x) + \gamma(y).$$

A $*$-pseudonorm is permitted to also take the standard value ∞.

(10.1.2) DEFINITION *A convex infinitesimal relation on* **E** *is given by an infinitesimal gauge* IG [*where* IG *is a family of $*$-seminorms with* card(IG) < card($\mathscr{X}$)] *through the definition*

$x \overset{\mathrm{I}}{\approx} y$ *if and only if* $\gamma(x - y)$ *is infinitesimal for* $\gamma \in \mathrm{IG}$ *whenever* $x, y \in \mathbf{E}$.

We denote the infinitesimal vectors by

$$\inf_{\mathrm{IG}}(\mathbf{E}) = \{x \in \mathbf{E} : x \overset{\mathrm{I}}{\approx} 0\}.$$

A *bounded convex infinitesimal relation* on **E** is given by further specifying bounds with IG as follows. Let FG be a family of $*$-pseudonorms and suppose $\mathscr{B}$ is a family of internal absolutely convex sets $B \in \mathscr{B}$ such that for each $\varphi \in \mathrm{FG}$, $\sup[\varphi(b) : b \in B]$ is finite (10.1.17). We suppose $G = \mathrm{FG} \cup \mathrm{IG}$

has $\text{card}(G) < \text{card}(\mathscr{X})$ and $\text{card}(\mathscr{B}) < \text{card}(\mathscr{X})$. Let $Bd = \bigcup \mathscr{B} = \bigcup[B : B \in \mathscr{B}]$

$$x \overset{\mathscr{B}}{\approx} y \quad \text{if and only if} \quad x \overset{\mathrm{I}}{\approx} y \text{ and } (x - y) \in Bd.$$

The bounded infinitesimal vectors are denoted

$$\inf_{\mathscr{B}}(\mathbf{E}) = \{x \in \mathbf{E} : x \overset{\mathscr{B}}{\approx} 0\} = \inf_{\mathrm{I}} \cap Bd.$$

Finally, *a constrained convex infinitesimal relation* on **E** is given by

$x \overset{\mathrm{G}}{\approx} y$ if and only if $\gamma(x - y)$ is infinitesimal for $\gamma \in \mathrm{IG}$ and $\varphi(x - y)$ is finite for $\varphi \in \mathrm{FG}$.

The infinitesimal vectors are denoted

$$\inf_{\mathrm{G}}(\mathbf{E}) = \{x \in \mathbf{E} : x \overset{\mathrm{G}}{\approx} 0\}.$$

When FG is a finite set, the bounded and constrained infinitesimals coincide letting $B_n = \{x \in \mathbf{E} : \max \varphi(x) < n\}$ and $\mathscr{B} = \{B_n : n \in {}^{\sigma}\mathbf{N}\}$.

(10.1.3) ***EXAMPLE:*** The norm infinitesimals of $\mathbf{E} = {}^*H^{\infty}(U)$ are given by

$$h \overset{u}{\approx} g \quad \text{if and only if} \quad h(z) \approx g(z) \quad \text{for all} \quad z \in {}^*U,$$

where $\mathrm{FG} = \varnothing$ and $\mathrm{IG} = \{\text{sup norm}\}$.

(10.1.4) ***EXAMPLES:*** (8.4.12), (8.4.14)–(8.4.16), and three cases of (8.4.38) on the space of all functions on X are examples where $\mathbf{E} = {}^*\mathbf{F}$ and $\mathrm{IG} = {}^{\sigma}G$, where G is an infinite family of standard seminorms. In these cases FG is empty. The reader should verify these remarks, especially the observation that $\mathrm{IG} \neq {}^*G$.

(10.1.5) ***EXAMPLE:*** Let $\mathbf{E} = {}^*\mathscr{C}^{\infty}({}^*\mathbf{R})$ the internal smooth functions on ${}^*\mathbf{R}$. The carrier (or support) of a function is the closed set

$$\text{carr}(f) = \text{cl}\{t : f(t) \neq 0\}.$$

The *distributional infinitesimals* are given by

$f \overset{\mathscr{D}}{\approx} g$ if and only if $f^{(m)}(t) \approx g^{(m)}(t)$ for finite derivatives, $m \in {}^{\sigma}\mathbf{N}$ and finite $t \in \mathcal{O}$ and $\text{carr}(f - g) \subseteq \mathcal{O}$.

The infinitesimal gauge consists of the family $\max[|f^{(m)}(t)| : -n \le t \le n;\ m, n \in {}^{\sigma}\mathbf{N}]$ and the finite gauge is $\sup[|\text{carr}(f)|]$.

The *finite points* are denoted $\mathscr{D} = \text{fin}_{\mathrm{G}}(\mathbf{E})$ and are simply the internal smooth functions with finite carrier that are finite in their standard derivatives.

Observe that the functionals "evaluation of the kth derivative"

$$f \to f^{(k)}(t)$$

are S-continuous for $\mathscr{D}$, $\mathscr{S}$, and $\mathscr{E}$ (8.4.14)–(8.4.15) when t and k are finite. When $k = 1$ and t is infinite, evaluation is discontinuous on $\mathscr{E}$, infinitesimal on $\mathscr{S}$, and zero on $\mathscr{D}$.

(10.1.6) ***EXAMPLE:*** Let $\ell_\Omega{}^\infty$ equal the space of internal sequences from $\{1, 2, \ldots, \Omega\}$ into $\mathbf{K}$. The Mackey infinitesimals are

$x \overset{M}{\approx} y$ if and only if $x(n) \approx y(n)$ for $n \in {}^\sigma\mathbf{N}$ and $x(\omega) \sim y(\omega)$ for $1 \le \omega \le \Omega$.

The reader should supply G. Notice that this space is internal and has $*$-finite dimension Ω.

(10.1.7) DEFINITION *The locally convex finite points will always be measured from the zero vector*

$$\text{fin}_G(\mathbf{E}) = \{x \in \mathbf{E} : \gamma(x) \sim 0 \text{ for all } \gamma \in G\}.$$

The G-locally convex infinitesimal hull of $\mathbf{E}$

$$\hat{\mathbf{E}} = \hat{\mathbf{E}}_G = \text{fin}_G(\mathbf{E})/\overset{G}{\approx}$$

is the linear space of finite mod infinitesimal vectors with the seminorms $\hat{p}(\hat{x}) = \widehat{p(x)}$ *for* $p \in \text{SG}$, *the set of S-continuous internal pseudonorms.*

The infinitesimal description of standard continuity is still easy; the epsilon–delta formulation is somewhat awkward [compare (5.2.1) and (8.4.23)].

(10.1.8) PROPOSITION *Let* $p\colon \mathbf{E} \to {}^*[0, \infty]$ *be an internal pseudonorm. The following conditions are equivalent:*

(IC) *$x \overset{G}{\approx} 0$ implies $p(x) \approx 0$.*

(SC) *For each standard $\varepsilon \in {}^\sigma\mathbf{R}^+$ and each map $b\colon \text{FG} \to {}^\sigma\mathbf{R}^+$, there exists a standard $\delta \in {}^\sigma\mathbf{R}^+$, a standardly finite set $\gamma_1, \ldots, \gamma_m \in \text{IG}$, and a standardly finite set $\varphi_1, \ldots, \varphi_n \in \text{FG}$ such that if $\gamma_i(x) < \delta$, $1 \le i \le m$, and $\varphi_j(x) < b(\varphi_j)$, $1 \le j \le m$, then $p(x) < \varepsilon$.*

PROOF: Suppose $x \overset{G}{\approx} 0$ implies $p(x) \approx 0$ and both ε and b are given. Each $*$-finite extension of $(\text{IG}, \text{FG}, {}^\sigma\mathbf{R}^+, b)$, $(\mathscr{I}, \mathscr{F}, \mathscr{R}, \beta)$ that satisfies $\gamma(x) < \min(\mathscr{R})$ for $\gamma \in \mathscr{I}$ and $\varphi(x) < \beta(\varphi)$ for $\varphi \in \mathscr{F}$, forces $x \overset{G}{\approx} 0$ so $p(x) < \varepsilon$. By Luxemburg's $*$-finite family lemma, there is a properly finite family satisfying the condition.

Conversely, if $x \overset{G}{\not\approx} 0$ and $p(x) > \varepsilon \in {}^{\sigma}\mathbf{R}^{+}$, then for every standard $\delta \in {}^{\sigma}\mathbf{R}^{+}$, letting $b(\varphi) = \mathrm{st}(\varphi(x)) + 1$, we have $\gamma(x) < \delta$ for $\gamma \in \mathrm{IG}$, $\varphi(X) < b(\varphi)$ and yet $p(x) > \varepsilon$ so (SC) cannot hold either.

Some simple observations are as follows:

(10.1.9) ***OBSERVATION:*** $\mathrm{fin}(\mathbf{E})$ *is an* $\mathcal{O}$*-module and a* ${}^{\sigma}\mathbf{K}$*-vector space.*

(10.1.10) ***OBSERVATION:*** $\hat{\mathbf{E}}$ *is a* $\hat{\mathbf{K}}$*-vector space* ($\hat{\mathbf{K}} = \mathbf{R}$ or C, of course), moreover you can do the linear algebra with $\mathcal{O}$ in $\mathrm{fin}(\mathbf{E})$ and ignore infinitesimal differences. For example, if $\hat{x}_1, \ldots, \hat{x}_n$ are linearly independent on $\hat{\mathbf{E}}$ over $\hat{\mathbf{K}}$, then

(a) $x_1, \ldots, x_n$ are $*$-linearly independent over $*\mathbf{K}$ in $\mathbf{E}$,
(b) $\sum \lambda_j x_j$ is finite iff $\sum |\lambda_j|$ is finite, and
(c) $\lambda_j \approx \lambda_j' \sim 0$ implies $\sum \lambda_j x_j \approx \sum \lambda_j' x_j$, in particular,

$$\widehat{(\sum \lambda_j x_j)} = \sum \hat{\lambda}_j \hat{x}_j .$$

(10.1.11) ***OBSERVATION:*** *If* $\hat{T}$ *is a linear map* from $\hat{\mathbf{K}}^n$ into $\hat{\mathbf{E}}$, *there is an internal linear* $T\colon \mathbf{K}^n \to \mathbf{E}$ *such that the hull of* T *equals* $\hat{T}$. Let $\{e_1, \ldots, e_n\}$ be a standard basis for $\mathbf{K}^n$ and suppose $\hat{T}(\hat{e}_j) = \hat{x}_j$ are linearly independent. If x_j in $\mathbf{E}$ represents $\hat{x}_j$ in $\hat{\mathbf{E}}$, let $T(e_j) = x_j$ and extend by internal linearity. The hull of T is $\hat{T}$. If the dimension of the image of $\hat{T}$ is $m < n$, write $\hat{T} = \hat{S}\hat{R}$, where $\hat{R}\colon \hat{\mathbf{K}}^n \to \hat{\mathbf{K}}^m$ is onto and $\hat{S}\colon \hat{\mathbf{K}}^m \to \hat{\mathbf{E}}$ is injective. Apply the last construction to $\hat{S}$ and note that $\hat{R}$ is standard and the hull of itself.

(10.1.12) ***OBSERVATION:*** If $\mathbf{X}$ is a subspace of $\hat{\mathbf{E}}$ whose Hamel dimension is less than $\mathrm{card}(\mathscr{X})$, there is an internal $*$-finite dimensional space $\mathbf{F} \subseteq \mathbf{E}$ whose hull contains $\mathbf{X} \subseteq \hat{\mathbf{F}}$. Hulls of $*$-finite-dimensional spaces are called *hyperfinite dimensional* so *every standard size subspace of* $\hat{\mathbf{E}}$ *is contained in a hyperfinite-dimensional subspace.* Let $\{x_\alpha\}$ be a set of vectors in $\mathbf{E}$ such that $\{\hat{x}_\alpha\}$ is a Hamel basis for $\mathbf{X}$. Extend $\{x_\alpha\}$ to a $*$-finite set of vectors using the remark following (7.6.1). Let $\mathbf{F}$ be the internal linear span of that $*$-finite set. Since $\mathbf{F}$ is internal our hull constructions will apply to it, in particular, *hyperfinite-dimensional subspaces are closed.*

A simple case of what we have in mind is as follows. Let $\ell^2(*\mathbf{N}) = \mathbf{E}$, the Hilbert space of square summable internal sequences (that is, the internal series $\sum_{k=0}^{\infty} |a_k|^2$ is in $*\mathbf{R}$). The inner product is $\langle a, b\rangle = \sum_{k=0}^{\infty} a_k b_k$ and finite vectors are those for which $\sum_{k=0}^{\infty} |a_k|^2 \in \mathcal{O}$. A Hilbert basis for this space is given by $*\{e_n : n \in \mathbf{N}\}$ where $e_0 = (1, 0, 0, \ldots)$, $e_1 = (0, 1, 0, \ldots)$, $e_n = (\delta_k{}^n)$. The span of the vectors $e_0, \ldots, e_\Omega$ is a $*$-finite-dimensional space $\ell_\Omega{}^2$ and internally it can be treated just like any finite-dimensional space (for example, an operator is an $\Omega \times \Omega$ matrix which has a Jordan form). A standard vector a in ${}^{\sigma}\ell^2$ satisfies $\sum_{k=0}^{\infty} |a_k|^2 \approx \sum_{k=0}^{\Omega} |a_k|^2$ by the infinitesimal limit condition of (5.1.7) so that projecting $\sum_{k=0}^{\infty} a_k e_k$ onto $\sum_{k=0}^{\Omega} a_k e_k$,

$$(a_0, a_1, a_2, \ldots) \xmapsto{P_\Omega} (a_0, a_1, \ldots, a_\Omega, 0, 0, 0, \ldots),$$

only distorts the norm by an infinitesimal and ${}^{\sigma}\ell^2(\mathbf{N})$ is isometrically embedded in $\widehat{\ell_\Omega{}^2}$. (Of course, one can encounter difficulties in taking the hull because finiteness and infinitesimalness are external. The hull operation may not "commute" with internal linear algebra.)

(10.1.13) ***OBSERVATION:*** fin(**E**) *consists exactly of those vectors in* **E** *that are nearly annihilated by infinitesimal scalar multiplication.* That is, $x \in \mathbf{E}$ is finite iff for every $\varepsilon \approx 0$, $\varepsilon x \overset{G}{\approx} 0$. This is a fairly obvious but important observation. If $\gamma(x) \sim 0$ for $\gamma \in G$, then $\gamma(\varepsilon x) = |\varepsilon|\gamma(x) \approx 0$. Conversely, if $\gamma(x) \not\sim 0$ for some $\gamma \in G$, then either $\gamma(x) = \infty$ for $\gamma \in \mathrm{FG}$ and then $\gamma(\varepsilon x) = \infty$ for all $\varepsilon \neq 0$, or $\gamma(x) \neq \infty$, but is infinite in $*K$. Then $\varepsilon = (\gamma(x))^{-1/2}$ makes $\gamma(\varepsilon x) = (\gamma(x))^{1/2} \not\sim 0$.

(10.1.14) ***OBSERVATION:*** Now suppose *p is an S-continuous $*$-pseudo-norm* and x is finite. Then $|\varepsilon| p(x) = p(\varepsilon x) \approx 0$ and the scalar $p(x)$ is finite, in other words *$\hat{p}(\hat{x})$ is a seminorm on* $\hat{\mathbf{E}}$ $(\varepsilon \approx 0)$.

A useful lemma on the infinitesimal gauge is as follows:

(10.1.15) LEMMA *If $\gamma(x) \approx 0$ for all $\gamma \in$* IG, *then there is an infinite scalar Ω such that we still have $\gamma(\Omega x) \approx 0$ for $\gamma \in$* IG. This follows from polysaturation since the internal sets

$$A(n, \gamma) = \{m \in *\mathbf{N} : n < m \text{ and } \gamma(mx) < 1\}$$

for $n \in {}^{\sigma}\mathbf{N}$ and $\gamma \in \mathrm{IG}$ form a "small" family (note card(IG) $<$ card($\mathscr{X}$) by hypothesis) with the finite intersection property. An element of the intersection satisfies our assertion; see (7.6.1).

(10.1.16) ***OBSERVATION:*** The finite points in a locally convex space have intrinsic meaning unlike the case of an arbitrary uniform space. In fact, this is easy to see in the case where $\mathrm{FG} = \varnothing$, for then *one gauge* IG_1 *is finer than another* IG_2 *if and only if* $\mathrm{fin}_1(\mathbf{E}) \subseteq \mathrm{fin}_2(\mathbf{E})$ *if and only if* $x \overset{1}{\approx} y$ *implies* $x \overset{2}{\approx} y$. This is a simple consequence of the homogeneity (H). If $x \in \mathrm{fin}_1(\mathbf{E})$ implies $x \in \mathrm{fin}_2(\mathbf{E})$ and $y \overset{1}{\approx} 0$, then by (10.1.15) $\Omega y \overset{1}{\approx} 0$ for an infinite Ω in $\mathbf{K}$. Surely Ωy is in $\mathrm{fin}_1(\mathbf{E})$, so also in $\mathrm{fin}_2(\mathbf{E})$ and $(1/\Omega)(\Omega y) = y \overset{2}{\approx} 0$. Conversely, if $x \overset{1}{\approx} 0$ implies $x \overset{2}{\approx} 0$, we use (10.1.13) to show that $\mathrm{fin}_1(\mathbf{E}) \subseteq \mathrm{fin}_2(\mathbf{E})$. For ε, an infinitesimal scalar, $\varepsilon x \overset{1}{\approx} 0$ implies $\varepsilon x \overset{2}{\approx} 0$ and x is nearly 2-annihilated as in $\mathrm{fin}_2(\mathbf{E})$.

(10.1.17) *OBSERVATION:* The *absolutely convex envelope* of a set A is given by forming all the sums of the form

$$\sum_{j=1}^{n} \lambda_j a_j, \qquad \sum_{j=1}^{n} |\lambda_j| \leq 1,$$

with $a_j \in A$ for $1 \leq j \leq n$ and $\lambda_j \in \mathbf{K}$. The *convex envelope* requires $\lambda_j \geq 0$ and $\sum_{j=1}^{n} \lambda_j = 1$. The *hyperconvex envelope* of A consists of all the $*$-finite sums

$$\sum_{j=1}^{\Omega} \lambda_j a_j, \quad a_j \in A, \qquad \sum_{i=1}^{\Omega} \lambda_j = 1, \quad \lambda_j \geq 0,$$

that is, all the *internal* convex combinations from inside A. Notice that A need not be internal itself. The *absolutely hyperconvex envelope* of A consists of all the $*$-finite sums

$$\sum_{j=1}^{\Omega} \lambda_j a_j, \qquad \sum_{j=1}^{\Omega} |\lambda_j| \leq 1, \qquad a_j \in A.$$

(10.1.18) *OBSERVATION:* The *infinitesimals form an absolutely hyperconvex set*. If $\{x_j : 1 \leq j \leq \Omega\}$ is an internal set of G-infinitesimals,

$$\Gamma = \sup[\gamma(x_j) : 1 \leq j \leq \Omega] \quad \text{is infinitesimal}$$

and

$$\Phi = \sup\,[\varphi(x_j) : 1 \leq j \leq \Omega] \quad \text{is finite.}$$

Therefore we have,

$$\gamma\left(\sum_{j=1}^{\Omega} \lambda_j x_j\right) \leq \sum_{j=1}^{\Omega} |\lambda_j| \gamma(x_j) \leq \Gamma \cdot \sum_{j=1}^{\Omega} |\lambda_j| \approx 0 \qquad \text{and} \qquad \varphi\left(\sum_{j=1}^{\Omega} \lambda_j x_j\right) \leq \Phi.$$

(10.1.19) *OBSERVATION:* *The finite vectors also are absolutely hyperconvex.*

(10.1.20) HULL COMPLETENESS THEOREM

(a) *If B is an internal set for which* $\sup[\varphi(x) : x \in B]$ *is finite for each* φ *in* FG, *then the locally convex hull* $\hat{B} \subseteq \hat{\mathbf{E}}$ *is a complete subset of* $\hat{\mathbf{E}}$.

(b) *If* FG *contains only a finite set of $*$-pseudonorms, in particular, if* FG $= \varnothing$, *then* $\hat{\mathbf{E}}$ *is complete.*

PROOF: Part (a) follows from (8.4.24) once we observe that (10.1.8) with $b(\varphi) = \text{st}\{\sup[\varphi(x) : x \in B]\} + 1$ says IG generates the uniformity on $\hat{B}$.

Part (b) reduces to the case of a single $*$-pseudonorm by taking the maximum of the finite set FG. Each set of the form

$$\Phi_n = \{x \in \mathbf{E} : \varphi(x) < 2^n\}$$

is internal and satisfies the hypotheses of part (a) so $\hat{\Phi}_n$ forms a sequence of absolutely convex complete sets that satisfy:

(1) $\hat{\Phi}_n + \hat{\Phi}_n \subseteq \hat{\Phi}_{n+1}$,
(2) an absolutely convex subset V of $\hat{\mathbf{E}}$ is a neighborhood of zero in $\hat{\mathbf{E}}$ if and only if $V \cap \hat{\Phi}_n$ is a neighborhood of zero in $\hat{\Phi}_n$ for each n.

The Raĭkov–Köthe completeness theorem says $\hat{\mathbf{E}}$ is complete (see *Mathematical Reviews* **21**, 3747). Some improvement on part (b) is possible in various special settings.

The remainder of the section focuses on the case of a standard locally convex space $\mathbf{E} = {}^*\mathbf{F}$ with IG $= {}^\sigma N$, where N is the family of standard continuous seminorms on $\mathbf{F}$ and where the finite guage is empty, FG $= \varnothing$. Some references here are Robinson [1966], Luxemburg [1969], Young [1972], and especially Henson and Moore [1972]. In this setting we are specializing Section 8.4, where $\mathbf{E}$ is a uniform space with semimetrics $\gamma(x - y)$. We shall simply write $x \approx y$ for $x \overset{G}{\approx} y$ when only one linear gauge is involved.

(10.1.21) ***OBSERVATION:*** *The finite points* $\mathrm{fin}_G({}^*\mathbf{F})$ *of a standard space form a chromatic set*, since "$\gamma(x) < K$" is a standard sentence true for x where $K = \mathrm{st}(\gamma(x)) + 1$. If $y \in \mu_{\mathscr{D}}\{x\}$, then y must also satisfy the sentence.

(10.1.22) DEFINITION *The discrete union monad of* $\mathrm{fin}_N({}^*F)$ *is called the set of bounded points* $\mathrm{bd}_N({}^*\mathbf{F}) = \nu_{\mathscr{D}}(\mathrm{fin}_N({}^*\mathbf{F})) = \bigcup[{}^*B : {}^*B \subseteq \mathrm{fin}_N({}^*\mathbf{F})]$. This agrees with the conventional terminology for bounded subsets of linear spaces since if ${}^*B \subseteq \mathrm{fin}({}^*\mathbf{F})$, then the set $\{q(b) : B \in {}^*B\}$ is bounded by every infinite scalar and hence by some finite scalar. Conversely, if $\sup[q(b) : b \in B] = f(q)$ exists in $\mathbf{R}$ for each standard q, then ${}^*B$ contains only finite points.

(10.1.23) ***REMARK:*** *The bounded points form a* $\mathscr{O}$*-module* that contains all the standard points.

(10.1.24) PROPOSITION *The following are equivalent:*

(a) *The topology on* $\mathbf{F}$ *can be defined by a single seminorm.*
(b) $\mathrm{bd}_N({}^*\mathbf{F}) = \mathrm{fin}_N({}^*\mathbf{F})$, *the finite points are monadic.*
(c) *The infinitesimals are bounded* $\mathrm{inf}_N({}^*\mathbf{F}) \subseteq \mathrm{bd}_N({}^*\mathbf{F})$.

PROOF: That (a) implies (b) implies (c) is obvious. That (c) implies (a) comes from Cauchy's principle and the well-known fact that (a) is equivalent to there being a bounded neighborhood of zero. (The gauge of an absolutely convex subneighborhood is the seminorm.)

(10.1.25) PROPOSITION *The following are equivalent for a standard locally convex space* $(\mathbf{F}, N)$:

(S) *Every bounded set is relatively compact [respectively, precompact].*
(M) $\mathrm{bd}_N({}^*\mathbf{F}) \subseteq \mathrm{ns}_N({}^*\mathbf{F})$ [*respectively,* $\mathrm{pns}_N({}^*\mathbf{F})$].

PROOF: Apply (8.3.11) [respectively, (8.4.34)].

(10.1.26) DEFINITION *A duality between two linear spaces* $\mathbf{F}$ *and* $\mathbf{G}$ *is an internal bilinear form* $\langle , \rangle : \mathbf{F} \times \mathbf{G} \to \mathbf{K}$ *separating the points of each, that is, for each* $f \in \mathbf{F}$ *there is a* $g \in \mathbf{G}$ *so that* $\langle f, g \rangle \neq 0$ *and for each* $g \in \mathbf{G}$ *there is an* $f \in \mathbf{F}$ *so that* $\langle f, g \rangle \neq 0$.

One consequence of the Hahn–Banach theorem is that a locally convex space $\mathbf{F}$ is in duality with the space of continuous linear functionals on $\mathbf{F}$ which we denote by $\mathbf{F}'$. The prime depends on the topology of $\mathbf{F}$, so we could write $(\mathbf{F}, N)'$.

The spaces $L^1(0, 1)$ and $L^\infty(0, 1)$ mentioned above are in duality under the pairing $f \in L^1$, $g \in L^\infty \mapsto \int_0^1 f(x)g(x)\,dx$.

The space of sequences of finite support $\mathbf{F} = \{(a_1, \ldots, a_m, 0, 0, \ldots)\}$ is in duality with itself: $(a_1, \ldots, a_m, 0, 0, \ldots), (b_1, \ldots, b_n, 0, 0, \ldots) \mapsto \sum_{i=1}^{\infty} a_i b_i$.

We could also consider internal dualities, for example, K^ω, for $*$-finite ω or the space $\mathcal{D}$ of (10.1.5) in duality with ${}^*C^\infty$ by $\int_{-\infty}^{\infty} f(x)g(x)\,dx$.

The remainder of the section deals with a standard dual pair $\langle \mathbf{F}, \mathbf{G} \rangle$ and the nonstandard extension ${}^*\mathbf{F}$, ${}^*\mathbf{G}$, ${}^*\langle \cdot, \cdot \rangle = \langle \cdot, \cdot \rangle$.

(10.1.27) DEFINITION *Let* A *be a subset of* ${}^*\mathbf{F}$.

(i) *The infinitesimal polar of* A *is*

$$A^{\mathrm{i}} = \{g \in {}^*\mathbf{G} : \langle a, g \rangle \approx 0 \text{ for all } a \in A\}.$$

(a) *The absolute polar of* A *is*

$$A^{\mathrm{a}} = \{g \in {}^*\mathbf{G} : |\langle a, g \rangle| \leq 1 \text{ for all } a \in A\}.$$

(f) *The finite polar of* A *is*

$$A^{\mathrm{f}} = \{g \in {}^*\mathbf{G} : \langle a, g \rangle \in \mathcal{O} \text{ for all a} \in A\}.$$

Similar definitions apply when B is a subset of $\mathbf{G}$, for example,

$$B^{\mathrm{i}} = \{f \in {}^*\mathbf{F} : \langle f, b \rangle \approx 0 \text{ for all } b \in B\}.$$

(10.1.28) ***REMARK:*** Using this terminology we can say that the *S-continuous linear functionals* are the infinitesimal polar of the monad of zero $[\mu_N(0))]^{\mathrm{i}}$ in the pair $\langle {}^*\mathbf{F}, {}^*\mathbf{F}'\rangle$. They are also the finite polar of the finite points of $^*\mathbf{F}$, $[\mathrm{fin}({}^*\mathbf{F})]^{\mathrm{f}}$. It is therefore natural to consider $(\mathrm{fin}\ {}^*\mathbf{F})^{\mathrm{f}}$ mod$(\mathrm{fin}\ {}^*\mathbf{F})^{\mathrm{i}}$ as a "hull" of the space of functionals.

Since if $x \in \mu(0)$ there is an infinite λ so that λx is still infinitesimal, we see that $\mu(0)^{\mathrm{f}}$ is also the set of S-continuous linear functionals. By simple inclusion $\mu(0)^{\mathrm{i}} = \mu(0)^{\mathrm{a}} = \mu(0)^{\mathrm{f}} = \mathrm{fin}({}^*\mathbf{F})^{\mathrm{f}} = (\mathrm{SC}_N({}^*\mathbf{F}) \cap {}^*\mathbf{F}')$.

In a standard space without constraints the infinitesimals are uniform infinitesimals and a topological $\bigcap$-monad.

(10.1.29) TECHNICAL LEMMA (Basic properties of polars)

(a) $A^{\mathrm{i}} \subseteq A^{\mathrm{a}} \subseteq A^{\mathrm{f}}$.
(b) A^s *is hyper absolutely convex for* $s = \mathrm{i}, \mathrm{a}, \mathrm{f}$.
(c) $A \subseteq B$ *implies* $A^s \supseteq B^s$, $s = \mathrm{i}, \mathrm{a}, \mathrm{f}$.
(d) A^{i} *and* A^{f} *are* $\mathcal{O}$*-modules.*
(e) *If A is a $\mathcal{O}$-module, then* $A^{\mathrm{i}} = A^{\mathrm{a}}$.
(f) $A \subseteq A^{\mathrm{aa}} \subseteq A^{\mathrm{ii}} \subseteq A^{\mathrm{ff}} = A^{\mathrm{if}}$.
(g) $A^{sss} = A^s$, $s = \mathrm{i}, \mathrm{a}, \mathrm{f}$.

PROOF: (a–e) and (g) are left as exercises.

(f): $A \subseteq A^{\mathrm{aa}}$ is clear. $A^{\mathrm{i}} \subseteq A^{\mathrm{a}}$ so, by (c), $A^{\mathrm{aa}} \subseteq A^{\mathrm{ia}}$ and, by (d) and (e), $A^{\mathrm{aa}} \subseteq A^{\mathrm{ia}} \subseteq A^{\mathrm{ii}}$. Also, $A^{\mathrm{ii}} \subseteq A^{\mathrm{if}}$ and by two applications of (c), $A^{\mathrm{ff}} \subseteq A^{\mathrm{if}}$. If $x \notin A^{\mathrm{ff}}$, then for some $y \in A^{\mathrm{f}}$, $a = \langle x, y\rangle$ is infinite and so is $\sqrt{a}$. Now we have $(1/\sqrt{a})y \in A^{\mathrm{i}}$ and $\langle x, (1/\sqrt{a})y\rangle = 1/\sqrt{a}$, a infinite so $x \notin A^{\mathrm{if}}$ and $A^{\mathrm{if}} = A^{\mathrm{ff}}$.

We will consider locally convex structures on the spaces **F** and **G** of our dual pair. First, Σ-convergences, see (8.4.38–8.4.45).

(10.1.30) DEFINITION A *polar topology on* **G** *is described as follows. Let π be a union monad in* $^*\mathbf{F}$ *with the following properties:*

(a) *If $f \in \pi$, then $\langle f, g\rangle$ is finite for standard* $g \in {}^\sigma\mathbf{G}$.
(b) ${}^\sigma\pi = {}^\sigma\mathbf{F}$.

Infinitesimals associated with π in $^*\mathbf{G}$ *are given by*

$$g \overset{\pi}{\approx} h \quad \textit{if and only if} \quad \langle f, g\rangle \approx \langle f, h\rangle \quad \textit{for} \quad f \in \pi$$

or

$$g \overset{\pi}{\approx} 0 \quad \textit{if and only if} \quad g \in \pi^{\mathrm{i}}.$$

The π-finite points are given by $\mathrm{fin}_\pi({}^*\mathbf{G}) = \pi^{\mathrm{f}}$.

(10.1.31) ***EXAMPLE:*** Let $\pi = {}^\sigma\mathbf{F}$. This is called the *weak topology* $\sigma(\mathbf{G}, \mathbf{F})$ because it is the coarsest so that elements of ${}^\sigma\mathbf{F}$ are S-continuous as functionals on $^*\mathbf{G}$.

(10.1.32) ***EXAMPLE:*** Let $\pi = \bigcup[{}^*H : H$ is absolutely convex and $\sigma(\mathbf{F}, \mathbf{G})$-compact]. This is called the *Mackey topology* $\tau(\mathbf{G}, \mathbf{F})$ and is the finest so that continuous functionals are in $\mathbf{F}$.

(10.1.33) ***EXAMPLE:*** Let $\pi = \mathrm{bd}_{\sigma(\mathbf{F},\mathbf{G})}({}^*\mathbf{F})$, the discrete union monad of the points finite in the sense of property (a), that is, the weakly bounded points. This is called the *strong topology* $\beta(\mathbf{G}, \mathbf{F})$, and may not be compatible with the dual pair in the sense that strongly continuous functionals need not have representations in $\mathbf{F}$.

When $\mathbf{G} = (\mathbf{E}, N)$, for a countable family N and $\mathbf{F} = (\mathbf{E}, N)'$, then $\tau(\mathbf{E}, \mathbf{E}')$ equals N. A metrizable space carries the Mackey topology of its dual. If $(\mathbf{E}, N)$ is a complete metrizable space $\beta(\mathbf{E}, \mathbf{E}') = N$.

(10.1.34) ***EXAMPLE:*** Suppose $\mathbf{G}$ already has a convex structure described by a family of seminorms N. Let $\pi = [\mathrm{fin}_N({}^*\mathbf{G})]^{\mathrm{f}}$, which is a $\bigcup$-monad. Provided standard elements of ${}^\sigma\mathbf{F}$ are N-S-continuous so that (b) is satisfied, we have $g \overset{\pi}{\approx} h$ iff $g \overset{N}{\approx} h$. Standardly, uniform convergence on equicontinuous sets reproduces compatible topologies; see (10.1.36).

Various other polar topologies are interesting in special contexts.

(10.1.35) EXERCISES

(a) Show that polar topolgies are locally convex and linear by showing that the infinitesimals are hyperconvex and that addition and scalar multiplication are S-continuous.

(b) The point of condition (a) is that ${}^\sigma\mathbf{G}$ consists of finite points. What is the point of condition (b)?

(c) What is $\mathrm{fin}({}^*\mathbf{F})^{qr}$ equal to for $qr = $ ii, ia, if, ff when $\mathbf{G} = \mathbf{F}'$?

(10.1.36) **THEOREM** *Let* $(\mathbf{F}, N)$ *be a locally convex space with continuous dual* $\mathbf{F}'$. *Then the S-continuous linear functionals* $\mu_N(0)^{\mathrm{i}} = \mu_N(0)^{\mathrm{a}} = \mu_N(0)^{\mathrm{f}} = \mathrm{fin}_N({}^*\mathbf{F})^{\mathrm{f}}$ *form the equicontinuous union monad, and the polar topology of all S-continuous linear functionals is the same as that induced by the original gauge* $\mu_N(0) = \mu_N(0)^{\mathrm{ii}}$.

PROOF: First, we show that the S-continuous linear functionals form a monad. Let $\mathcal{U}$ be a base for neighborhoods of zero in $\mathbf{F}$, so $\{U^{\mathrm{a}} : U \in \mathcal{U}\}$ forms a base for equicontinuous sets in $\mathbf{F}'$. Suppose $g \in {}^*\mathbf{F}' \backslash \bigcup[{}^*U^{\mathrm{a}} : U \in \mathcal{U}]$. For each $U \in \mathcal{U}$, $A(U) = \{f : f \in {}^*U$ and $|\langle f, g\rangle| \geq 1\}$ is then nonempty and the collection has the finite intersection property, so there is a p in all

$A(U)$'s. Now there is also an infinite integer ω so that ωp is still infinitesimal. Thus g is not S-continuous, since $|\langle \omega p, g\rangle| \geq \omega$.

Second, we show that $\mu(0) = \mu(0)^{\mathrm{ii}}$. Suppose $f \not\approx 0$ so $q(f) \not\approx 0$ for some standard seminorm q. The functional $\varphi(\lambda f) = \lambda q(f)$ satisfies

$$|\varphi(\lambda f)| \leq q(\lambda f)$$

so by the internal Hahn–Banach theorem φ has an extension subordinate to q on all of $\mathbf{F}$. Thus, φ is an S-continuous functional and $\varphi(f) \not\approx 0$ so $f \notin \mu(0)^{\mathrm{ii}}$.

The reader should notice that compatible topologies have the same closed absolutely convex sets; in fact, *the closed absolutely convex envelope of* $B \subseteq \mathbf{E}$ *is given by* B^{aa}. This has an easy infinitesimal proof which we leave as an exercise; it is simply the Hahn–Banach theorem and the fact that compatible continuities are the same for linear maps. The reader may even wish to give a proof of the Hahn–Banach theorem based on finite extendability and the associated concurrence.

(10.1.37) THEOREM *Let* $\langle \mathbf{F}, \mathbf{G}\rangle$ *be a dual pair. Then* $\mathrm{pns}_{\sigma(\mathbf{F}, \mathbf{G})}({}^*\mathbf{F}) = \mathrm{fin}_{\sigma(\mathbf{F}, \mathbf{G})}({}^*\mathbf{F})$, *in particular, a* $\sigma(\mathbf{F}, \mathbf{G})$*-bounded set is totally bounded. The weak hull of* $\mathbf{F}$ *is the algebraic dual* $\mathbf{G}^{\#}$ *of all linear functionals on* $\mathbf{G}$,

$$\hat{\mathbf{F}}_{\sigma(\mathbf{F}, \mathbf{G})} \cong (\mathbf{G}^{\#}, \sigma(\mathbf{G}^{\#}, \mathbf{G})).$$

PROOF: For the first statement we must show that each element of $({}^{\sigma}\mathbf{G})^{\mathrm{f}}$ is contained in an infinitesimal intersection monad. We need the following lemma:

If linearly independent $b_1, \ldots, b_n$ *in* $\mathbf{G}$ *and* $\lambda_1, \ldots, \lambda_n$ *in* $\mathbf{K}$ *are specified, there is an* $f \in \mathbf{F}$ *so that*

$$\langle f, b_j\rangle = \lambda_j.$$

This lemma is simple linear algebra and the fact that dual pairs separate points.

Let a be a $\sigma(\mathbf{F}, \mathbf{G})$-finite point and consider a basic neighborhood of zero given by $b_1, \ldots, b_n$ in ${}^{\sigma}\mathbf{G}$ and ε in ${}^{\sigma}\mathbf{R}^+$. The neighborhood is given by $\{x : |\langle x, b_j\rangle| < \varepsilon,\ 1 \leq j \leq n\}$ and we conclude the proof by using (8.4.29). Apply the lemma above to $\lambda_j = \mathrm{st}\langle a, b_j\rangle$ obtaining $f \in {}^{\sigma}\mathbf{F}$ with $\langle a, b_j\rangle \approx \langle f, b_j\rangle$, $1 \leq j \leq m$. We see that $a \in \{x \in {}^*\mathbf{F} : \langle x - f, b_j\rangle < \varepsilon, 1 \leq j \leq m\}$.

The second part follows from the fact that we can actually represent linear functionals, that is, for $\varphi \in \mathbf{G}^{\#}$ we have $f \in {}^*\mathbf{F}$ with $\varphi(g) = \langle f, g\rangle$, for $g \in {}^{\sigma}\mathbf{G}$. The representability comes from the $*$-transform of the above algebraic lemma by embedding a standard algebraic basis in a $*$-finite set and using φ to define the scalars.

Weakly bounded sets are totally bounded since now the bounded points, defined as the discrete monad of the finite points, and the totally bounded points, defined as the discrete monad of the pre-near-standard points coincide.

(10.1.38) ALAOGLU'S THEOREM *Let* $\mathbf{G} = \mathbf{F}'$. *Suppose* U *is an* N-*neighborhood of zero in* $\mathbf{F}$. *Then* U^{a} *is* $\sigma(\mathbf{F}', \mathbf{F})$ *compact.*

PROOF: It suffices to show each $g \in {}^*U^{\mathrm{a}}$ is $\sigma(\mathbf{F}', \mathbf{F})$-near-standard. Define a standard h by $\langle f, h\rangle = \mathrm{st}\langle f, g\rangle$ for $f \in {}^{\sigma}\mathbf{F}$. We know h is in $\mathbf{F}'$ because it has a value less than one in ${}^{\sigma}U$ and thus U. We know U^{a} is closed, hence it is compact.

(10.1.39) MACKEY–ARENS' THEOREM *Let* $\langle \mathbf{F}, \mathbf{G}\rangle$ *be a dual pair. Any locally convex Hausdorff topology on* $\mathbf{F}$ *which is compatible with the duality, that is, so* $(\mathbf{F}, N)' = \mathbf{G}$, *is a polar topology generated by a monad of absolutely convex* $\sigma(\mathbf{G}, \mathbf{F})$-*compact subsets of* $\mathbf{G}$. *In particular, the Mackey topology described in* (10.1.32) *is the finest compatible one.*

In particular, this says the S-continuous functionals of a compatible topology are contained in the weakly compact points $\mathrm{cpt}_{\sigma(\mathbf{G}, \mathbf{F})}({}^*\mathbf{G}) \subseteq \mathrm{bd}_{\sigma(\mathbf{G}, \mathbf{F})}({}^*\mathbf{G})$ so the Mackey or any compatible topology is coarser than the strong. [Sometimes strictly, for example, $\mathbf{F} = L^1$, $\mathbf{G} = L^{\infty}$, and the strong on $\mathbf{L}^{\infty}$ equals the norm topology which is not compatible.]

PROOF: First, $\mu(0)^{\mathrm{i}} = \mu(0)^{\mathrm{a}}$, since it is a $\mathcal{O}$-module. By the last result $v(U^{\mathrm{a}} : U$ a neighborhood of zero$)$ is contained in the weakly compact points. Finally, by (10.1.36), $\mu(0) = \mu(0)^{\mathrm{ii}}$.

Conversely, suppose φ is a standard Mackey continuous linear functional on $\mathbf{F}$. We must show that φ can be represented in $\mathbf{G}$. The continuity condition means that there is an absolutely convex weakly compact set $C \subseteq \mathbf{G}$ so that $|\varphi(f)| \leq 1$ on C^{a}. We claim that this inequality means there is an $h \in {}^*C$ so that $\varphi(f) \approx \langle f, h\rangle$ for $f \in {}^{\sigma}\mathbf{F}$. By weak near standardness of all of ${}^*C$, there is a $g \in {}^{\sigma}C$ so that $\langle f, h\rangle \approx \langle f, g\rangle$ for $f \in {}^{\sigma}\mathbf{F}$. Since φ and g are standard, $\varphi(f) = \langle f, g\rangle$ on $\mathbf{F}$.

We conclude by justifying the claim. First, φ can be weakly represented in ${}^*\mathbf{G}$ from (10.1.37), but we need to show that it is nearly represented in ${}^*C$. An $h \in {}^*C$ exists with $\varphi(f) \approx \langle f, h\rangle$ for $f \in {}^{\sigma}\mathbf{F}$, once we show that the following relation is concurrent:

$$\{((f, \varepsilon), g) : f \in \mathbf{F},\ \varepsilon \in \mathbf{R}^+,\ g \in C \text{ and } |\varphi(f) - \langle f, g\rangle| \leq \varepsilon\}.$$

If this is not concurrent, there are $f_1, \ldots, f_n$ and ε all standard for which $|\varphi(f_j) - \langle f_j, g\rangle| > \varepsilon$ for all $g \in C$.

By the Hahn–Banach theorem, in $\mathbf{F}^{\#}$ we can separate the $\sigma(\mathbf{F}^{\#}, \mathbf{F})$-closed absolutely convex set C from φ by a $\sigma(\mathbf{F}^{\#}, \mathbf{F})$-continuous functional ψ. Thus $\psi(\varphi) > 1$ while $|\psi(C)| \leq 1$. Since ψ is continuous it is in $\mathbf{F}$ (and not just $\mathbf{F}^{\#\#}$) contradicting the inequality since $\psi \in \mathbf{F}$ implies $\psi \in C^{a}$. This proves Mackey–Arens' theorem.

Let $\langle \mathbf{F}, \mathbf{G}\rangle$ be a dual pair and π a union monad which induces a polar topology on $^*\mathbf{G}$. By condition (a) in the definition of polar topology $\pi \subseteq \text{fin}_{\sigma(\mathbf{F}, \mathbf{G})}(^*\mathbf{F})$ and since it is a monad $\pi \subseteq \text{bd}_{\sigma(\mathbf{F}, \mathbf{G})}(^*\mathbf{F})$. Besides the infinitesimal relation which π induces on $^*\mathbf{G}$ through $g \underset{^*\mathbf{G}}{\overset{\pi}{\approx}} h$ if and only if $(g - h) \in \pi^{i}$, π induces a bounded infinitesimal relation on $^*\mathbf{F}$ through $f \underset{^*\mathbf{F}}{\overset{\pi}{\approx}} k$ if and only if $(f - k) \in \pi$ and $\langle f, g\rangle \approx \langle k, g\rangle$ for each $g \in {}^{\sigma}\mathbf{G}$. S-continuity with respect to these bounded infinitesimals in $^*\mathbf{F}$ precisely characterizes the points of the completion of $\mathbf{G}$ in the polar topology. By (10.1.36) we may assume that π consists of absolutely convex bounds so that definition (10.1.2) applies to $\overset{\pi}{\approx}$ on $^*\mathbf{F}$. The S-continuity with respect to the π-bounded infinitesimals on $^*\mathbf{F}$ will be referred to here as $\overset{\pi}{\approx}$-continuous: $f \in [\pi \cap \mu_{\sigma(\mathbf{F}, \mathbf{G})}(0)]$ implies $\psi(f) \approx 0$. (The other π-continuity is on $^*\mathbf{G}$.)

(10.1.40) GROTHENDIECK'S COMPLETENESS THEOREM *Let $\langle \mathbf{F}, \mathbf{G}\rangle$ and π be as in the preceding paragraph. The π-completion of $\mathbf{G}$ is the hull of*

$$\text{pns}_{\pi}(^*\mathbf{G}) = [\pi \cap {}^{\sigma}\mathbf{G}^{i}]^{i} = [\pi \cap \mu_{\sigma(\mathbf{F}, \mathbf{G})}(0)]^{i}.$$

Moreover, every $\overset{\pi}{\approx}$-continuous internal linear function on $^*\mathbf{F}$, $\psi: {}^*\mathbf{F} \to \mathbf{K}$, is π-represented by a $g \in \text{pns}_{\pi}(^*\mathbf{G})$,

$$\psi(f) \approx \langle f, g\rangle \qquad \text{for all} \quad f \in \pi,$$

so $(\mathbf{G}, \pi)$ is complete if and only if every standard functional weakly continuous on π-equicontinuous sets in $\mathbf{F}$ is already represented in $\mathbf{G}$.

PROOF: First we show that $\text{pns}_{\pi}(^*\mathbf{G}) \subseteq [\pi \cap {}^{\sigma}\mathbf{G}^{i}]^{i}$. If $h \in \text{pns}$, then for each standard ε and $^*P \subseteq \pi$, there exists a $g \in {}^{\sigma}\mathbf{G}$ such that $|\langle h - g, p\rangle| < \varepsilon$ for $p \in {}^*P$. If p is in $^*P \cap {}^{\sigma}\mathbf{G}^{i}$, then $\langle g, p\rangle \approx 0$ and

$$|\langle h, p\rangle| \leq |\langle h - g, p\rangle| + |\langle g, p\rangle| \lessapprox \varepsilon.$$

But ε was arbitrary, so $\langle h, p\rangle \approx 0$.

Second, suppose ψ is an internal $\overset{\pi}{\approx}$-continuous linear functional on $^*\mathbf{F}$, in particular, we could have $\psi(f) = \langle f, h\rangle$ for $h \in [\pi \cap {}^{\sigma}\mathbf{G}^{i}]^{i}$. We will show that given $P = P^{aa}$ with $^*P \subseteq \pi$ and $\varepsilon \in {}^{\sigma}\mathbf{R}^{+}$, there is a standard $g \in {}^{\sigma}\mathbf{G}$ so that $|\psi(p) - \langle p, g\rangle| < \varepsilon$ for $p \in {}^*P$. In case $\psi = \langle \cdot, h\rangle$, this shows that h is pre-near-standard. Otherwise we apply saturation to obtain an $h \in {}^*\mathbf{G}$ such that

$$\psi(p) \approx \langle p, h \rangle \qquad \text{for} \quad p \in \pi.$$

In the latter case the same argument applied to $\langle \cdot, h \rangle$ shows that it is π-pre-near-standard in ${}^*\mathbf{G}$. There is no loss of generality in taking $P = P^{aa}$.

The π-continuity of ψ means that whenever V is an internal subset of $\mu_{\sigma(\mathbf{F}, \mathbf{G})}(0)$ we have

$$|\psi(p)| \leq \varepsilon/2 \qquad \text{for} \quad p \in {}^*P \cap V.$$

This must therefore hold on a standard $\sigma(\mathbf{F}, \mathbf{G})$-neighborhood by Cauchy's principle, that is, there exist linearly independent standard vectors $g_1, \ldots, g_n \in {}^\sigma\mathbf{G}$ such that $|\langle p, g_j \rangle| \leq 1$, $1 \leq j \leq n$, and $p \in {}^*P$ implies $|\psi(p)| \leq \varepsilon/2$. Moreover, we may as well assume that for each j, $\langle p, g_j \rangle \neq 0$ for some $p \leq P$, otherwise $p \in {}^*P$ implies $|\langle p, g_j \rangle| \leq 1$ and the weak constraint is unnecessary to conclude that $|\psi(p)| \leq \varepsilon/2$.

Since P is a closed absolutely convex set it defines a seminorm q on the space $\mathbf{P} = \{\lambda {}^*P : \lambda > 0\}$ given by

$$q(z) = \inf(\lambda : z \in \lambda P).$$

We know that if $|\langle p, g_j \rangle| \leq 1$ for $1 \leq j \leq n$ and $q(p) \leq 1$, that $|\psi(p)| < \varepsilon/2$ so on the anihilator space $\mathbf{A} = \{p \in \mathbf{P} : \langle p, g_j \rangle = 0 \text{ for } 1 \leq j \leq n\}$ we have $|\psi(p)| \leq \frac{1}{2}\varepsilon q(p)$. By the $*$-Hahn–Banach theorem there is an extension $\varphi : \mathbf{P} \to \mathbf{K}$ of $\psi | \mathbf{A}$ satisfying $|\varphi(p)| \leq \frac{1}{2}\varepsilon q(p)$ on $\mathbf{P}$ and $\varphi(a) = \psi(a)$ on $\mathbf{A}$. The linear function $(\psi - \varphi)$ on $\mathbf{P}$ is zero on $\mathbf{A}$, but whenever $\langle x, g_j \rangle = 0$, $1 \leq j \leq n$, implies $(\psi - \varphi)(x) = 0$, $\psi - \varphi = \sum_{j=1}^n \lambda_j g_j$ on $\mathbf{P}$ by a $*$-theorem of linear algebra. The g_j's are linearly independent standard $\mathbf{G}$-vectors and $|\psi(p) - \langle p, \sum_{j=1}^n \lambda_j g_j \rangle| = |\varphi(p)| \leq \frac{1}{2}\varepsilon q(p)$. We claim that the λ_j's must be finite scalars and once we show this,

$$\left| \psi(p) - \left\langle p, \sum_{j=1}^{n} \kappa_j g_j \right\rangle \right| \leq \varepsilon \qquad \text{for} \quad p \in {}^*P,$$

letting $\kappa_j = \mathrm{st}(\lambda_j)$. The κ_j's and g_j's are standard, so that the assertion we secondly set out to prove is verified taking $g = \sum_{j=1}^n \kappa_j g_j$.

By the way we have selected the g_j's we may pick a dual set $p_i \in P$ so that $\langle p_i, g_j \rangle = \delta_{ij} \cdot a$ where the a is standard, nonzero, and less than one. Also, in that case $|\psi(p_i)| \leq \varepsilon/2$ by the choice of the g_j's and $|\psi(p_i) - \langle p_i, \sum_{j=1}^n \lambda_j g_j \rangle| = |\varphi(p_i)| \leq \varepsilon/2$ forcing $|\lambda_j| \leq \varepsilon/a$, in particular finite. This proves every internal $\overset{\pi}{\approx}$-continuous functional on ${}^*\mathbf{F}$ is π-represented by a π-pns point in ${}^*\mathbf{G}$.

The standard formulation at the end of the theorem follows easily from the fact that a space is complete if and only if each pre-near-standard point is near standard. We leave it as a translation exercise.

(10.1.41a) APPLICATION *If K is a π-totally bounded subset of* **G**, *then the closed absolutely convex envelope $C = K^{aa}$ is also totally bounded.*

PROOF: By (8.4.34), $K \subseteq [\pi \cap {}^\sigma\mathbf{G}^i]^i$ and (10.1.36) we may assume π is closed under standard scalar multiplication so $[\pi \cap {}^\sigma\mathbf{G}^i]^i = [\pi \cap {}^\sigma\mathbf{G}^i]^a$. Consequently $K^{aa} \subseteq [\pi \cap {}^\sigma\mathbf{G}^i]^{aaa} = [\pi \cap {}^\sigma\mathbf{G}^i]^a = [\pi \cap {}^\sigma\mathbf{G}^i]^i$ and thus C is totally bounded.

This proof, in fact the whole theorem, is quite abstract. In a Banach space it simplifies considerably since the bounds can be given by one dual norm constraint.

Let $(\mathbf{E}, \|\cdot\|)$ be a normed space and $(\mathbf{E}', |||\cdot|||)$ the dual normed space with dual norm $|||x'||| = \sup(|\langle x, x'\rangle| : \|x\| \leq 1)$. In this case the bounded infinitesimals in $\mathbf{E}'$ associated with the norm of **E** are "$x' \overset{\mathscr{B}}{\approx} y'$ if and only if $|||x' - y'|||$ is finite and $\langle x, x'\rangle \approx \langle x, y'\rangle$ for each $x \in {}^\sigma\mathbf{E}$." By homogeneity of $|||\cdot|||$ weak continuity on the unit ball $\{|||x'||| \leq 1\}$ suffices for linear maps. The points in the completion of **E** are those $x \in {}^*\mathbf{E}$ which are $\sigma(\mathbf{E}', \mathbf{E})$-continuous on the unit ball of $\mathbf{E}'$.

(10.1.41b) MAZUR'S THEOREM *If* **E** *is a Banach space, the closed convex envelope of a norm compact set K is also compact.*

PROOF: Let C be the convex envelope of K. Each $c \in {}^*C$ has the form $c = \sum_{j=1}^{\Omega} c_j k_j$, $k_j \in {}^*K$, $0 \leq c_j$, and $\sum_{j=1}^{\Omega} c_j = 1$. Since ${}^*K$ is compact each k_j is near standard, that is, there is an $h_j \in {}^\sigma K$ such that $\|k_j - h_j\| \approx 0$. We take an arbitrary $z \in {}^*\mathbf{E}'$ with $|||z||| \leq 1$. Now $\varepsilon_j = |\langle k_j - h_j, z\rangle| = \|k_j - h_j\| \cdot |||z||| \approx 0$, so if z is a weak infinitesimal as well, in particular, $\langle h_j, z\rangle \approx 0$ for $h_j \in {}^\sigma K$, then $\langle \sum_{j=1}^{\Omega} c_j k_j, z\rangle \approx 0$ and c is norm pre-near standard by (10.1.40) and the remark preceding our theorem. Since **E** is complete, c is near standard (8.4.28). Since ${}^*C$ contains all near-standard points its standard part is compact (8.3.11).

(10.1.42) EXERCISES

(a) Let $(\mathbf{E}, N)$ be a complete locally convex space and suppose M is a finer gauge on **E** with $(\mathbf{E}, N)' = (\mathbf{E}, M)'$. Show that $(\mathbf{E}, M)$ is complete.

(b) Let **G** be a dense subspace of $(\mathbf{E}, N)$ with the induced gauge. Show that $\text{pns}({}^*\mathbf{G}) = \text{pns}({}^*\mathbf{E}) \cap {}^*\mathbf{G}$, $\text{fin}({}^*\mathbf{G}) = \text{fin}({}^*\mathbf{E}) \cap {}^*\mathbf{G}$, and $\hat{\mathbf{G}} = \hat{\mathbf{E}}$. Every standard hull is the hull of a complete standard space.

(10.1.43) KREIN'S THEOREM *Let* $(\mathbf{E}, N)$ *be a complete locally convex space and let* $\mathbf{E}' = (\mathbf{E}, N)'$ *be the continuous dual. If* K *is a* $\sigma(\mathbf{E}, \mathbf{E}')$*-compact subset of* $\mathbf{E}$, *then the* $\sigma(\mathbf{E}, \mathbf{E}')$*-closed convex envelope* C *of* K *is also* $\sigma(\mathbf{E}, \mathbf{E}')$*-compact.*

PROOF: Every point of $^*K$ is weakly near standard, if $k \in {}^*K$, there is an $h \in {}^\sigma K$ such that $\langle h, y\rangle \approx \langle k, y\rangle$ for every $y \in {}^\sigma\mathbf{E}'$ by (8.3.7). By Grothendeick's theorem if C is not compact, there is a $p \in {}^*C$ so that the standard functional $\langle w, y\rangle = \text{st}\langle p, y\rangle$ for $y \in {}^\sigma\mathbf{E}'$ is not $\sigma(\mathbf{E}', \mathbf{E})$-continuous on U^a for some neighborhood U in $\mathbf{E}$, that is, for some $\varepsilon \in {}^\sigma\mathbf{R}^+$, if $\delta \in {}^\sigma\mathbf{R}^+$ and F is a finite subset of $\mathbf{E}$, there is a $y \in U^a$ such that $|\langle f, y\rangle| < \delta$ for $f \in F$ and $|\langle w, y\rangle| > \varepsilon$. This will lead to a contradiction.

Now we define a sequence $\{y_n\}$ of elements of U^a and a sequence $\{D_n\}$ of countable subsets of K. Each D_n will be enumerated (repetition allowed) $d(n, 1), d(n, 2), \ldots$. First, choose $y_1 \in U^a$ so that $|\langle w, y_1\rangle| > \varepsilon$. Choose $D_1 \subseteq K$ so that for all $x \in K$ and $\delta \in {}^\sigma\mathbf{R}^+$, there is a $d \in D_1$ such that $|\langle x - d, y_1\rangle| < \delta$. Then D_1 can be built from finite sets for $\delta = 1, \frac{1}{2}, \frac{1}{3}, \ldots$, since K is compact. Proceed defining y's and D's by induction: $y_{n+1} \in U^a$ satisfies $|\langle w, y_{n+1}\rangle| > \varepsilon$, but for each i, $j \geq 1$ with $i + j \leq n + 1$, $|\langle d(i, j), y_{n+1}\rangle| \leq 2^{-n}$. Choose a subset D_{n+1} of K such that for each $x \in K$ and $\delta \in {}^\sigma\mathbf{R}^+$, there is a d in D_{n+1} with $|\langle x - d, y_j\rangle| < \delta$ for $1 \leq j \leq n + 1$. The sequences $\{y_n\}$ and $\{D_n\}$ satisfy: (1) $|\langle w, y_n\rangle| > \varepsilon$, (2) if $d \in \bigcup D_n = D$, $\lim|\langle d, y_n\rangle| = 0$, (3) if $x \in K$, $n \in \mathbf{N}$, and $\delta \in {}^\sigma\mathbf{R}^+$, then for some $d \in D_n$, $|\langle x - d, y_j\rangle| < \delta$ for $1 \leq j \leq n$.

Next, we show that $\lim\langle x, y_n\rangle = 0$ for every x in K. If $\lim\langle k, y_n\rangle \neq 0$, then $\langle k, y_n\rangle \not\approx 0$ for some infinite $\Omega \in {}^*\mathbf{N}_\infty$. However, y_Ω is $\sigma(\mathbf{E}', \mathbf{E})$-near-standard to $y_0 \in {}^\sigma\mathbf{E}'$ by Alaoglu's theorem (10.1.38) since $y_\Omega \in {}^*U^a$. By condition (2) on D above we know $\langle {}^*d, y_\Omega\rangle \approx 0$ for $d \in D$, therefore $\langle {}^*d, y_0\rangle = 0$ for all $d \in {}^\sigma D$. By transfer of condition (3) on our sequences, there is a $d \in {}^*D$ such that $\langle {}^*x - d, y_j\rangle \approx 0$ for $1 \leq j \leq \Omega$. Let d_0 be the $\sigma(\mathbf{E}, \mathbf{E}')$-standard part of this $d \in {}^*K$. Then $\langle x - d_0, y_j\rangle = 0$ for all finite j and thus for all j. Consequently $\langle x, y_\Omega\rangle = \langle d_0, y_\Omega\rangle \approx \langle d_0, y_0\rangle \approx \langle d, y_0\rangle \approx 0$, contradicting the statement that $\lim\langle k, y_n\rangle \neq 0$. Hence $\lim\langle k, y_n\rangle = 0$ for all $k \in K$.

Since all the y_n's are in U^a they are uniformly bounded on K and the following lemma tells us that $\lim[\text{st}\langle p, y_n\rangle] = 0$, that is, $|\langle w, y_n\rangle| < \varepsilon/2$ contrary to the statement that w is not $\sigma(\mathbf{E}', \mathbf{E})$-continuous on U^a. This proves the theorem modulo:

(10.1.44) LEMMA $(\mathbf{E}, K$, and C as above) *If* $\{f_n\}$ *is a sequence of uniformly bounded continuous linear functionals and* $\lim f_n(x) = 0$ *for each* $x \in K$, *then* $\lim_n(\text{st}[f_n(z)]) = 0$ *for each* $z \in {}^*C$.

Scaling if necessary, we assume $|f_n(K)| \leq 1$. By compactness, each $k \in {}^*K$ satisfies $\lim_n(\mathrm{st}[f_n(k)]) = 0$, since $k \approx h \in {}^\sigma K$, $f_n(h)$ tends to zero, and f_n is continuous.

Let $p \in {}^*C$ be given by $p = \sum_{j=1}^{\omega} \lambda_j q_j$, $q_j \in K$, $0 \leq \lambda_j$, $\sum_{j=1}^{\omega} \lambda_j = 1$. We will show that $|f_n(p)| < \delta$ for sufficiently large n and an arbitrary standard $\delta \in {}^\sigma\mathbf{R}^+$. Let this δ be fixed and define

$$A_n = \{k : 1 \leq k \leq \omega \text{ and } |f_n(q_k)| < \delta/2\}.$$

We know that for each $k \in I = \{j : 1 \leq j \leq \omega\}$ there is a finite m such that $k \in A_n$ for all $n > m$.

We define an internal finitely additive probability measure on the internal subsets of I by

$$\mu(A) = \sum_{k \in A} \lambda_k .$$

Now we claim $\lim_n[\mathrm{st}(\mu(A_n))] = 1$. The claim proves the lemma because once $\mu(A_n) > 1 - \delta/2$,

$$\begin{aligned} |f_n(p)| &\leq \sum_{k \in A_n} \lambda_k |f_n(q_k)| + \sum_{k \notin A_n} \lambda_k |f_n(q_k)| \\ &\leq \sum_{k \in A_n} \lambda_k\left(\frac{\delta}{2}\right) + \sum_{k \notin A_n} \lambda_k < \frac{\delta}{2} + \frac{\delta}{2} = \delta. \end{aligned}$$

The claim can be justified by observing that if $\lim_n[\mathrm{st}(\mu(A_n))] \neq 1$, there is a subsequence B_m such that $\mu(B_m) \leq 1 - \delta$ for all m. From this subsequence we can further extract a subsubsequence H_i such that $\bigcup[H_i : 1 \leq i \leq n] \neq I$ for each finite n, so the union of the $\bigcup[H_i : i \in {}^\sigma\mathrm{N}] \neq I$ either, by saturation applied to the finite union property. This is a contradiction because no infinite family of A_n's omits a $k \in I$ as we remarked after the definition of A_n.

(10.1.45) ***APOLOGY:*** There is a good deal more to be said about this proof of Henson and Moore for the last lemma. What underlies it is the fact that an internal $*$-finitely additive probability measure on an internal algebra of sets in an $\aleph_1$-saturated model induces a unique external countably additive measure on the external sigma algebra generated by the internal algebra. This book regrettably lacks a chapter on "$*$-finite probability" (Loeb [1974b].)

(10.1.46) ***REMARK:*** *Differential calculus in Banach spaces* is no more difficult than for functions of finitely many real variables, in fact, *all the theorems and proofs of Section 5.7 hold when* **E** *and* **F** *are Banach spaces* provided that we add that $Df_a(\cdot)$ is a continuous linear map in the definitions, i.e., let $S\mathrm{Lin}^n(\mathbf{E}, \mathbf{F})$ denote the symmetric *continuous* n-linear maps. (Theorem (5.7.1) only concerns matrices of course and (5.7.8) holds in a Hilbert space.) The feature of the norm which makes this true is that the spaces of continuous n-linear maps inherit a natural compatible norm structure.

Two clear options present themselves for a general definition. For each scalar infinitesimal

$$f(a + \varepsilon h) - f(a) = \varepsilon \cdot Df_a(h) + \varepsilon \cdot \eta \qquad \text{for all} \quad h \in bd(^*\mathbf{E}),$$

or

$$f(a + \varepsilon k) - f(a) = \varepsilon \cdot Df_a(k) + \varepsilon \cdot \eta \qquad \text{for all} \quad k \in \text{fin}(^*\mathbf{E}),$$

where η is infinitesimal in $^*\mathbf{F}$. One is still faced with the question of what continuous differentiability should mean, for the space of continuous linear maps Lin(**E**, **F**) *cannot* be given a compatible locally convex topology which even makes evaluation continuous! Evaluation is the map $\mathbf{E} \times \text{Lin}(\mathbf{E}, \mathbf{F}) \to \mathbf{F}$ given by $(e, L(\cdot)) \mapsto L(e)$. This is the source of a plethora of different definitions. Indeed there may not even be a single best answer with completely general **E** and **F**. We propose to examine the problem by asking what "infinitesimally close" should mean for n-linear maps $Df_a^{\,n}(\cdot)$ and $Df_b^{\,n}(\cdot)$ when $a \approx b$. Our answer cannot be topological, but that should not matter as long as it is natural and compatible with the linear operations. We want it to provide a smooth useful calculus most of all. This will be taken up elsewhere.

10.2 HILBERT SPACES

Let **H** be a complex Hilbert space in our superstructure $\mathscr{X}$ with inner product $\langle\ ,\ \rangle$ and the associated norm $\|x\|^2 = \langle x, x\rangle$. These two functions extend to $^*\mathbf{H}$ and have the same formal properties, in particular, $|\langle x, y\rangle| \le \|x\| \cdot \|y\|$, so that $\langle\ ,\ \rangle$ takes finite values on norm finite vectors. The hull of $^*\mathbf{H}$, $\hat{\mathbf{H}} = \text{fin}(^*\mathbf{H})/\text{inf}(^*\mathbf{H})$, has $\text{st}\langle x, y\rangle = \langle \hat{x}, \hat{y}\rangle$ as an inner product and is complete by the hull completeness theorem, that is, $(\hat{\mathbf{H}}, \langle\ ,\ \rangle)$ is a Hilbert space which contains a copy of **H** isometrically embedded as $^\sigma\mathbf{H}$.

The space **H** serves as its own dual space, that is, each continuous linear functional $\varphi\colon \mathbf{H} \to \mathbf{C}$ is represented through the inner product by some $y \in \mathbf{H}$, $\varphi(x) = \langle x, y\rangle$ for all $x \in \mathbf{H}$ (see W. Rudin [1974, Chapter 4]). The theory of the last section now applies to the dual pair $\langle \mathbf{H}, \mathbf{H}\rangle$, the unit ball is weakly compact by Alaoglu's theorem, for example.

(10.2.1) ***OBSERVATION:*** *A weak infinitesimal in this setting is a vector $y \in {}^*\mathbf{H}$ that is nearly perpendicular to $^\sigma\mathbf{H}$, that is, $\langle x, y\rangle \approx 0$ for $x \in {}^\sigma\mathbf{H}$.*

(10.2.2) ***REMARK:*** By Alaoglu's theorem *every norm finite point of* $^*\mathbf{H}$ *is weakly near standard*, in fact, if $y \in \text{fin}(^*\mathbf{H})$, the map

$$\varphi(x) = \text{st}\langle x, y\rangle, \qquad x \in {}^\sigma\mathbf{H}$$

is continuous and standard, therefore it is represented by some $z \in \mathbf{H}$, $\langle x, y\rangle \approx \langle x, z\rangle$ for $x \in {}^\sigma\mathbf{H}$.

For simplicity, we shall restrict our discussion to a separable space $\mathbf{H}$, that is, one for which an orthonormal *sequence* $\{e_n : n \in \mathbf{N}\}$ exists satisfying

$$a = \sum_{n=1}^{\infty} \langle a, e_n\rangle e_n$$

where the limit is in the sense of the norm. We denote $\langle a, e_n\rangle = a_n$. The sequences $\{a_n\}$ are in the classical space $\ell^2(\mathbf{N})$ mentioned in (8.3.13) and (10.1.12) above and the Riesz–Fischer theorem (W. Rudin [1974, Chapter 4]) asserts that this representation is a Hilbert space isomorphism. The representation has the extension

$$a_\Omega = \langle a, e_\Omega\rangle, \qquad \Omega \in {}^*\mathbf{N},$$

where $^*\{e_n : n \in \mathbf{N}\}$ is denoted $\{e_n : n \in {}^*\mathbf{N}\}$, in particular, if Ω is infinite, e_Ω is in the extended basis, for example, see (10.1.12).

(10.2.3) THEOREM *A point $a \in {}^*\mathbf{H}$ is norm near standard if and only if $\|a\|^2 = \sum_{k=1}^{\infty} |a_k|^2$ is finite and for each infinite $\Omega \in {}^*\mathbf{N}_\infty$, $\sum_{k=1}^{\Omega} |a_k|^2 \approx \|a\|^2$.*

PROOF: Left as an exercise. *Hint:* See (5.1.7) and Robinson [1966, Theorem 7.2.2]. Clearly, the standard part should be given externally by $b_n = \hat{a}_n = \text{st}(a_n)$; one simply has to push the nearness of a_n and b_n out to infinite subscripts.

(10.2.4) ***REMARK:*** The vectors e_Ω for infinite Ω are nearly perpendicular to ${}^\sigma\mathbf{H}$ by the convergence of the standard sequence $\sum_{k=1}^{\infty} |a_k|^2$, (5.1.7), namely $\langle a, e_\Omega\rangle = a_\Omega$ and $\sum_{k=1}^{\Omega} |a_k|^2 - \sum_{k=1}^{(\Omega-1)} |a_k|^2 \approx 0$, so $a_\Omega \approx 0$. Generalizing this slightly we see that *a vector b with $\|b\|$ finite is nearly perpendicular to ${}^\sigma\mathbf{H}$ if and only if $b_n \approx 0$ for $n \in {}^\sigma\mathbf{N}$.* First,

$$|\langle a, b\rangle| \le \left|\sum_{k=1}^{n} a_k b_k\right| + \left|\sum_{k=n}^{\infty} a_k b_k\right|.$$

By Robinson's sequential lemma, the first term is infinitesimal out to some infinite $\Omega = n$. The second term

$$\left|\sum_{k=\Omega}^{\infty} a_k b_k\right|^2 \le \left(\sum_{k=\Omega}^{\infty} |a_k|^2\right)\left(\sum_{k=\Omega}^{\infty} |b_k|^2\right)$$

by $*$-transform of Schwarz' inequality. Finally

$$\sum_{k=\Omega}^{\infty} |b_k|^2 \leq \|b\|^2$$

is finite, and

$$\sum_{k=\Omega}^{\infty} |a_k|^2 \approx 0$$

by standard convergence (5.1.7). The product of an infinitesimal and a finite number is infinitesimal. The converse is left as an exercise.

(10.2.5) ***OBSERVATION:*** Since the weak and norm locally convex structures do not coincide there are norm-infinite weak infinitesimals; explicitly let $\omega \ll \Omega$ be as in (8.3.13) and (8.4.44) and (7.6.3).

EXERCISE (Puritz): *Show that ωe_Ω is nearly perpendicular to ${}^\sigma\mathbf{H}$ and has infinite norm ω.*

(10.2.6) INFINITESIMALS IN HILBERT SPACE We shall have use for three infinitesimal relations on $*\mathbf{H}$.

Norm infinitesimals:

$$x \overset{n}{\approx} y \qquad \text{if and only if} \qquad \|x - y\| \approx 0.$$

Bounded weak infinitesimals:

$x \overset{B}{\approx} y$ if and only if $\|x - y\|$ is finite and $\langle x, z\rangle \approx \langle y, z\rangle$ for $z \in {}^\sigma\mathbf{H}$.

Weak infinitesimals:

$$x \overset{w}{\approx} y \qquad \text{if and only if} \qquad \langle x, z\rangle \approx \langle y, z\rangle \quad \text{for } z \in {}^\sigma\mathbf{H}.$$

As we saw in (10.1.12), ${}^\sigma\mathbf{H}$ can be externally embedded in the linear subspace S, generated by $\{e_1, \ldots, e_\Omega\}$, and we call the hull $\hat{S}$ (which contains $\widehat{{}^\sigma\mathbf{H}}$ isometrically) a *hyperfinite-dimensional subspace* of $\hat{\mathbf{H}}$. Bernstein and Robinson [1966] used the fact that S is formally finite dimensional to solve an invariant subspace problem for polynomially compact operators. Robinson's book [1966] also has an account of their solution. Lomonozov [1973] subsequently gave a very elegant solution to a more general problem using Schauder's fixed point theorem and a clever nonlinearization. The Bernstein–Robinson methods still give the easiest proof that a compact operator has an invariant subspace and we think their techniques deserve further attention by experts. Of the other work along these lines, we mention Bernstein [1972] and Moore [1974]. We shall sketch a little of the theory of "hyperfinite extensions of bounded operators" in the remainder of the section, omitting most of the details, referring the reader to Moore [1974] or Robinson [1966] where necessary.

(10.2.7) DEFINITION The hull $\hat{S}$ of a $*$-finite-dimensional subspace $S \subseteq {}^*\mathbf{H}$ which contains $\widehat{{}^\sigma\mathbf{H}} \subseteq \hat{S}$ will be called a *hyperfinite-dimensional extension* of $\mathbf{H}$. We shall call S a *$*$-finite-dimensional extension* of $\mathbf{H}$ even though it may not contain $\mathbf{H}$.

Let $T: \mathbf{H} \to \mathbf{H}$ be a standard bounded linear operator on $\mathbf{H}$. An internal linear operator $A: S \to S$ such that the norm of A

$$\|A\| = \sup[\|As\| : \|s\| \le 1, s \in S]$$

is a finite scalar will be called a *finite operator on* S. The hull of a finite operator A on S will be called a *hyperfinite extension* of T provided that the hull $\hat{A}$ restricted to $\widehat{{}^\sigma\mathbf{H}}$ equals T. Each $*$-finite-dimensional subspace S has an internal $*$-orthogonal projection $P_S: {}^*\mathbf{H} \to S$ by transfer [for example, P_Ω of (10.1.12)] and the hull of the operator

$$T_S = P_S T P_S = P_S T \quad \text{restricted to} \quad S,$$

is called the *standard hyperfinite extension* of T. (Check that T_S is finite, agrees with T on $\widehat{{}^\sigma\mathbf{H}}$, and $\|\hat{T}_S\| = \|T\|$.)

We denote the *adjoint* of $T: \mathbf{H} \to \mathbf{H}$ by $T^\#$ so $\langle Tx, y\rangle = \langle x, T^\# y\rangle$ and $\|T\| = \|T^\#\|$. An operator is *Hermetian* or *self-adjoint* if $T = T^\#$. Observe that $(\hat{T}_S)^\# = [(T^\#)_S]\hat{}$ [with the help of (10.2.8)].

The basic idea in the hyperfinite extension of operators is to study internally the finite-dimensional internal operator A with linear algebra and show that the hull operation preserves what you want to know about T. Notice that $\hat{S}$ is a Hilbert space by the hull completeness theorem.

(10.2.8) LEMMA *Let S produce a hyperfinite-dimensional extension of* $\mathbf{H}$. *Let A and B be finite operators on S and $\lambda \in \mathcal{O}$ a finite scalar. Then*

(a) $\|\hat{A}\| = \mathrm{st}({}^*\|A\|)$.
(b) $(A + B)\hat{} = \hat{A} + \hat{B}$.
(c) $(\lambda A)\hat{} = \hat{\lambda}\hat{A} = \mathrm{st}(\lambda) \cdot \hat{A}$.
(d) $(AB)\hat{} = \hat{A}\hat{B}$.
(e) $(\hat{A})^\# = (A^\#)\hat{}$.
(f) *If P is a $*$-orthogonal projection on S, then $\hat{P}$ is an orthogonal projection on $\hat{S}$.*

PROOF: Left as an exercise.

(10.2.9) LEMMA *The following are equivalent:*

(a) *T is self-adjoint.*

(b) *There exists a self-adjoint hyperfinite extension of T.*
(c) *Every standard hyperfinite extension $\hat{T}_S$ is self-adjoint.*

PROOF: Certainly (c) implies (b). Since a hyperfinite extension agrees with T on $\widehat{{}^\sigma\mathbf{H}}$, (b) implies (a). If T is self-adjoint, (10.2.8) says $(\hat{P}_S \hat{T} \hat{P}_S)^{\#} = \hat{P}_S{}^{\#} \hat{T}^{\#} \hat{P}_S{}^{\#} = P_S T P_S$, so T_S is self-adjoint, and (a) implies (c).

A standard operator T is said to be *compact* if it maps norm bounded sets to relatively norm-compact sets, that is, $T(B)$ has compact closure when $\sup[\|b\| : b \in B] < \infty$; see (8.3.13) and (8.4.49).

(10.2.10) THEOREM *The following are equivalent for a standard linear operator T on $\mathbf{H}$:*

(a) *T is compact.*
(b) *T has a compact hyperfinite extension.*
(c) *Every standard hyperfinite extension $\hat{T}_S$ is compact.*
(d) $\hat{T}_S(\hat{S}) \subseteq \widehat{{}^\sigma\mathbf{H}}$.
(e) *T maps norm-finite points of $^*\mathbf{H}$ to norm-near-standard points.*
(f) *If $x \overset{B}{\approx} 0$, then $T(x) \overset{n}{\approx} 0$.*

PROOF: The equivalence of (a) and (e) is a simple exercise on the characterization of compactness (8.3.7) in terms of near-standard points (8.4.49).

(e) ⇒ (f): Since T is compact, it is bounded so when $y \overset{w}{\approx} 0$, $Ty \overset{w}{\approx} 0$, since for $z \in {}^\sigma\mathbf{H}$, $\langle Ty, z\rangle = \langle y, T^{\#}z\rangle \approx 0$, since $T^{\#}z \in {}^\sigma\mathbf{H}$. Also if y is finite, Ty is finite, $\|Ty\| \leq \|T\| \cdot \|y\|$. By these two remarks plus (10.2.3) and (10.2.4), $x \overset{B}{\approx} 0$ implies $Tx \overset{n}{\approx} 0$.

(f) ⇒ (e): By (10.2.2) every finite x is weakly near standard $x = z + y$ with $z \in {}^\sigma\mathbf{H}$ and $y \overset{w}{\approx} 0$. $Tx = Tz + Ty$ and $Ty \overset{n}{\approx} 0$ so $Tz \overset{n}{\approx} Tx$ and Tz is standard.

(c) ⇒ (b): By specialization.

(b) ⇒ (a): The restriction of a compact operator is compact.

(a) ⇒ (c): Let B be the unit ball in $\mathbf{H}$ and $\varepsilon \in {}^\sigma\mathbf{R}^+$. By hypothesis, there exist $\{x_1, \ldots, x_n\}$ in $\mathbf{H}$ such that $T(B) \subseteq \bigcup_{i=1}^n \{x \in \mathbf{H} : \|x - x_i\| < \varepsilon\}$. Thus $^*T(^*B) \subseteq \bigcup_{i=1}^n \{x \in {}^*\mathbf{H} : \|x - x_i\| < \varepsilon\}$ by the standard definition principle and therefore

$$T_S(^*B \cap S) \subseteq \bigcup_{i=1}^{n} \{x \in S : \|x - x_i\| < \varepsilon\}$$

and

$$T_S((^*B \cap S)\hat{\ }) \subseteq \bigcup_{i=1}^{m} \{\hat{x} \in \hat{S} : \|x - x_i\| \leq \varepsilon\}.$$

The unit ball in $\hat{S}$ equals $({}^*B \cap S)\hat{}$.

(e) $\Rightarrow$ (d): The hull of the near-standard points equals ${}^\sigma\mathbf{H}$.

(d) $\Rightarrow$ (a): The last part follows from the following interesting extension of Moore.

(10.2.11) THEOREM *Let $\hat{S}$ be a hyperfinite-dimensional subspace of $\hat{\mathbf{H}}$. A finite operator A on S has a compact hull, that is, $\hat{A}$ is compact on $\hat{S}$ if and only if $\hat{A}(\hat{S})$ is separable.*

PROOF: If $\hat{A}$ is compact, $\hat{A}(\hat{S})$ is separable; see (8.3.13). Conversely, if $\hat{A}$ is noncompact, there is a sequence $\{\hat{x}_n\}$ on $\hat{S}$ with $\|\hat{x}_n\| \le 1$ and $\|\hat{A}\hat{x}_m - \hat{A}\hat{x}_n\| > \varepsilon$ for $m \neq n$. Let $\{x_n : 1 \le n \le \omega\}$ be an internal extension of choices for the finite x_n's. This exists by (7.6.2). In fact, we may assume $\|x_n\| \le 1$ and $\|Ax_m - Ax_n\| > \varepsilon$ for $m \neq n$. Now $\{j : 1 \le j \le \omega\}$ has uncountable cardinality by construction of ${}^*\mathscr{X}$ in Section 7.5 so $\hat{A}(\hat{S})$ is inseparable.

Robinson [1966, Theorem 7.2.6] gives a simple proof that a compact operator has an ε-approximation by an operator of finite rank. The idea is simply to standardize the notion that the $*$-finite-dimensional operator T_S approximates T. We leave it as an exercise.

An internal operator A on S is $*$-finite dimensional and therefore has a Jordan normal form, every point in the spectrum has an eigenvector. Moore [1974] shows quite easily that

(10.2.12) THEOREM *If A is a finite operator on a $*$-finite-dimensional extension S of $\mathbf{H}$, then the spectrum of $\hat{A}$ equals the point spectrum of $\hat{A}$.*

Also

(10.2.13) THEOREM *The point spectrum of a standard hyperfinite extension of a bounded operator contains the spectrum of the original operator.*

Robinson [1966] proves

(10.2.14) THEOREM *The eigenvectors of a standard hyperfinite extension of a compact self-adjoint operator associated with noninfinitesimal eigenvalues are near-standard vectors. The standard part of the eigenvector for a finite eigenvalue is an eigenvector for the standard operator.*

A self-adjoint operator with only infinitesimal eigenvalues is an infinitesimal operator, so a compact operator has an eigenvalue.

We close with Moore's [1974] proof of the spectral theorem for a bounded self-adjoint operator. Let T be a bounded self-adjoint operator on $\mathbf{H}$ and let S be any $*$-finite-dimensional extension. Then T_S is $*$-self-adjoint on S and $\hat{T}_S$ is a self-adjoint extension of T to $\hat{S}$. If V is a finite-dimensional inner product space and Q is a self-adjoint linear transformation of V into itself, there exists an orthonormal basis $\{e_1, e_2, \ldots, e_n\}$ of V and $\{\lambda_1, \lambda_2, \ldots, \lambda_n\} \subseteq \mathbf{R}$ such that

(i) $\lambda_1 \leq \lambda_2 \leq \cdots \leq \lambda_n$ and
(ii) $Q(\sum_{i=1}^{n} a_i e_i) = \sum_{i=1}^{n} a_i \lambda_i e_i$ for all choices of $a_1, a_2, \ldots, a_n$ in $\mathbf{C}$.

Hence there exists a $*$-orthonormal basis $\{\psi_1, \psi_2, \ldots, \psi_\omega\}$ for S and a $*$-finite set $\{\lambda_1, \lambda_2, \ldots, \lambda_\omega\} \subseteq {}^*\mathbf{R}$ such that

(i) $\lambda_i \leq \lambda_{i+1}$ for $i = 1, 2, \ldots, \omega - 1$ and
(ii) ${}^*T_S(\sum_{i=1}^{\omega} a_i \psi_i) = \sum_{i=1}^{\omega} a_i \lambda_i \Psi_i$ for any internal $*$-finite sequence $\{a_1, a_2, \ldots, a_\omega\} \subseteq {}^*\mathbf{C}$.

In particular, $T_S \psi_i = \lambda_i \psi_i$ and since $\|T_S\| \leq \|T\|$ it follows that λ_i is finite for $i = 1, 2, \ldots, \omega$. For each real μ and $n \in {}^\sigma\mathbf{N}$ define $S(\mu, n)$ to be the $*$-span of $\{\psi_k : \lambda_k \leq \mu + n^{-1}\}$ and $F(\mu, n)$ to be the $*$-projection of S onto $S(\mu, n)$. Now $F(\mu, n) = 0$ if $\lambda_k > \mu + n^{-1}$ for all k, and for each μ the sequence $\{\hat{F}(\mu, n)\}$ is a monotone decreasing sequence of projections on $\hat{S}$. Define $E(\mu)$ to be the strong limit of $\hat{F}(\mu, n)$. Then $\{E(\mu) : \mu \in \mathbf{R}\}$ is the spectral resolution for $\hat{T}_S$. More precisely:

(10.2.15) THEOREM

(a) *$E(\mu) = 0$ if $\mu < \|\hat{T}\|$ and $E(\mu) = I$ if $\|\hat{T}\| < \mu$.*
(b) *$E(\mu)E(\alpha) = E(\alpha)E(\mu) = E(\min(\alpha, \mu))$ for all $\mu, \alpha \in \mathbf{R}$.*
(c) *$E(\mu)\hat{T}_S = \hat{T}_S E(\mu)$ all $\mu \in \mathbf{R}$.*
(d) *$\alpha(E(\beta) - E(\alpha)) \leq T_S(E(\beta) - E(\alpha)) \leq \beta(E(\beta) - E(\alpha))$ if $\alpha, \beta \in \mathbf{R}$ with $\alpha < \beta$.*
(e) *$\lim_{\mu \to \alpha^+} E(\mu) = E(\alpha)$.*

PROOF: (a) This is immediate since $\mathrm{st}(\|T_S\|) = \|\hat{T}_S\| = \|T\|$.

(b) Let $\mu, \alpha \in {}^\sigma\mathbf{R}$ and $n \in {}^\sigma\mathbf{N}$. Then

$$F(\mu, n)F(\alpha, n) = F(\alpha, n)F(\mu, n) = F(\min(\mu, \alpha), n).$$

Hence $\hat{F}(\mu, n)\hat{F}(\alpha, n) = \hat{F}(\alpha, n)\hat{F}(\mu, n) = \hat{F}(\min(\mu, \alpha), n)$. Taking limits we obtain (b).

(c) Let $\mu \in {}^\sigma\mathbf{R}$ and $n \in {}^\sigma\mathbf{N}$. Then $F(\mu, n)T_S = T_S F(\mu, n)$, hence $\hat{F}(\mu, n)\hat{T}_S = \hat{T}_S \hat{F}(\mu, n)$ and taking limits we obtain (c).

(d) Let $\alpha, \beta \in \mathbf{R}$ with $\alpha < \beta$ and $n \in \mathbf{N}$. Then

$$(\alpha + n^{-1})(F(\beta, n) - F(\alpha, n)) \leq T_S(F(\beta, n) - F(\alpha, n)) \leq (\beta + n^{-1})(F(\beta, n) - F(\alpha, n)).$$

Again (d) follows by applying the $\hat{}$ operation and taking limits.

(e) Finally if $\alpha \in {}^\sigma\mathbf{R}$, then $\{\hat{F}(\mu, n) : \mu \in \mathbf{R}, \mu > \alpha \text{ and } n \in {}^\sigma\mathbf{N}\}$ is cofinal with $\{F(\alpha, n) : n \in {}^\sigma\mathbf{N}\}$.

Now let P be the (external) projection of $\hat{S}$ onto ${}^{\sigma}\hat{\mathbf{H}}$. Since T_S leaves ${}^{\sigma}\mathbf{H}$ invariant, it is easy to show that P commutes with each projection $E(\mu)$. Hence if $G(\mu)$ is the restriction of $PE(\mu)$ to $\mathbf{H} \cong {}^{\sigma}\mathbf{H}$ for each $\mu \in \mathbf{R}$, then $\{G(\mu) : \mu \in \mathbf{R}\}$ is the spectral resolution for T.

Two more results of Moore [1974] are:

(10.2.16) THEOREM *An operator has a normal hyperfinite extension if and only if the operator is subnormal and SOME standard hyperfinite extension is normal if and only if the operator is normal.*

(10.2.17) THEOREM *A quasitriangular operator has a hyperfinite extension whose spectrum equals the spectrum of the operator.*

10.3 BANACH SPACES

In this section we shall examine the properties of the norm hull of a standard (B)-space $\mathbf{E}$ in our superstructure $\mathcal{X}$. For simplicity we choose the real scalars; for the most part only minor modifications are required to extend these results to the complex case. As noted in Chapter 3, all the "classical" Banach spaces are included in $\mathcal{X}$, for example, c_0, $\ell^p(\mathbf{N})$, $L^p(\mu)$ when μ is a standard measure, $C(X)$ when X is a standard compact space, and so on. (Of course, we could allow $\mathbf{E}$ to be arbitrary and construct a superstructure over $\mathbf{R} \cup \mathbf{E}$.)

For convenience [unlike (10.1.41)] we denote the norm on $\mathbf{E}$ by $p(x) = \|x\|$. Associated with $\mathbf{E}$ are the spaces:

$^*\mathbf{E}$ the nonstandard extension of $\mathbf{E}$ in $^*\mathcal{X}$ with the extended norm
$$p(x) = {}^*p(x),$$

$\mathbf{E}'$ the continuous dual with dual norm in $\mathcal{X}$
$$q(y) = p'(y) = \sup[|\langle x, y\rangle| : p(x) \le 1],$$

$(^*\mathbf{E})' = {}^*(\mathbf{E})'$ the internal continuous dual in $^*\mathcal{X}$ with extended dual norm
$$q(y) = \sup[|\langle x, y\rangle| : p(x) \le 1,\ x \in {}^*\mathbf{E}],$$

$\hat{\mathbf{E}}$ the norm hull of $^*\mathbf{E}$ with the hull norm
$$\hat{p}(\hat{x}) = \mathrm{st}(p(x)) \qquad \text{for} \quad x \in \mathrm{fin}_p({}^*\mathbf{E}),$$

$(\mathbf{E}')^\wedge$ the norm hull of $(^*\mathbf{E}', q)$ with hull of the dual norm
$$\hat{q}(\hat{y}) = \mathrm{st}(q(y)),\ y \in \mathrm{fin}_q({}^*\mathbf{E}'),$$

$(\hat{\mathbf{E}})'$ the continuous dual of the external Banach space $\hat{\mathbf{E}}$ with dual norm
$$\hat{p}'(\varphi) = \sup[|\varphi(\hat{x})| : \hat{p}(\hat{x}) \le 1].$$

Observe that the hulls are Banach spaces by the hull completeness theorem. The external dual $(\hat{\mathbf{E}})'$ is complete since the scalars are. It is a special feature of having a single norm that allows us to take the hull of the internal dual space, itself as a normed space, and consider it in duality with the hull of the original space (otherwise another natural factorization would be finite mod infinitesimal maps). We need not apologize for having to keep track of the difference between the "dual of the hull" and "the hull of the dual"—there are many simple natural connections and a beautiful result of Henson and Moore states:

E *is superreflexive* (in the sense of R. C. James [1972]) *if and only if* $(\hat{\mathbf{E}})' = (\mathbf{E}')\hat{}$ *if and only if* $(\hat{\mathbf{E}})$ *is a reflexive Banach space.*

We believe that this result is a prime example of a simple infinitesimal condition reflecting complicated limiting behavior in the original space. The section comes from Henson and Moore [1972, 1974c, 1974d].

We begin our study by showing that $(\mathbf{E}')\hat{}$ *is norm embedded in* $(\hat{\mathbf{E}})'$.

(10.3.1) PROPOSITION *If* $\hat{x} \in \hat{\mathbf{E}}$ *and* $\hat{y} \in (\mathbf{E}')\hat{}$, *then* $|\langle \hat{x}, \hat{y} \rangle| \le \hat{p}(\hat{x})\hat{q}(\hat{y})$. *Moreover, there is a* $\hat{z} \in \hat{\mathbf{E}}$ *such that* $\langle \hat{z}, \hat{y} \rangle = \hat{q}(\hat{y})$ *and* $\hat{p}(\hat{z}) = 1$, *in other words,* $\hat{q}(\hat{y}) = \hat{p}'(\hat{y})$ *for* $\hat{y}$ *viewed as an element of* $(\hat{\mathbf{E}})'$.

PROOF: Select $x \in \mathrm{fin}(^*\mathbf{E})$ and $y \in \mathrm{fin}(^*\mathbf{E}')$ with the hulls $\hat{x}$ and $\hat{y}$, respectively. By $*$-transform of the definition of q, $|\langle x, y \rangle| \le p(x)q(y)$, take standard parts of both sides to obtain the first assertion. There is a $z \in {}^*\mathbf{E}$ of p-norm one for which $\langle z, y \rangle \approx q(y)$ by the $*$-definition of q, taking standard parts shows $\hat{p}'(\hat{y}) \ge \hat{q}(\hat{y})$. This proves the proposition.

(10.3.2) RETRACTION THEOREM *Let* $\mathbf{X}$ *be a subspace of* $\hat{\mathbf{E}}$ *with Hamel dimension less than* $\mathrm{card}(\mathscr{X})$ *and let* $\varphi: \hat{\mathbf{E}} \to \mathbf{R}$ *be a* $\hat{p}$*-bounded linear functional in* $(\hat{\mathbf{E}})'$. *There is a* $\hat{y} \in (\mathbf{E}')\hat{}$ *such that* $\hat{q}(\hat{y}) \le \hat{p}'(\varphi)$ *and for* $\hat{x} \in \mathbf{X}$,

$$\langle \hat{x}, \hat{y} \rangle = \varphi(\hat{x}).$$

Functionals on standard-size subspaces can be retracted to $(\mathbf{E}')\hat{}$.

PROOF: Let $H = \{\hat{x}_\alpha\}$ be a Hamel basis for $\mathbf{X}$. We know that any finite set $\{\hat{x}_1, \ldots, \hat{x}_n\}$ of independent hulled elements in H come from $*$-linearly independent x_j's in $^*\mathbf{E}$, say $F = \{x_1, \ldots, x_n\}$. Without restriction, scale φ so that $\hat{p}'(\varphi) = 1$. Take $\lambda_j = \varphi(\hat{x}_j)$ and define a linear functional on the span of F by

$$\sum \alpha_j x_j \mapsto \sum \alpha_j \lambda_j.$$

On the span of F this functional has norm less than or infinitesimally near one, so by $*$-Hahn–Banach there is a y in $^*\mathbf{E}'$ with $q(y) \lessapprox 1$ and $\langle x_j, y \rangle = \lambda_j$ for $1 \le j \le n$.

For each finite subset $F \subseteq H$ and each $\delta \in {}^\sigma\mathbf{R}^+$, let $E(F, \delta) = \{y \in {}^*\mathbf{E}' : \langle x_\alpha, y\rangle = \varphi(x_\alpha), x_\alpha \in F, q(y) \le 1 + \delta\}$. By the preceding remarks these δ-extensions have the finite intersection property, so by polysaturation a nonempty intersection. This proves the retraction to $\mathbf{X}$ in $(\mathbf{E}')\hat{}$.

(10.3.3) EXERCISE (Henson and Moore): For the dual pair $(\hat{\mathbf{E}}, (\mathbf{E}')\hat{})$ show that B is $\sigma(\hat{\mathbf{E}}, (\mathbf{E}')\hat{})$-bounded if and only if B is $\hat{p}$-bounded.

(10.3.4) DEFINITION *Let $(\mathbf{F}, n)$ be a normed space and $\lambda \ge 1$. A linear map $T: \mathbf{F} \to \mathbf{E}$ is called a λ-embedding if*

$$n(x) \le p(Tx) \le \lambda \cdot n(x)$$

for all $x \in \mathbf{F}$. If such a T exists, we say that $(\mathbf{F}, n)$ is λ-embedded in $(\mathbf{E}, p)$ by T. The reader should check that $(\mathbf{F}, n)$ is λ-embedable in $(\mathbf{E}, p)$ if and only if there is an invertible linear map $Q: \mathbf{F} \to \mathbf{G} \subseteq \mathbf{E}$ such that $\|Q\| \cdot \|Q^{-1}\| \le \lambda$.

(10.3.5) DEFINITION *Let $(\mathbf{F}, n)$ be a normed space and $\lambda \ge 1$. We say $\mathbf{F}$ is finitely λ-representable in $(\mathbf{E}, p)$ if for each finite-dimensional subspace $F \subseteq \mathbf{F}$ and each $\varepsilon > 0$, (F, n) is $(\lambda + \varepsilon)$-embedable in $(\mathbf{E}, p)$.* We will simply say *finitely representable* when $\lambda = 1$.

Hyperfinite-dimensional spaces provide a link between these two notions (10.1.12).

(10.3.6) THEOREM *Let $(\mathbf{F}, n)$ and $(\mathbf{E}, p)$ be normed spaces in $\mathscr{X}$ and $\lambda \ge 1$. The following are equivalent.*

(a) *$(\mathbf{F}, n)$ is λ-embedable in $(\hat{\mathbf{E}}, \hat{p})$.*
(b) *$(\mathbf{F}, n)$ is finitely λ-representable in $(\hat{\mathbf{E}}, \hat{p})$.*
(c) *$(\mathbf{F}, n)$ is finitely λ-representable in $(\mathbf{E}, p)$.*

PROOF: (a)$\Rightarrow$(b): Clear.
(b)$\Rightarrow$(c): In (10.1.11) we showed how to "internalize" finite-dimensional maps. The norm estimates for $(\lambda + \varepsilon/2)$-embedding into $\hat{\mathbf{E}}$ are only slightly distorted in the process so the internal map is a $(\lambda + \varepsilon)$-embedding of the finite-dimensional subspace. Apply Leibniz' principle.
(c)$\Rightarrow$(a): Apply the $*$-transform of the condition to a $*$-finite-dimensional extension $\mathbf{G}$ of $\mathbf{F}$, ${}^\sigma\mathbf{F} \subseteq \mathbf{G} \subseteq {}^*\mathbf{F}$, and the internal map $T: \mathbf{G} \to {}^*\mathbf{E}$ is a $(\lambda + \varepsilon)$-embedding with ε selected infinitesimal. Then T restricted to ${}^\sigma\mathbf{F} \cong \mathbf{F}$ is a λ-embedding.

(10.3.7) EXERCISE: Let $\hat{\mathbf{H}}$ be a hyperfinite-dimensional extension of $\mathbf{E}$. Show that (a)–(c) of the theorem are equivalent to λ-embedability in $\hat{\mathbf{H}}$.

(10.3.8) ℓ^2 IS EMBEDDED IN EVERY STANDARD BANACH NORM HULL $(\hat{\mathbf{E}}, \hat{p})$ This embedding is developed by the fundamental theorem of Dvoretsky [1960] which states that ℓ^2 is finitely representable in every infinite-dimensional normed space over **R**.

One of the important properties of a Banach space that can be phrased in terms of which spaces are finitely representable is R. C. James' notion of superreflexivity.

(10.3.9) DEFINITION $(\mathbf{E}, p)$ *is superreflexive if only reflexive Banach spaces are finitely representable in* $(\mathbf{E}, p)$.

A reflexive Banach space is one whose second dual $\mathbf{E}''$ is isomorphic to **E** itself, that is, **E** represents each continuous functional on $\mathbf{E}'$. Hilbert spaces are reflexive since the first dual is already isomorphic to the space. The L^p-spaces for $1 < p < \infty$ are reflexive and the dual "is" L^q with $(1/p) + (1/q) = 1$. The first dual of L^1 is L^∞, but not every functional on L^∞ is represented again on L^1, for example, with counting measure on **N**, ℓ^∞ has "evaluation at $\Omega \in {}^*\mathbf{N}_\infty$" [see Section 9.1] which cannot be represented in ℓ^1. Irreflexive Banach spaces are characterized by the fact that the unit ball is not $\sigma(\mathbf{E}, \mathbf{E}')$-compact. Notice in $\mathbf{E}'$ that $\sigma(\mathbf{E}', \mathbf{E})$ and $\sigma(\mathbf{E}', \mathbf{E}'')$ are distinct in this case. (See Alaoglu's and Krein's theorems above.)

(10.3.10) THEOREM *The following are equivalent:*

(a) $(\mathbf{E}, p)$ *is superreflexive.*
(b) $(\hat{\mathbf{E}}, \hat{p})$ *is reflexive.*
(c) $(\hat{\mathbf{E}}, \hat{p})$ *is superreflexive.*
(d) $(\mathbf{E}')\hat{} = (\hat{\mathbf{E}})'$.

PROOF: James [1972] shows that superreflexivity is equivalent to "(J): For some $r \in \mathbf{R}$, $0 < r < 1$, and some positive integer n, there do not exist finite sequences $\{x_1, \ldots, x_n\}$ in **E** and $\{y_1, \ldots, y_n\}$ in $\mathbf{E}'$ which satisfy:

$$p(x_i) = q(y_j) = 1 \qquad \text{for} \quad 1 \le i, j \le n,$$

$$r = \langle x_i, y_j \rangle \qquad \text{if} \quad 1 \le j \le i \le n,$$

$$0 = \langle x_i, y_j \rangle \qquad \text{if} \quad 1 \le i < j \le n."$$

Clearly, if this condition fails in **E** the hull of the finite sequences force the failure in $\hat{\mathbf{E}}$, so if $\hat{\mathbf{E}}$ is superreflexive, **E** is. This shows (c) $\Rightarrow$ (a).

(a) $\Rightarrow$ (b): Next, suppose **E** is superreflexive and $\hat{\mathbf{E}} = \mathbf{F}$ has been constructed. Let $\mathscr{X}$ be the superstructure based on $\mathbf{F} \cup \mathbf{R}$ and apply (10.3.6). The isometric embedding $\mathbf{E} \to \mathbf{F}$ (σ of the first extension shows $\hat{\mathbf{E}}$ is isometrically embedded in $\hat{\mathbf{F}}$ $(\hat{\hat{\mathbf{E}}})$. Therefore ${}^\sigma\mathbf{F}$ is embedable in $\hat{\mathbf{E}}$ and finitely representable in **E**. By definition then $\hat{\mathbf{E}}$ must be reflexive, since only reflexive spaces are finitely represented in superreflexive spaces.

EXERCISE: Give a proof of this part of the theorem that does not use double extensions.

(b)$\Rightarrow$(d): If $(\hat{\mathbf{E}})'$ properly contains $(\mathbf{E}')\hat{}$, the Hahn–Banach theorem says there is a separating functional in $(\hat{\mathbf{E}})''$, zero on $(\mathbf{E}')\hat{}$ but not identically zero. This requires the completeness of $(\mathbf{E}')\hat{}$. Such a functional cannot be in $\hat{\mathbf{E}}$, since then it would be identically zero. Consequently $(\hat{\mathbf{E}})'' \neq \hat{\mathbf{E}}$.

(d)$\Rightarrow$(J)$\equiv$(c): If (J) fails, take $r \approx 1$ and $\Omega \in {}^*\mathbf{N}_\infty$ and internal sequences $\{x_i : 1 \le i \le \Omega\}$, $\{y_j : 1 \le j \le \Omega\}$ from (J). By Alaoglu's theorem the closed unit ball of $(\hat{\mathbf{E}})'$ is $\sigma(\hat{\mathbf{E}}', \hat{\mathbf{E}})$-compact, so there is a limit point $\varphi = \hat{y} \in (\mathbf{E}')\hat{}$ of $\{\hat{y}_j : j \in {}^\sigma\mathbf{N}\}$. Therefore, for each n, $1 \le n \le \Omega$ and $m \in {}^\sigma\mathbf{N}$,

$$\langle \hat{x}_n, \hat{y}_n \rangle = \begin{cases} 0, & m > n, \\ 1, & m \le n; \end{cases}$$

and since $\varphi = y$ is a weak-star limit,

$$\langle \hat{x}_n, \hat{y} \rangle = \begin{cases} 0, & n \in {}^\sigma\mathbf{N}, \\ 1, & n \text{ infinite and } \le \Omega. \end{cases}$$

This cannot happen since

$$n \mapsto \langle x_n, y \rangle$$

is an internal map and Robinson's sequential lemma applies.

(10.3.11) COROLLARY *Let $1 < p < \infty$ and $(1/p) + (1/q) = 1$. The dual of $\widehat{L^p}$ is $\widehat{L^q}$ for standard measures.*

PROOF: Uniformly convex spaces are superreflexive and L^p-spaces are uniformly convex. This comes from the classical inequality of Clarkson. Of course, the dual of L^p is L^q—see any graduate text on analysis.

(10.3.12) ***EXAMPLE:*** Let $\{p_n | n \in \mathbf{N}\}$ be a strictly increasing sequence of real numbers such that $1 < p_n$ for each n and $p_n < \infty$. For each n let q_n be defined by $(1/p_n) + (1/q_n) = 1$. Let $\mathbf{E}_n$ be the sequence space ℓ^{p_n} over $\mathbf{R}$ and let p_n denote the ℓ^{p_n} norm for each n in $\mathbf{N}$. Then $(\mathbf{E}_n', P_n')$ is ℓ^{q_n} with the ℓ^{q_n} norm.

Let $\mathbf{E}$ be the subspace of the product $\pi\mathbf{E}_n$ $(n \in \mathbf{N})$ consisting of all $x = (x_1, x_2, \ldots)$ such that

$$P(x) = \left(\sum_{n=1}^{\infty} [P_n(x_n)]^2 \right)^{1/2} < \infty.$$

Then $(\mathbf{E}, P)$ is a reflexive Banach space and the dual space of $(\mathbf{E}, P)$ may be identified with $(\mathbf{E}', P')$, where $\mathbf{E}'$ is the subspace of $\pi\mathbf{E}_n'$ $(n \in \mathbf{N})$ consisting of all $y = (y_1, y_2, \ldots)$ such that

$$P'(x) = \left(\sum_{n=1}^{\infty} [P_n'(y_n)]^2\right)^{1/2} < \infty.$$

The pairing between $\mathbf{E}$ and $\mathbf{E}'$ is given by

$$\langle x, y \rangle = \sum_{n=1}^{\infty} \langle x_n, y_n \rangle.$$

Now let $r < 1$ be an element of $\mathbf{R}$ and let n be in $\mathbf{N}$. Pick p_k such that $rn^{1/p_k} < 1$. For each j in N let e_j be the R-valued sequence which has the entry 1 in the jth place and zeros elsewhere. Then each e_j belongs to $\mathbf{E}_k$ and to $\mathbf{E}_k'$. For each i with $1 \le i \le n$ define x_i in $\mathbf{E}_k$ by

$$x_i = 1^{-1/p_k}(e_1 + \cdots + e_i).$$

Then

$$P_k(x_i) = 1, \qquad P_k'(e_j) = 1 \qquad \text{for} \quad 1 \le i, j \le n,$$
$$r < \langle x_i, e_j \rangle \qquad \text{if} \quad 1 \le j \le i \le n,$$

and

$$0 = \langle x_i, e_j \rangle \qquad \text{if} \quad 1 \le i < j \le n.$$

Since $(\mathbf{E}_k, P_k)$ and $(\mathbf{E}_k', P_k')$ are embedded (as paired spaces) in (E, P) and (E', P'), respectively, it follows that (E, P) does not satisfy condition (J).

Luxemburg [1972b] showed that there is a norm-preserving function F from $\mathbf{E}''$ the second dual of a Banach space $\mathbf{E}$ into $\hat{\mathbf{E}}$ that preserves the pairing between $\mathbf{E}'$ and $\mathbf{E}''$. This answered a question of Robinson [1964]. Luxemburg's solution was based on Helly's theorem. Henson and Moore [1974d] showed a linear map exists by using

(10.3.13) LOCAL REFLEXIVITY PRINCIPLE *$\mathbf{E}$ is a Banach space, V a finite-dimensional subspace of $\mathbf{E}'$, and U a finite-dimensional subspace of $\mathbf{E}''$ with $\delta > 0$. For each U, V, δ there exists a linear injection $T: U \to \mathbf{E}$ with*

$$Tx = x \qquad \text{for} \quad x \in \mathbf{E} \cap U,$$
$$\varphi(Tu) = u(\varphi) \qquad \text{for} \quad u \in U \quad \text{and} \quad \varphi \in V,$$

and

$$\|T\| \, \|T^{-1}\| < 1 + \delta.$$

PROOF: See the work of Lindenstrauss and Rosenthal [1969].

Taking U a $*$-finite-dimensional extension of ${}^{\sigma}\mathbf{E}''$ and V a $*$-finite-dimensional extension of $\mathbf{E}'$ and δ infinitesimal we have an internal linear injection $T: U \to {}^*\mathbf{E}$ with norm nearly one. The hull $\hat{T}$ restricted to ${}^{\sigma}\mathbf{E}''$ proves:

(10.3.14) THEOREM *There is a linear isometry*

$$T: {}^{\sigma}\mathbf{E}'' \to \hat{\mathbf{E}}, \qquad Tx = x \qquad \text{for all} \quad x \in {}^{\sigma}\mathbf{E}$$

$$\varphi(Te) = e(\varphi) \qquad \text{for all} \quad e \in {}^{\sigma}\mathbf{E}'' \qquad \text{and} \quad \varphi \in {}^{\sigma}\mathbf{E}'.$$

Each $\hat{x} \in \hat{\mathbf{E}}$ can be considered as a functional on ${}^{\sigma}\mathbf{E}'$ by restriction, that is, there is a linear projection $R: \hat{\mathbf{E}} \to {}^{\sigma}\mathbf{E}''$ such that for each $\hat{x} \in \hat{\mathbf{E}}$ and $y \in {}^{\sigma}\mathbf{E}'$, $\langle \hat{x}, y \rangle = (Rx)(y)$.

(10.3.15) EXERCISE (Henson and Moore): $RT: {}^{\sigma}\mathbf{E}'' \to {}^{\sigma}\mathbf{E}''$ is the identity. The map $TR: \hat{\mathbf{E}} \to {}^{\sigma}\mathbf{E}''$ has norm one. Show that the space ${}^{\sigma}\mathbf{E}$ is complemented in $\hat{\mathbf{E}}$ if and only if $\mathbf{E}$ is complemented in $\mathbf{E}''$.

(10.3.16) EXERCISE (Henson and Moore): If $\hat{\mathbf{H}}$ is a hyperfinite dimensional extension of ${}^{\sigma}\mathbf{E}$ in $\hat{\mathbf{E}}$, show that there is a linear functional preserving embedding Q: $\mathbf{E}'' \to \hat{\mathbf{H}}$.

(10.3.17) EXERCISE: Show that all generalized limits arise as one of Robinson's summability methods—a sum with an ℓ^1-finite internal sequence; see Section 6.4.

The following results can be found in the work of Henson and Moore [1974c, d] using Cozart and Moore [1974].

(10.3.18) THEOREM *The following are equivalent:*

(a) *$(\mathbf{E}, p)$ is B-convex.*
(b) *$(\hat{\mathbf{E}}, \hat{p})$ is B-convex.*
(c) *$(\hat{\mathbf{E}}, \hat{p})$ does not contain an isometric copy of ℓ^1.*
(d) *$(\hat{\mathbf{E}}, \hat{p})$ does not contain a linear homeomorphic copy of ℓ^1.*

(10.3.19) THEOREM *If $1 \le p < \infty$ and (L, P) is an abstract L^p space in $\mathscr{X}$, then (L, P) is Riesz isometrically embedable in $\hat{\ell}^p$. If (L, P) is an abstract M-space in $\mathscr{X}$, (L, P) is Riesz isometrically embedable in $\hat{c}_0$. $(\mathbf{E}, P)$ is a $\mathscr{L}_p$-space if and only if $(\hat{\mathbf{E}}, \hat{P})$ is an $\mathscr{L}_p$-space $(1 \le p \le \infty)$.*

10.4 DISTRIBUTIONS

In this section we sketch the way the generalized functions of ${}^*\mathscr{C}^\infty(\mathbf{R})$ act on the spaces $\mathscr{D}$, $\mathscr{S}$, $\mathscr{E}$ [of (8.4.14), (8.4.15), and (10.1.5)] as "generalized functions" or "distributions" in the conventional sense. The various distributions we obtain will all be $*$-smooth point functions. We use the partial duality for $\varphi, f \in {}^*\mathscr{C}^\infty$

$$\langle \varphi, f \rangle = \int_{-\infty}^{\infty} \varphi(t) f(t)\, dt$$

when the absolutely convergent $*$-Lebesgue integral exists in ${}^*\mathbf{R}$, the symbol $\langle \varphi, f \rangle$ is undefined otherwise. When carr(f) is bounded in ${}^*\mathbf{R}$, $\langle \varphi, f \rangle$ exists in ${}^*\mathbf{R}$ for all $\varphi \in {}^*\mathscr{C}^\infty$. Similarly if $\varphi(t) = e^{-t^2}$ and f has at most polynomial growth $\langle \varphi, f \rangle$ exists in ${}^*\mathbf{R}$ by the $*$-transform of the classical estimates for convergence.

Recall the definitions of the basic spaces:

$$\mathscr{D} = \{\varphi \in {}^*\mathscr{C}^\infty : \varphi \overset{\mathscr{D}}{\sim} 0\}$$

$$\varphi \overset{\mathscr{D}}{\approx} \psi \quad \text{if and only if} \quad \begin{cases} \varphi^{(k)}(t) \approx \psi^{(k)}(t) & \text{for} \quad k \in {}^\sigma\mathbf{N} \\ \text{and} & \\ \varphi(t) = \psi(t) & \text{for infinite} \quad t \in {}^*\mathbf{R}. \end{cases}$$

(10.4.1) DEFINITION *A function $f \in {}^*\mathscr{C}^\infty$ will be called a finite distribution if $\varphi \in \mathscr{D}$ implies $\langle \varphi, f \rangle$ is a finite scalar; it will be called an infinitesimal distribution if $\varphi \in \mathscr{D}$ implies $\langle \varphi, f \rangle$ is infinitesimal.* The space of finite distributions is denoted by D', the infinitesimal distributions by d'.

$$\mathscr{S} = \{\varphi \in {}^*\mathscr{C}^\infty : \varphi \overset{\mathscr{S}}{\sim} 0\}$$

$$\varphi \overset{\mathscr{S}}{\approx} \psi \quad \text{if and only if} \quad (1 + |t|^n)[\varphi^{(k)}(t) - \psi^{(k)}(t)] \approx 0$$

for all $t \in {}^*\mathbf{R}$, where $n, k \in {}^\sigma\mathbf{N}$.

(10.4.2) DEFINITION *A function $f \in {}^*\mathscr{C}^\infty$ is called a finite tempered distribution if $\varphi \in \mathscr{S}$ implies $\langle \varphi, f \rangle$ is finite; it is called an infinitesimal tempered distribution if $\varphi \in \mathscr{S}$ implies $\langle \varphi, f \rangle$ is infinitesimal.* The finite tempered distributions are denoted S', the infinitesimal tempered distributions s':

$$\mathscr{E} = \{\varphi \in {}^*\mathscr{C}^\infty : \varphi \overset{\mathscr{E}}{\sim} 0\}$$

$$\varphi \overset{\mathscr{E}}{\approx} \psi \quad \text{if and only if} \quad \varphi^{(k)}(t) \approx \psi^{(k)}(t) \quad \text{for } t \text{ finite and } k \in {}^\sigma\mathbf{N}.$$

(10.4.3) DEFINITION *A function $f \in {}^*\mathscr{C}^\infty$ is called a finite compact distribution if $\varphi \in \mathscr{E}$ implies $\langle \varphi, f \rangle$ is finite; it is called an infinitesimal* *compact distribution if* $\varphi \in \mathscr{E}$ *implies* $\langle \varphi, f \rangle$ *is infinitesimal.* The finite compact distributions are denoted E', the infinitesimal ones by e'.

(10.4.4) EXERCISE: Let f be an infinitely SU-differentiable function with values in a locally convex space **F**. How would you define vector-valued distributions $D'(\mathbf{F})$, etc.? How much of the remaining theory can you generalize? How would you change things if the domain of the $*$-smooth functions was a manifold?

(10.4.5) PROPOSITION *Differentiation is an S-continuous linear operation on each of the spaces* D', S', E' *as well as on* $\mathscr{D}$, $\mathscr{S}$, $\mathscr{E}$.

PROOF: Integration by parts says

$$\int_{-\infty}^{\infty} \varphi(t) f^{(\prime)}(t)\, dt = -\int_{-\infty}^{\infty} \varphi^{(\prime)}(t) f(t)\, dt,$$

and if φ is in $\mathscr{D}$, $\mathscr{S}$, or $\mathscr{E}$, so is $\varphi^{(\prime)}$. Therefore, if $f \in d'$, s', or e' so is $f^{(\prime)}$, likewise $f \in D'$, S', or E' implies $f^{(\prime)}$ in the respective space. On $\mathscr{D}$, $\mathscr{S}$, $\mathscr{E}$, simply check the definitions.

(10.4.6) A DELTA FUNCTION Let

$$c(t) = \begin{cases} \exp(-1/(1 - |t|^2)), & |t| < 1 \\ 0, & \text{otherwise.} \end{cases}$$

This is a simple variation on Cauchy's flat function; it is $\mathscr{C}^\infty$. Let ε be a positive infinitesimal and let

$$d(t) = c(t/\varepsilon).$$

The function d is $*\mathscr{C}^\infty$ and carried on the interval $[-\varepsilon, \varepsilon]$. Let $k = \int_{-\infty}^{\infty} d(t)\, dt$ and let

$$\delta(t) = d(t)/k = (1/k)c(t/\varepsilon).$$

The delta function δ is $*\mathscr{C}^\infty$, positive, and has integral one.

(10.4.7) EXERCISE: Show that δ is not in $\mathscr{D}$. *Hint:* Estimate $\delta(0)$ or $\delta'(t)$.

(10.4.8) PROPOSITION *The delta function* $\delta(t)$ *is a finite compact distribution and for each* $\varphi \in \mathscr{E}$, $n \in {}^\sigma\mathbf{N}$,

$$\langle \varphi, \delta^{(n)} \rangle \approx (-1)^n \varphi^{(n)}(0).$$

Moreover, it is nearly unique, that is, if $f \in D'$ *(respectively,* S', E'*) satisfies* $\langle \varphi, f \rangle \approx \varphi(0)$ *for each* $\varphi \in \mathscr{D}$ *(respectively,* $\mathscr{S}$, $\mathscr{E}$*), then* $(\delta - f) \in d'$ *(respectively,* s', e'*).*

PROOF: First, $\langle \varphi, \delta \rangle$ is finite since

$$\left| \int_{-\infty}^{\infty} \varphi(t)\delta(t)\, dt \right| \leq \max[|\varphi(t)| : |t| \leq \varepsilon] \cdot \int_{-\varepsilon}^{\varepsilon} \delta(t)\, dt$$

and the max is attained and finite for φ in $\mathscr{D}$, $\mathscr{S}$, or $\mathscr{E}$. Finiteness of $\varphi'(t)$ for $|t| \leq \varepsilon$ means that $\max[|\varphi(t) - \varphi(s)| : |t|, |s| \leq \varepsilon]$ is infinitesimal. Hence

$$\min[\varphi(t) : |t| \leq \varepsilon] \leq \langle \varphi, \varepsilon \rangle \leq \max[\varphi(t) : |t| \leq \varepsilon]$$

and $\langle \varphi, \delta \rangle \approx \varphi(0)$. Integration by parts and the fact that differentiation is continuous on $\mathscr{D}$ (respectively, $\mathscr{S}$, $\mathscr{E}$) shows $\langle \varphi, \delta^{(n)} \rangle \approx (-1)^n \varphi^{(n)}(0)$.

Finally, if $\langle \varphi, f \rangle \approx \varphi(0)$ for $\varphi \in \mathscr{D}$ (respectively, $\mathscr{S}$, $\mathscr{E}$), then $\langle \varphi, f - \delta \rangle \approx 0$ and $(f - \delta) \in d'$ (respectively, s', e').

(10.4.9) EXERCISE: Let ε be a positive infinitesimal and $g(t) = (\varepsilon\pi)^{-1/2} \exp(-t^2/\varepsilon)$. Which of the spaces D', S', E' does g belong to? Which of the spaces d', s', e' does $(g - \delta)$ belong to? Give other examples of delta functions. Is $\delta * \delta$ infinitesimally close to δ? In which senses?

(10.4.10) δ-MOLLIFICATION LEMMA *Let μ be a $*$-Borel measure with finite total variation on standard bounded intervals, for example, a standard locally finite Borel measure. The δ-mollification of μ*

$$\delta * \mu(x) = \int_{-\infty}^{\infty} \delta(x - t)\, d\mu(t)$$

is a finite distribution and $\langle \varphi, \delta * \mu \rangle \approx \int_{-\infty}^{\infty} \varphi(t)\, d\mu(t)$ for $\varphi \in \mathscr{D}$. When $d\mu(t) = f(t)\, dt$ for $f \in D'$, $(\delta * f - f) \in d'$.

PROOF:

$$\langle \varphi, \delta * \mu \rangle = \int \varphi(x) \int \delta(x - t)\, d\mu(t)\, dx = \iint \varphi(x)\delta(x - t)\, dx\, d\mu(t)$$

$$\approx \int \varphi(t)\, d\mu(t),$$

by $*$-Fubini's theorem and the uniform infinitesimal estimate of the error between $\langle \varphi(x), \delta(x - t) \rangle$ and $\varphi(t)$ coming from the fact that they are internal. The mollification is $*$-smooth by the $*$-transform of Leibniz' rule.

(10.4.11) EXERCISE: Which measures have δ-mollifications in E' and S'? If $\varphi \in \mathscr{D}$, show that $\delta * \varphi \overset{\mathscr{D}}{\approx} \varphi$. If $\psi \in \mathscr{E}$, is $\delta * \psi \overset{\mathscr{E}}{\approx} \psi$?

Almost everything converges distributionally:

(10.4.12) PROPOSITION *Let f be a finite distribution and g an internally Lebesgue measurable function. Any of the following conditions imply that* $(f - \delta * g) \in d'$:

(a) $f(t) \approx g(t)$ a.e. dt *for finite intervals* $[a, b]$.

(b) $f(t) \sim g(t)$ a.e. dt *on finite intervals* $[a, b]$ *and* $f(t) \approx g(t)$ *except on sets of infinitesimal measure on* $[a, b]$.

(c) *For finite* $p \geq 1$, $\int_a^b |f(t) - g(t)|^p \, dt \approx 0$ *for finite a and b.*

PROOF: Left as an exercise.

(10.4.13) PROPOSITION *If* $\{f_n\}$ *is a sequence of finite distributions converging weakly, that is, for each* $\varphi \in {}^\sigma\mathscr{D}$,

$$\lim_n [\mathrm{st}\langle \varphi, f_n \rangle] = L(\varphi) \text{ exists,}$$

then there is a finite distribution f_Ω *such that* $f_n \to f_\Omega$, *that is, for each* $\varepsilon \in {}^\sigma\mathbf{R}^+$, *each finite interval* $[a, b]$, *and each bound* $b\colon {}^\sigma\mathbf{N} \to {}^\sigma\mathbf{R}^+$ *there exists an* $m \in {}^\sigma\mathbf{N}$ *such that if* $n > m$ *and* $\sup(|\varphi^{(k)}(t)| : t \in [a, b]) \leq b(k)$, *then*

$$|\langle \varphi, f_n - f_\Omega \rangle| < \varepsilon.$$

PROOF: Finite distributions are S-continuous on the normed subspaces of $\mathscr{D}$ given by $\sup[|\varphi^k(t)| : k < N]$ on functions with support in $[-M, M]$, $M \in {}^\sigma\mathbf{N}$. The standard part $\mathrm{st}\langle \varphi, f_n \rangle = F_n(\varphi)$ is a continuous functional on the standard space by the general S-continuity theorem. By the uniform boundedness principle the sequence F_n is uniformly bounded on this subspace, say,

$$\sup\{F_n(\varphi) : \sup[|\varphi^{(k)}(t)| : k < N] \leq 1 \;\&\; \mathrm{carr}(\varphi) \subseteq [-M, M]\} \leq B(M, N).$$

The bound $B(M, N) + 1$ will then apply to all the f_n's. Extend $B(M, N) + 1 = C(M, N)$ to an internal function and f_n to an internal sequence of ${}^*\mathscr{C}^\infty$-functions satisfying

$$|\langle \varphi, f_\omega \rangle| \leq C(M, N)$$

whenever

$$\sup|\varphi^{(k)}(t)| \leq 1, \quad k \leq N \qquad \text{and} \qquad \mathrm{carr}(\varphi) \subseteq [-M, M].$$

Extend the standard function $L(\varphi)$ to the nonstandard model and consider the set of indices

$$J(\varphi, \varepsilon) = \{j \in {}^*\mathbf{N} : m < j, |\langle \varphi, f_j \rangle - L(\varphi)| < \varepsilon\},$$

which is internal by the internal definition principle. When φ is standard in ${}^\sigma\mathscr{D}$ and $\varepsilon \in {}^\sigma\mathbf{R}^+$, $J(\varphi, \varepsilon) \supseteq \{j \in {}^\sigma\mathbf{N} : j > m\}$ for the m coming from the Weierstrassian definition of $\lim_n [\mathrm{st}\langle \varphi, f_n \rangle] = L(\varphi)$. Therefore J contains an infinite subscript. This family has the finite intersection property and if $\Omega \in \bigcap [J(\varphi, \varepsilon) : \varphi \in {}^\sigma\mathscr{D}, \varepsilon \in {}^\sigma\mathbf{R}]$, then f_Ω is the distribution we seek.

The remaining details are left as an exercise with the hint that the infinitesimal hull of $\mathscr{D}$, $\hat{\mathscr{D}}$, is easily seen to be the standard Schwartz test space—the standard part ψ of $\varphi \in \mathscr{D}$ satisfies $\psi \overset{\mathscr{D}}{\approx} \varphi$.

As an application of this theorem suppose c_n is a sequence of finite numbers for which $\lim_n [\operatorname{st}(c_n/n^k)] = 0$. Then an internal extension of $\{c_n\}$ defines a $*$-finite trigonometric distribution

$$\sum_{n=0}^{\Omega} c_n \exp(int)$$

since $\sum_{n=1}^{m} c_n(i/n)^k \exp(int)$ converges absolutely and can be differentiated k times in D' (the constant term is a finite distribution already).

(10.4.14) PROPOSITION *The finite distributions D' form an S-continuous $\mathscr{E}$-module; if $\varphi \overset{\mathscr{E}}{\approx} \psi \in \mathscr{E}$ and $(f - g) \in d'$ for $f, g \in D'$, then $F(t) = \varphi(t)f(t)$ and $G(t) = \psi(t)g(t)$ are in D' and $(F - G) \in d'$.*

PROOF: Let $h(t) \in \mathscr{D}$. Then $\varphi(t)h(t) \overset{\mathscr{D}}{\approx} \psi(t)h(t)$ by Leibniz' product rule and $\langle \varphi(t)h(t), f(t)\rangle \approx \langle \psi(t)h(t), f(t)\rangle \approx \langle \psi(t)h(t), g(t)\rangle$.

(10.4.15) EXERCISE: Are the spaces S' and E' also $\mathscr{E}$-modules?

(10.4.16) DEFINITION *We say a finite distribution f vanishes distributionally on a finite interval (a, b) provided that whenever $\varphi \in \mathscr{D}$ has* $\operatorname{carr}(\varphi) \subseteq (a, b)$, *then* $\langle \varphi, f \rangle \approx 0$. The *distributional carrier* of f is the *standard* set given in $\mathbf{R}$ by

$$\operatorname{CARR}(f) = \bigcap [\mathbf{R} \backslash (\hat{a}, \hat{b}) : f \text{ vanishes distributionally on } (a, b)].$$

We define

$$g(t) = \begin{cases} f(t) & \text{for } \ t \in {}^*\operatorname{CARR}(f) \\ 0 & \text{for } \ t \notin {}^*\operatorname{CARR}(f) \end{cases}$$

and $F(t) = (\delta * g)(t)$. Then $\operatorname{st}[\operatorname{carr}(F)] = \operatorname{CARR}(f)$ and $(F - f) \in d'$, in other words *we can replace f with an infinitesimally nearby distribution whose point carrier is nearly the distributional carrier of f.*

(10.4.17) *REMARK:* A finite compact distribution $g \in E'$ must have finite point carrier, since there is no restriction on the size of $\mathscr{E}$-test functions at infinite points. Consequently, $\operatorname{st}[\operatorname{carr}(g)] \supseteq \operatorname{CARR}[g]$ are *compact sets* in $\mathbf{R}$.

(10.4.18) ***REMARK:*** A finite distribution $f \in D'$ with CARR(f) compact in **R** can be replaced by an $F \in D'$ with $(F - f) \in d'$ and st(carr(F)) = CARR(f). The d'-equivalent F is clearly also in E' and satisfies

$$|\langle \varphi, f \rangle| \leq k \sup[|\varphi^{(m)}(t) : |t| \leq b \ \& \ m \leq n]$$

for finite constants k, b, and n. This follows from the general theorem on S-continuity of F as a function on $\mathscr{E}$.

(10.4.19) DEFINITION *When there is a finite order of differentiation n such that*

$$|\langle \varphi, f \rangle| \leq k \sup[|\varphi^{(m)}(t)| : m \leq n]$$

for all $\varphi \in \mathscr{D}$, the distribution f is said to have finite order. The preceding paragraph says compactly carried finite distributions have finite order.

(10.4.20) EXERCISE: A finite distribution of finite order is d'-close to a kth-derivative of an S-continuous finite distribution. This exercise may be harder than those above, its interest to us is only the contact it provides with (1) the theorem in Sobolev–Schwartz' theory which says a standard distribution of finite order is $D^k g$ for a continuous function g—the pointwise standard part of our S-continuous function; (2) the basic construction and analogous result in Mikusinski's theory. The conventional purpose of this theorem is to say (locally) that distributions are what arise from extending derivatives to continuous functions. We are relying on the canonical nonstandard extension $^*\mathscr{C}^\infty$ instead. The standard local representation theorem implies that every standard distribution is nearly represented in D' since $\delta * f$ for a continuous function is in D' and gives nearly the same values. The hull $\hat{\mathscr{D}} = \mathscr{D}/(\overset{\mathscr{D}}{\approx})$ is Schwartz' test space and the hull of a functional $\Phi = \langle \cdot, f \rangle$ when $f \in D'$ is a Schwartz' distribution by the general theorem on hulls of S-continuous functions.

As we have seen, one has almost complete freedom convergence-wise when working with distributions. This is the reason for the success of the conventional theory. Still, a conventional distribution answer or an approximate $^*\mathscr{C}^\infty$-function answer is usually not very interesting—one wants a "regular" answer to the problem.

(10.4.21) SOBOLEV'S LEMMA (simplest case) *Let $f \in {}^*\mathscr{C}^\infty$ and such that whenever a and b are finite, $\int_a^b |f(t)|^2\, dt$ and $\int_a^b |f'(t)|^2\, dt$ are finite. Then f is a finite distribution and a finite-valued S-continuous function.*

The standard part of this distributional answer will be a standard continuous function.

PROOF: First if $\varphi \in \mathscr{D}$, $\operatorname{carr}(\varphi) \subseteq [a, b]$ for some finite a and b. Then

$$\left|\int_a^b \varphi(t) f(t)\, dt\right| \leq \int_a^b |\varphi(t) f(t)|\, dt$$

$$\leq \left(\int_a^b |\varphi(t)|^2\, dt\right)^{1/2} \left(\int_a^b |f(t)|^2\right)^{1/2},$$

and therefore $\langle \varphi, f \rangle$ is finite.

Next, define

$$g(x) = \int_a^x |f'(t)|\, dt$$

so g is an increasing function with $g'(x) = |f'(x)|$ by the $*$-fundamental theorem of calculus. Since

$$f(x) - f(a) = \int_a^x f'(t)\, dt$$

we know

$$|f(x) - f(a)| \leq |g(x) - g(a)| = |g(x)|.$$

Again, by Schwarz' inequality

$$|g(x)| \leq \left(\int_a^x 1^2\, dt\right)^{1/2} \left(\int_a^x |f'(t)|^2\, dt\right)^{1/2} \leq (x - a)^{1/2} \cdot \mathcal{O}.$$

Thus we see that f is S-continuous.

Finally, if $f(a)$ is infinite, by S-continuity f is infinite of the same sign on a finite interval contrary to finiteness of $\int_a^b |f(t)|^2\, dt$.

This concludes our sketch of distributions as $^*\mathscr{C}^\infty$-generalized functions. We have regrettably not dealt with Fourier transforms and their applications, but hope the picture is clear enough so that the interested reader can develop this aspect of the theory.

To date, no complete nonlinear theory of distributions has emerged to solve even the restricted square root of the delta-function problem arising in quantum mechanics. Thurber and Katz have used "fractional powers of delta functions" (see Hurd and Loeb [1974]) with infinitesimal techniques. We hope these methods will lead to more progress in such problems, although we know they cannot make the difficulties vanish.

10.5 MIXED SPACES

In this section we investigate some further examples of bounded and constrained infinitesimal relations. Recall that in Grothendeick's completeness theorem and in Schwartz' test space $\mathscr{D}$ we encountered natural examples of bounded and constrained infinitesimals. Observe also that (10.1.8) can be phrased in terms of a kind of inductive limit on bounded sets.

Except for pointing out the basic infinitesimal formulation of general mixed topologies we shall deal only with a special kind where the constraint is given by a single norm and even then focus primarily on special examples. The reader is referred to the work of Collins [1975] and Cooper [1975] for a more complete treatment and to Cooper [1971] for a brief clear introduction to the subject. Classically, mixed topologies arise from notions of bounded convergence; see (10.5.5).

Let **E** be a linear space with two locally convex linear uniformities u and v satisfying:

(1) u is finer than v, and

(2) $(\mathbf{E}, u)$ is a (DF)-space, that is, there is a fundamental sequence (B_n) of bounded sets which has the property that whenever (U_n) is a sequence of closed absolutely convex neighborhoods of zero so that $U = \bigcap_{n=1}^{\infty} U_n$ absorbs bounded sets of **E**, then U is also a neighborhood of zero.

(3) There is a base (B_n) of absolutely convex bounded sets such that $B_n + B_n \subseteq B_{n+1}$ and each B_n is v-closed.

(10.5.1) DEFINITION *The mixture of u and v, $m(u, v)$ is defined to be the finest locally convex linear uniformity on* **E** *that agrees with v on u-bounded sets.*

(10.5.2) DEFINITION *We define a bounded infinitesimal relation on* ***E**, *the mixed infinitesimals:*

$$x \overset{M}{\approx} y \quad \text{if and only if} \quad (x - y) \in bd_u(^*\mathbf{E}) \quad \text{and} \quad x \overset{v}{\approx} y.$$

(10.5.3) ***REMARK:*** Since there is a countable fundamental system of bounded sets in $\mathrm{bd}_u(^*\mathbf{E})$, the *mixed infinitesimal hull*

$$\hat{\mathbf{E}}_M = \mathrm{bd}_u(^*\mathbf{E})/\overset{M}{\approx}$$

with the gauge of M-continuous internal seminorms *is complete and contains* **E** with the mixed topology $m(u, v)$ embedded as ${}^\sigma\hat{\mathbf{E}}$. In particular, a standard seminorm is in the gauge of $m(u, v)$ if and only if it is M-continuous.

Standard seminorms arise as Minkowski functions of sets of the form

$$\bigcup_{n=1}^{\infty} [(V_1 \cap B_1) + \cdots + (V_n \cap B_n)],$$

where $\{B_n\}$ is a fundamental absolutely convex, v-closed sequence for u-bounded sets and $\{V_n\}$ is an arbitrary sequence of absolutely convex v-neighborhoods of zero. An internal seminorm q is v-continuous on $*B_j$ so there is a standard $*V_j$ such that $*V_j \cap *B_j$ is contained in $\{x \in *B_j : q(x) < \delta\}$ for standard δ. Thus we see that the topology on the mixed hull is generated by the standard M-continuous seminorms, in other words,

(10.5.4) PROPOSITION *$\hat{\mathbf{E}}_M$ is embedded in $\hat{\mathbf{E}}_m$ as the standard locally convex hull of* $\mathrm{bd}_m(*\mathbf{E})$ *with respect to the standard m-seminorms.*

(10.5.5) PROPOSITION *A sequence $\{x_n\} \subseteq \mathbf{E}$ converges in the mixed topology $x_n \xrightarrow{m} x_0$ if and only if $\{x_n\}$ is u-bounded and $x_n \xrightarrow{v} x_0$ if and only if $x_\omega \overset{M}{\approx} x_0$ for every infinite $\omega \in *\mathbf{N}_\infty$.*

If $x_\omega \overset{M}{\approx} x_0$ for all infinite ω, $*(x_n) \subseteq \mathrm{bd}_u(*\mathbf{E})$ and therefore is a bounded set. On bounded sets m-convergence is equivalent to v-convergence, which follows from the condition $x_\omega \overset{v}{\approx} x_0$.

Conversely, if p is an M-continuous standard seminorm and $x_n \xrightarrow{m} x_0$, then $p(x_n - x_0) \to 0$, in particular, $*(x_n)$ consists only of finite points and therefore forms a bounded set. Since m is finer than v, $x_n \xrightarrow{v} x_0$. (m and u have the same bounded sets.)

(10.5.6) PROPOSITION *The hull of an internal family Λ of linear maps from $*\mathbf{E}$ into a locally convex space $*\mathbf{F}$ is equicontinuous on the mixed hull $\hat{\mathbf{E}}_M$ if and only if each L in Λ is M-continuous in $*\mathbf{E}$.*

PROOF: For each $*B_j \subseteq \mathrm{bd}_u(*\mathbf{E})$ and seminorm q on $\mathbf{F}$,

$$\{x \in *B_j : (\forall L \in \Lambda)[q(L(x) < \varepsilon]\}$$

contains $\mu_v(0) \cap *B_j$ and is internal. By Cauchy's principle it contains a standard B_j-relative v-neighborhood V_j. The seminorm associated with $\bigcup_{n=1}^{\infty} [(V_1 \cap B_1) + \cdots + (V_n \cap B_n)]$ forces $q(L(x)) < \varepsilon$ on $\mathrm{bd}_u(*\mathbf{E})$. The converse is clear since on bounded sets m and v agree.

(10.5.7) ***REMARK:*** An internal set $K \subseteq \mathrm{bd}_u(*\mathbf{E}) \cap \mathrm{pns}_v(*\mathbf{E})$ has a *compact hull* K in $\mathbf{E}_M$. This follows from Cauchy's principle $K \subseteq *B \subseteq \mathrm{bd}_u(*\mathbf{E})$, and M, m, and v all agree on $*B$ so apply (8.4.34). In particular, a standard set is m-totally bounded if and only if it is u-bounded and v-totally bounded or m-compact if and only if u-bounded and v-compact.

(10.5.8) DEFINITION *The kittygory* MIXTOP *consists of the normed spaces* $(\mathbf{E}, \|\cdot\|)$ *in* $\mathscr{X}$ *together with a coarser locally convex topology v with the unit ball v-closed, that is, a standard object is a triple* $(\mathbf{E}, \|\cdot\|, v)$. *A morphism from* $(\mathbf{E}, \|\cdot\|, v)$ *to* $(\mathbf{F}, |||\cdot|||, w)$ *is a linear map that sends the unit ball of* **E** *into the unit ball of* **F** *and such that the restriction to the unit ball is v-w-weakly continuous.*

This is the special case of mixed topologies that will concern us most. The mixed topology defines a covariant functor into the locally convex spaces $(\mathbf{E}, \|\cdot\|, v) \to (\mathbf{E}, m(\|\cdot\|, v))$. Cooper shows that MIXTOP arises by "completing" Banach spaces in certain category-theory senses—projective limits in particular.

(10.5.9) PROPOSITION *When* **E** *is a Banach space with continuous dual* $\mathbf{E}'$, *the mixture of the dual norm and weak-star topology* $\sigma(\mathbf{E}', \mathbf{E})$ *is compatible with the duality. That is, every M-continuous internal linear functional is nearly represented in* **E**.

PROOF: Apply Grothendeick's completeness theorem (10.1.40).

In an irreflexive pair, such as (L^∞, L^1) one would like simple characterizations of the Mackey topology, the finest one which preserves the pair. In the case of L^∞ this means linear functionals are represented by countably additive rather than only finitely additive measures. The reason one wants the finest is that it imposes the weakest continuity requirement on the nonlinear functions defined on the space. Bewley [1972] gives an interesting example of a Mackey continuous utility function that is not weak-star continuous, for example. The Mackey–Arens' theorem (10.1.39) provides an extrinsic characterization in general. In some special cases bounded infinitesimals provide a natural intrinsic characterization. The author [1973] discovered this using ad hoc methods in the case of L^∞ and later learned it could be described by mixing. Cooper had previously (unpublished, but see [1975]) characterized the Mackey topology of (L^∞, L^1) by mixing. Our interest in the problem comes from the questions raised by Rubel [1971] concerning bounded pointwise approximation of holomorphic functions. We think this mixing provides a nice example of Rubel's "principle of the conservation of topologies" since "anything" you mix with the norm ends up as the Mackey topology.

Let μ be a standard finite measure. The *Mackey infinitesimals* of ${}^*L^\infty(\mu)$ are given by

$f \overset{M}{\approx} g$ *if and only if* $f(x) \sim g(x)$ *for* μ – a.e. x *and* $f(x) \approx g(x)$ *except on a set of infinitesimal* μ*-measure.*

Compare this with Example (10.1.6) with $\mu(S) = \sum_{s \in S} (1/s)^2$.

(10.5.10) THEOREM *The mixed hull $\widehat{L_M{}^\infty}$ equals L^∞ with the Mackey topology of (L^∞, L^1).*

PROOF: The principal ingredient is the Hahn–Vitali–Saks theorem in the form:

If $K \subseteq L^1(\mu)$ is weakly compact, then K is L^1-norm bounded and uniformly absolutely continuous.

The extension in $*L^\infty$ says that the mixed infinitesimals are finer than uniform convergence on weakly compact sets, see Stroyan [1973].

Compatibility follows from (10.1.40), but in this case a potentially useful lemma as follows can also be used:

LEMMA *If λ is a finite internal finitely additive measure* (on the $*$-algebra of μ) *and if $\mu(E) \approx 0$ implies $\lambda(E) \approx 0$, then* $\mathrm{st}[\lambda(*E)] = \Lambda(E)$ *defines a countably additive measure on the standard measurable sets.*

This much shows that L^∞ with the Mackey topology is embedded in $\widehat{L_M{}^\infty}$ as ${}^\sigma L^\infty$. The hull is a complete space, but in this case every finite point of $*L^\infty$ is M-near standard by Alaoglu's theorem (10.1.38) thus $\widehat{L_M{}^\infty} = {}^\sigma L^\infty$, in particular, *the Mackey topology on L^∞ is complete.*

Cooper extends this result to the case of a positive Radon measure on a locally compact space. One must add "except for sets of *locally* infinitesimal measure" to the end of the definition of Mackey infinitesimals.

(10.5.11) DEFINITIONS Let X be a locally compact Tychonoff space. Let $\mathrm{BC}(X)$ be the space of bounded continuous functions. Consider the following uniformities:

$$u \text{ given by the } \textit{uniform norm } \|f\|_\infty = \sup[|f(x)| : x \in X],$$

$$\kappa \text{ given by the } \textit{seminorms } |f|_K = \sup[|f(x)| : x \in K],$$

for K a compact subset of X, the *compact convergence uniformity,*

$$\beta \textit{ the strict uniformity} \text{ given by the seminorms}$$

$$|f|_\varphi = \sup[|f(x)\varphi(x)| : x \in X],$$

for $\varphi \in C_0(X)$, functions vanishing at infinity.

The infinitesimal relation of compact convergence is $f \overset{\kappa}{\approx} g$ if and only if $f(x) \approx g(x)$ for $x \in \mathrm{ns}(*X)$, the near-standard points of $*X$. We have used local compactness, $\mathrm{ns}(*X) = \bigcup [*K : K \subset\subset X]$.

The bounded infinitesimal relation of the mixture of u and κ is

$$f \overset{b}{\approx} g \text{ if and only if } f(x) \sim g(x) \text{ for } x \in {}^*X \text{ and } f(x) \approx g(x) \text{ for } x \in \text{ns}({}^*X).$$

Each of the β-seminorms $|\cdot|_\varphi$ for $\varphi \in C_0(X)$ is b-continuous, because if x is remote, $\varphi(x) \approx 0$, so $f \overset{b}{\approx} 0$ implies $f(x)\varphi(x) \approx 0$ for all $x \in {}^*X$. Moreover,

(10.5.12) THEOREM *The strict uniformity is the mixture of u and κ. The mixed hull* $\widehat{\text{BC}_b(X)} \neq (\text{BC}(X), \beta)$.

PROOF: See the work of Stroyan [1974] for a simple proof that strict seminorms generate the mixed topology using infinitesimals. We shall concentrate on properties of BC(X) that follow easily from the infinitesimal description. These go back to the reasons for which R. C. Buck introduced the strict topology.

The reason the mixed hull is larger than the complete standard space [see (10.5.7)] is that there are many dis-S-continuous finite functions. For example, take an infinitesimal compact neighborhood of a standard point x. By $*$-Urysohn's lemma there is a continuous function χ, $0 \leq \chi \leq 1$, $\chi = 0$ off some infinitesimal neighborhood of K and $\chi = 1$ on K.

(10.5.13) EXERCISE: Is $\widehat{\text{BC}_b(X)}$ a topological algebra? Is $\widehat{L_M^\infty}$?

(10.5.14) THEOREM *The continuous dual of* (BC(X), β) *is the space of countably additive finite total variation Borel measures* BM(X). *An internal measure is b-continuous if and only if its variation is a finite scalar, and it is nearly carried on the near-standard points, that is, remote measurable sets have infinitesimal variation.*

When X is not compact, the dual of BC(X) in the uniform norm consists of the countably additive measures on the Čech–Stone compactification. Such measures need not be countably additive on X. It should also be clear that another advantage of the strict topology is that approximation becomes easier.

PROOF ($\approx$ Buck): If L is a standard b-continuous linear functional, its restriction to CK(X), functions of kompact carrier, is norm continuous. By the Riesz representation L is an integral for CK(X) (see W. Rudin [1974]). Each bounded Borel measure must have σ-compact carrier, as the reader can verify, say $\bigcup K_n$. Let $a_n = |\mu|(K_n \backslash K_{n-1})$, so $\sum_{n=1}^{\infty} a_n < \infty$.

CK(X) is b-dense in BC(X) because there is (by $*$-Urysohn's lemma) a continuous function χ, one on $\bigcup_{n=0}^{\Omega} K_n$, and zero on the complement of a larger $*$-compact set. If $f \in \text{BC}({}^*X)$ with $f \overset{b}{\sim} 0$, then $\chi(x)f(x) \in \text{CK}({}^*X)$ with finite norm and $\chi f \overset{b}{\approx} f$.

We conclude by showing that $\sum_{n=1}^{\infty} a_n \in \mathcal{O}$ and $\sum_{n=\Omega}^{\infty} a_n \approx 0$ implies μ is b-continuous. If $f \overset{b}{\approx} g$, then $\sup[|f(x) - g(x)|] = h \in \mathcal{O}$ and $\{j : f(x) \approx g(x)$ for $x \in K_j\}$ contains an infinite subscript by Robinson's sequential lemma. Thus $\sup[|f(x) - g(x)| : x \in K_\Omega] = k \approx 0$, and

$$\left| \int f - g \, d\mu \right| \leq k \sum_{n=1}^{\Omega} a_n + h \sum_{\Omega}^{\infty} a_n \approx 0.$$

This proves the theorem.

Herz' theorem gives an important aspect of the more flexible approximation of β:

(10.5.15) THEOREM *Let G be a locally compact Abelian group and $0 \neq f \in \mathrm{BC}(G)$. Then there is a real λ such that $e^{i\lambda x}$ is in the strict spectrum of f, that is, the β-closed linear span of the translates of f.*

PROOF: See Collins [1975].

Let $H^\infty(U)$ denote the bounded holomorphic functions on the unit disk $|z| < 1$. Viewing $H^\infty(U)$ as a subalgebra of $\mathrm{BC}(X)$ we define

$h \overset{b}{\approx} k$ if and only if $h(z) \sim k(z)$ for all $z \in {}^*U$ and $h(z) \approx k(z)$ for $|z| \not\approx 1$.

(10.5.16) EXERCISE: Show that $h \overset{b}{\approx} k$ if and only if $h(z) \sim k(z)$ for $z \in {}^*U$ and $h(z) \approx k(z)$ for standard $z \in {}^\sigma U$.

An example of a b-infinitesimal is

$$f(z) = z^\Omega$$

for infinite $\Omega \in {}^*\mathbf{N}_\infty$. It is clearly not a norm infinitesimal.

(10.5.17) THEOREM *The mixed hull $\widehat{H_b{}^\infty(U)} = (H^\infty(U), \beta)$.*

PROOF: A finitely bounded holomorphic function is S-continuous (see Robinson's holomorphic function lemma [1966, Theorem 6.3.1]). Therefore a finite holomorphic function is b-close to its standard part.

Each bounded holomorphic function can be norm-identified with its a.e. radial limits on the unit circle, $H^\infty(U)$ is the subalgebra of $L^\infty(d\theta)$ of functions whose negative Fourier coefficients vanish. Hence M makes sense for H^∞:

$h \overset{M}{\approx} k$ if and only if $h(e^{i\theta}) \sim k(e^{i\theta})$ a.e. $d\theta$ and $h(e^{i\theta}) \approx k(e^{i\theta})$ except on sets of infinitesimal $d\theta$ measure.

An example of a holomorphic M infinitesimal is

$$g(z) = [\lambda(z)]^{\Omega}$$

where Ω is infinite in $^*\mathbf{N}_\infty$ and λ is the lens map shown in Fig. 10.5.1. The function g is not a norm-infinitesimal and M is strictly finer than b since the example f above is not an M-infinitesimal.

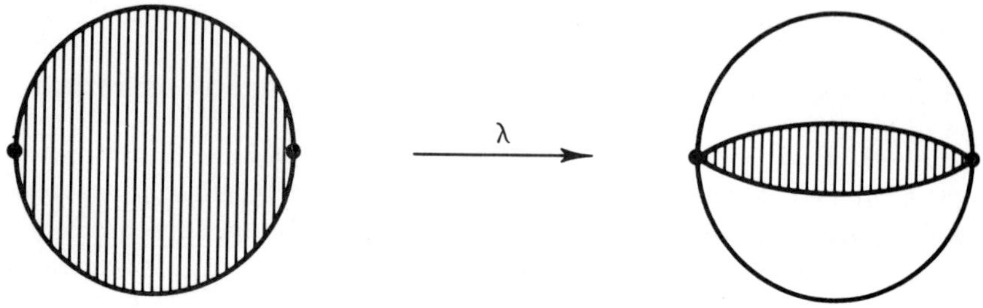

Fig. 10.5.1

(10.5.18) ***REMARK:*** Both b and M are compatible with Rubel's predual [1971].

10.6 (HM)-SPACES

In constructing the infinitesimal hull of $^*\ell^2(\mathbf{N})$ no two unit vectors $a_k = \delta_k{}^m$ and $b_k = \delta_k{}^n$ are identified since the distance between them is $\sqrt{2}$. Therefore, the hull depends on how big $^*\mathbf{N}$ is. If our reader recalls the construction of Section 7.5 it is apparent that $^*\mathbf{N}$ could have arbitrary high external cardinality—the hull depends on the model! The same is true for all infinite dimensional Banach spaces. This observation led Henson and Moore [1973] to investigate when a space has a single hull (over all enlargements). We propose to call these spaces "(HM)-spaces" since they do not coincide with any conventional classifications. The complete metrizable (HM)-spaces are exactly the (FM)-spaces, (HM) & (F) = (FM), but, in general, (HM)-spaces are not the Montel spaces. The (HM)-spaces include the nuclear spaces, the Schwartz spaces, and Bellenot [1975] showed that the property is self-dual between (F) and (DF) spaces.

Despite our remarks about different hulls for different models we shall study the kittygory of (HM)-spaces in the setting of our fixed nonstandard model from Section 7.5. In a sense these spaces are the "nearly finite-dimensional" infinite-dimensional spaces.

(10.6.1) DEFINITION *A standard locally convex space in $\mathscr{X}$, $(\mathbf{E}, N)$, is an* (HM)*-space provided*

$$\mathrm{pns}_N(^*\mathbf{E}) = \mathrm{fin}_N(^*\mathbf{E}).$$

In other words, when **E** is complete it is an (HM)-space when every finite vector is near a standard vector.

(10.6.2) PROPOSITION *The locally convex hull of a standard* (HM)*-space is its standard completion.*

PROOF: Left as an exercise.

If $\mathrm{pns}(^*\mathbf{E}) = \mathrm{fin}(^*\mathbf{E})$, then the discrete monad of each of these sets is the same, $\mathrm{pcpt}(^*\mathbf{E}) = \mathrm{bd}(^*\mathbf{E})$ and bounded sets are precompact.

(10.6.3) THEOREM *A metrizable locally convex space $\mathbf{E} \in \mathscr{X}$ is an* (HM)*-space if and only if every bounded set is totally bounded.*

PROOF: The "only if" is true in general, so let **E** be metrizable with a countable gauge of seminorms $\{p_k\}$ such that $p_k \leq p_{k+1}$.

Suppose $y \in \mathrm{fin}(^*\mathbf{E}) \backslash \mathrm{pns}(^*\mathbf{E})$. Let n_k be the smallest integer greater than $p_k(y)$. Scaling if necessary, we may assume $p_1(y - x) \geq 1$ for each $x \in {}^{\sigma}\mathbf{E}$, since y is not in the completion. Define a sequence in ${}^{\sigma}\mathbf{E}$ as follows. Let x_1 satisfy $p_1(x_1) \leq n_1$. If $x_1, \ldots, x_k$ have been chosen, with $p_1(x_i - x_j) \geq 1$ for $1 \leq i < j \leq k$ and $p_i(x_j) \leq n_i$ for $1 \leq i < j \leq k$, pick an x_{k+1} in ${}^{\sigma}\mathbf{E}$ with these same properties. This is possible, because $p_1(y - x_i) \geq 1$ and $p_i(y) \geq n_i$ for $1 \leq i \leq k$, that is, a $y \in {}^*\mathbf{E}$ exists, so by Leibniz' principle an x_{k+1} exists in **E**. The sequence $\{x_i\}$ is bounded but not totally bounded, so a metrizable non-(HM)-space has nonprecompact bounded sets.

Next we give an example of a locally convex space in which every bounded set is relatively compact, but which does not have invariant nonstandard hulls.

(10.6.4) ***EXAMPLE:*** Let **E** be the space of all **K**-valued functions g on **N** such that the set $\{k \in \mathbf{N} \mid g(k) \neq 0\}$ is finite. Let $\mathscr{F}$ be a free ultrafilter on **N** and let **F** be the space of all **K**-valued functions f on **N** such that for some M in $\mathscr{F}$ (M depending on f) f is bounded on M. Equip **E** with the topology $|\sigma|(\mathbf{E}, \mathbf{F})$ of uniform convergence on order intervals of **F**. This topology is generated by the family of seminorms

$$p_f(g) = \sum_{k=0}^{\infty} |g(k) f(k)|,$$

where f varies over **F**. It is well known and easy to show that the dual space $(\mathbf{E}, |\sigma|(\mathbf{E}, \mathbf{F}))'$ is **F**.

A subset B of **E** *is* $|\sigma|(\mathbf{E}, \mathbf{F})$*-bounded if and only if B is pointwise bounded and finite dimensional. Hence every* $|\sigma|(\mathbf{E}, \mathbf{F})$*-bounded subset of* **E** *is relatively* $|\sigma|(\mathbf{E}, \mathbf{F})$*-compact.*

PROOF: Clearly B is $|\sigma|(\mathbf{E}, \mathbf{F})$-bounded if it is pointwise bounded and finite dimensional. Suppose, conversely, that B is $|\sigma|(\mathbf{E}, \mathbf{F})$-bounded. Obviously B is pointwise bounded. Suppose the set $M = \{k \mid \text{for some } g \in B, g(k) \neq 0\}$ is not finite, and let M_1 and M_2 be disjoint, infinite subsets of M. Since $\mathscr{F}$ is an ultrafilter, either M_1 or M_2 is not in $\mathscr{F}$. We may assume $M_1 \notin \mathscr{F}$ and write $M_1 = \{n_1, n_2, \ldots\}$, where $n_1 < n_2 < \cdots$. Now define a function f with its support in M_1 such that p_f is not bounded on all of B. Since f is bounded on $\mathbf{N}\backslash M_1$, which is in $\mathscr{F}$, the function f is in **F**. Therefore B is not $|\sigma|(\mathbf{E}, \mathbf{F})$-bounded, which is a contradiction.

Choose p to be an element of $\mu(\mathscr{F})$, the monad of $\mathscr{F}$. Define e to be the internal $^*\mathbf{K}$-valued function on $^*\mathbf{N}$ that satisfies

$$e(k) = \begin{cases} 0 & \text{if } k \neq p \\ 1 & \text{if } k = p. \end{cases}$$

Then e is in $^*\mathbf{E}$, and since $p \in \mu(\mathscr{F})$, p is finite. On the other hand the constant function 1 is in **F** and for each $x \in {}^{\sigma}\mathbf{E}$, $p_1(x - e) \geq 1$ so that p is not pre-near-standard:

$$\mathrm{pns}_{|\sigma|(\mathbf{E}, \mathbf{F})}(^*\mathbf{E}) \neq \mathrm{fin}_{|\sigma|(\mathbf{E}, \mathbf{F})}(^*\mathbf{E}).$$

(10.6.5) THEOREM *Every Schwartz space is an* (HM)*-space.*

PROOF: Henson and Moore [1973].

(10.6.6) THEOREM *If* $(\mathbf{E}, N)$ *is an* (F)*-space or a* (DF)*-space, the following are equivalent*:

(a) $(\mathbf{E}, N)$ *is an* (HM)*-space.*
(b) $(\mathbf{E}', \beta(\mathbf{E}', \mathbf{E}))$ *is an* (HM)*-space.*

PROOF: Bellenot [1975].

10.7 POSTSCRIPT TO CHAPTER 10

Two of the subjects in mathematics that we are most fond of are functional analysis and the general theory of infinitesimals. In our opinion the importance of both of these subjects lies in the applications they have in the mathematical sciences. Of course, functional analysis is old and well tested—abstractly, concretely, in mathematical and scientific applications. Infinitesimals in this setting are relatively new and untested. This chapter tries to show the role of infinitesimals in part of functional analysis. It falls far short of incorporating functional analysis within the theory of infinitesimals, which was not our intention anyway.

We hope research on infinitesimals in abstract functional analysis will draw the attention of more experts. Even more, we hope that researchers in the mathematical sciences will test infinitesimal methods directly in their models—both in pure and applied mathematics. We believe that this will be fruitful in areas ranging from differential geometry to economics (Brown and Robinson [1972]). For example, it could be helpful in models where "white noise" acts simply because the basic model is closer to the scientific hypotheses.*

**Note added in proof:* Robert M. Anderson has written a very interesting paper on Brownian motion as an infinitesimal random walk. His work will appear in the *Israel Journal of Mathematics*.

REFERENCES

Almgren, F. J.

[1966] "Plateau's Problem." Benjamin, New York.

Barwise, K. J.

[1974] Back and forth thru infinitary logic, "Studies in Model Theory," M. Morley (ed.), Vol. 8. Amer. Math. Soc., Providence, Rhode Island.

Behrens, M. F.

[1969] Untitled preprint; also see Behrens [1972].

[1970] The corona conjecture for a class of infinitely connected domains, *Bull. Amer. Math. Soc.* **76**, 387–391.

[1971] The maximal ideal space of algebras of bounded analytic functions on infinitely connected domains, *Trans. Amer. Math. Soc.* **161**, 359–379.

[1972] Analytic and meromorphic function in the open unit disk. Preprint.

[1947a] Analytic sets in $\mathscr{M}(D)$. In "Victoria Symposium on Nonstandard Analysis," A. Hurd and P. Loeb (eds.). Springer-Verlag, Berlin and New York.

[1974b] Boundary values for meromorphic functions defined in the open unit disk. In "Victoria Symposium on Nonstandard Analysis," A. Hurd and P. Loeb (eds.). Springer-Verlag, Berlin and New York.

[1974c] A local inverse function theorem. In "Victoria Symposium on Nonstandard Analysis," A. Hurd and P. Loeb (eds.). Springer-Verlag, Berlin and New York.

Bellenot, S. F.

[1975] On nonstandard hulls of convex spaces. Preprint.

Bernstein, A. R.

[1972] The spectral theorem—a non-standard approach. *Z. Math. Logik Grundlagen Math.* **18**, 419–434.

Bernstein, A. and Loeb, P.

[1974] A non-standard integration theory. In "Victoria Symposium on Nonstandard Analysis," A. Hurd and P. Loeb (eds.). Springer-Verlag, Berlin and New York.

Bernstein, A. and Robinson, A.

[1966] Solution of an invariant subspace problem of K. T. Smith and P. R. Halmos. *Pacific J. Math.* **16**, 421–431.

Bewley, T. J.

[1972] Existence of equilibria in economies with infinitely many commodities. *J. Econom. Theory.* **4**, 514–540.

Bhatia, N. and Szegö, G.

[1970] "Stability Theory of Dynamical Systems." Springer-Verlag, Berlin and New York.

Birkhoff, G. D.

[1927] "Dynamical Systems," Colloq. Pub. Vol. 9. Amer. Math. Soc. Providence, Rhode Island; Also revised edition edited by Moser.

Brown, D. J. and Robinson, A.

[1972] A limit theorem on the cores of large standard exchange economies. *Proc. Nat. Acad. Sci. U.S.A.*, **69**, 1258–1260.

Carathéodory, C.
[1932] "Conformal Representation." Cambridge University Press, London and New York.
[1954] "Theory of Functions of a Complex Variable." Chelsea, Bronx, New York.
Carleson, L.
[1962] Interpolation by bounded analytic functions and the corona problem. *Ann. of Math.* **76**, 547–559.
Chang, C. C. and Keisler, H. J.
[1973] "Model Theory." North-Holland Publ., Amsterdam.
Choquet, G.
[1968] "Construction d'ultrafiltres sur **N**." *Bull. Sci. Math.* **92** (2), 41–48; Deux Classes remarquable d'ultrafiltres sur **N**. *Bull. Sci. Math.* **92** (2), 143–153.
Collins, J.
[1976] Strict, weighted, and mixed topologies, and applications. *Advances in Math.* **19**, (2), 207–237.
Cooper, J. B.
[1971] The strict topology and spaces with mixed topologies. *Proc. Amer. Math. Soc.* **30** (3), 583–592.
[1975] The mixed topology and applications. Preprint.
Cozart, D. and Moore, L.
[1974] The nonstandard hull of a normed Riesz space. *Duke Math. J.* **41** (2), 263–275.
Dixmier, J.
[1951] Sur certaines espaces consideres par M. H. Stone. *Summa Brasil Math.* **2**, 151–182.
Dvoretsky, A.
[1960] Some results on convex bodies and Banach spaces. *Proc. Internat. Symp. Linear Spaces Jerusalem*, 1960, 123–160.
Fenstad, J. and Nyberg, A.
[1969] Standard and non-standard methods in uniform topology. Logic Colloquium 1969, 353–359.
Goldberg, R. R.
[1964] "Methods of Real Analysis." Ginn (Blaisdell), Boston, Massachusetts.
Gottschalk, W. and Hedlund, G.
[1955] "Topological Dynamics." Amer. Math. Soc., Providence, Rhode Island.
Haddad, L.
[1972] Comments on nonstandard topology. Preprint.
Hahn, H.
[1907] Uber die nichtarchimedische groszensysteme. *S.-B. Wiener Akad. Math.-Natur. Kl.* **116** (Abt. IIa), 601–655.
Henson, C. W.
[1972a] The nonstandard hulls of a uniform space. *Pacific J. Math.* **43** (1), 115–137.
[1972b] On the nonstandard representation of measures. *Trans. Amer. Math. Soc.* **172**, 437–446.
[1973] The monad system of the finest compatible uniform structure. Preprint, to appear *Proc. Amer. Math. Soc.*
[1974] The isomorphism property in nonstandard analysis and its use in the theory of Banach spaces. *J. Symbolic Logic.* **39** (4), 717–731.
Henson, C. and Moore, L.
[1972] The nonstandard theory of topological vector spaces. *Trans. Amer. Math. Soc.* **172**, 405–435.
[1973] Invariance of the nonstandard hulls of locally convex spaces. *Duke Math. J.* **40**, 193–205.

[1974a] Invariance of the nonstandard hulls of a uniform space. In "Victoria Symposium on Nonstandard Analysis," A. Hurd and P. Loeb (eds.). Springer-Verlag, Berlin and New York.

[1974b] Semi-reflexivity of the nonstandard hulls of a locally convex space. In "Victoria Symposium on Nonstandard Analysis," A. Hurd and P. Loeb (eds.). Springer-Verlag, Berlin and New York.

[1974c] Subspaces of the nonstandard hull of a normed space. *Trans. Amer. Math. Soc.* **197**, 131–143.

[1974d] Nonstandard hulls of the classical Banach spaces, *Duke Math. J.* **41** (2), 277–284.

Hirsch, M. and Smale, S.

[1974] "Differential Equations, Dynamical Systems, and Linear Algebra." Academic Press, New York.

Hoffman, K.

[1967] Bounded analytic functions and Gleason parts, *Ann. of Math.* **86** (2), 74–111.

Hurd, A.

[1971a] Non-standard analysis of dynamical systems I: Limit motions, stability. *Trans. Amer. Math. Soc.* **160**, 1–26.

[1971b] Local conditions for equivalence of compact dynamical systems, *Amer. J. Math.* **93**, 742–752.

Hurd, A. and Loeb, P.

[1974] "Victoria Symposium on Nonstandard Analysis." Springer-Verlag, Berlin and New York.

James, R. C.

[1972] Super-reflexive Banach spaces, *Canad. J. Math.* **24**, 896–904.

Johnson, D. R.

[1957] Bibliography of nonstandard analysis. *Lecture Notes in Math.*, mimeographed, Univ. of Pittsburgh, Pittsburg, Pennsylvania.

Keisler, H. J.

[1967] Ultraproducts which are not saturated. *J. Symbolic Logic* **32**, 23–46.

[1976] "Elementary Calculus: An Approach Using Infinitesimals." Prindle, Weber & Schmidt, Boston, Massachusetts.

Kerr-Lawson, A.

[1965] A filter description of the homomorphisms of H^∞. *Canad. J. Math.* **17**, 734–757.

Koch, A. and Mikkelsen, Ch. J.

[1974] Topos-theoretic factorization of nonstandard extensions. In "Victoria Symposium on Nonstandard Analysis," A. Hurd and P. Loeb (eds.). Springer-Verlag, Berlin and New York.

Kugler, L.

[1969] Nonstandard analysis of almost periodic functions. In "Applications of Model Theory to Algebra, Analysis and Probability," W. A. J. Luxemburg (ed.). Holt, New York.

Laugwitz, D.

[1959] Eine Einführung der δ-Funktionen, *S.-B. Bayerische Akad. Wiss.* **4**, 41–59.

[1961] Anwendungen unendlich kleiner Zahlen I & II. *J. Reine Angew. Math.* **207**, 53–60; **208**, 22–34.

[1974] On Abraham Robinson's sequential lemma. Preprint.

Laugwitz, D. and Schmieden, C.

[1958] Eine Erweiterung der Infinitesimalrechnung, *Math. Z.* **69**, 1–39.

Leibniz, G. (see bibliography in Robinson [1966]).

[1974] "Leibniz in Paris 1672–1676," by J. E. Hofmann. Cambridge Univ. Press, London

and New York.
[1920] "The Early Mathematical Manuscripts of Leibniz," by J. M. Child, Open Court Publ., La Salle, Illinois.

Lindenstrauss, J. and H. P. Rosenthal
[1969] The $\mathscr{L}_p$-spaces. *Israel J. Math.* **7**, 325–349.

Loeb, P.
[1972] A non-standard representation of measurable spaces L_∞ and $L_\infty{}^*$. In "Contributions to Non-Standard Analysis," W. A. J. Luxemburg and A. Robinson (eds.) North-Holland Publ., Amsterdam.
[1973] A combinatorial analog of Lyapunov's theorem for infinitesimally generated atomic vector measures. *Proc. Amer. Math. Soc.* **39**, 585–586.
[1974] A nonstandard representation of Borel measures and σ-finite measures. In "Victoria Symposium on Nonstandard Analysis," A. Hurd and P. Loeb (eds.). Springer-Verlag, Berlin and New York.
[1975] Conversion from non-standard to standard measure spaces and applications in probability theory. *Trans. Amer. Math. Soc.*, **211**, 113–122.

Lomonozov, V. I.
[1973] Invariant subspaces of a family of operators commuting with a compact operator, *Funkcional Anal. i Prilozen.* **7**, 3.

Łos, J.
[1955] Quelques remarques, theoremes, et problemes sur les classes definissables d'algebras. In "Mathematical Interpretations of Formal Systems," Skolem *et al.* (eds.). North-Holland, Amsterdam.

Luxemburg, W. A. J.
[1962] "Non-standard Analysis, Lectures on A. Robinson's Theory of Infinitesimal and Infinitely Large Numbers." Caltech Bookstore, Pasadena, revised 1964.
[1969] "Applications of Model Theory to Algebra, Analysis and Probability." Holt, New York.
[1972a] A non-standard analysis approach to Fourier analysis. In "Contributions to Non-Standard Analysis," W. A. J. Luxemburg and A. Robinson (eds.). North-Holland Publ., Amsterdam.
[1972b] On some binary relations occurring in analysis. In "Contributions to Non-Standard Analysis." W. A. J. Luxemburg and A. Robinson (eds.), North-Holland Publ., Amsterdam.

Luxemburg, W. A. J. and Robinson, A.
[1972] "Contributions to Non-Standard Analysis." North-Holland Publ., Amsterdam.

Luxemburg, W. A. J. and Taylor, R. F.
[1970] Almost commuting matrices are near commuting matrices. *Proc. Royal Acad., Amsterdam, ser. A*, **73**, 96–98.

Machover, M. and Hirschfeld, J.
[1969] "Lectures on Non-Standard Analysis," Springer-Verlag, Berlin and New York.

Moore, L. C.
[1974] Hyperfinite extensions of bounded operators on a separable Hilbert space. (To appear, *Trans. Amer. Math. Soc.*).

Nemytskii, V. and Stepanov, V.
[1960] "Qualitative Theory of Differential Equations." Princeton Univ. Press, Princeton, New Jersey.

Newman, J. R.
[1956] "The World of Mathematics," Vol. 1, Simon and Schuster, New York.

Puritz, C. W.

[1972] Skies, constellations and monads. In "Contributions to Non-Standard Analysis," W. A. J. Luxemburg and A. Robinson (eds.). North-Holland Publ., Amsterdam.

Raikov, D. A.

[1959] A criterion for completeness in locally convex spaces, *Uspehi Mat. Nauk* **14** (85), 223–229.

Robinson, A.

[1961] Non-standard analysis, *Proc. Roy. Acad. Amsterdam Ser. A*, **64**, 432–440.

[1964] On generalized limits and linear functionals, *Pacific J. Math.* **14**, 269–283.

[1966] "Non-Standard Analysis." North-Holland Publ., Amsterdam.

[1969] Compactification of groups and rings, and non-standard analysis, *J. Symbolic Logic*, **34** (4), 576–588.

[1973a] Metamathematical problems. *J. Symbolic Logic*, **38** (3), 500–516.

[1973b] Function theory on some non-archimedean fields. *Amer. Math. Monthly*, **80** (6).

Robinson, A. and Zakon, E.

[1969] A set-theoretical characterization of enlargements. In "Applications of Model Theory to Algebra Analysis and Probability," W. A. J. Luxemburg (ed.). Holt, New York.

Rosenthal, P.

[1969] *American Math. Monthly Res. Problem*, **76**, p. 925.

Rubel, L. A.

[1971] Bounded convergence of analytic functions, *Bull. Amer. Math. Soc.* **77**, 13–24 and bibliography.

Rudin, M. E.

[1975] "Lectures on Set Theoretic Topology," *Amer. Math. Soc. Regional Conference Series*, no. 23. Providence, Rhode Island.

Rudin, W.

[1956] Homogeneity problems in the theory of Čech compactifications, *Duke Math. J.* **23**, 409–419.

[1974] "Real and Complex Analysis," 2nd ed. McGraw-Hill, New York.

Stroyan, K.

[1972a] Uniform continuity and rates of growth of meromorphic functions, in "Contributions to Non-Standard Analysis," W. A. J. Luxemburg and A. Robinson (eds.). North-Holland Publ., Amsterdam.

[1972b] Additional remarks on the theory of monads. In "Contributions to Non-Standard Analysis," W. A. J. Luxemburg and A. Robinson (eds.). North-Holland Publ., Amsterdam.

[1973] A characterization of the Mackey uniformity $m(L^\infty, L^1)$ for finite measures. *Pacific J. Math.* **49**, 223–228.

[1974] A nonstandard characterization of mixed topologies. In "Victoria Symposium on Nonstandard Analysis," A. Hurd and P. Loeb (eds.). Springer-Verlag, Berlin and New York.

Wattenberg, F.

[1971] Non-standard topology and extensions of monad systems to infinite points. *J. Symbolic Logic* **36**, 463–476.

[1973] Monads of infinite points and finite product spaces. *Trans. Amer. Math. Soc.* **176**, 351–368.

Whitney, H.

[1957] "Geometric Integration Theory," Princeton Univ. Press, Princeton, New Jersey.

Young, L.
[1972] Functional analysis—a non-standard treatment with semifields. In "Contributions to Non-Standard Analysis," W. A. J. Luxemburg and A. Robinson (eds.). North-Holland Publ., Amsterdam.

Zalcman, L.
[1975] A heuristic principle in complex function theory. *Amer. Math. Monthly* **82**, 813–817.

INDEX

A

Accordion, Schwarz', 124
Additive function, 266
Adequate filter, 183
Alaoglu's theorem, 279
Almost periodic function, 259
Almost periodic point of a flow, 156
Archimedean order, 67
Ascoli's theorem, 217

B

Back-and-forth, 193
Banach space, 292
Banach–Mazur limits, 154
Birkhoff's theorem, 157
Blaschke product, 246
Bloch's principle, 219
Bohr group, 262
Bolzano's theorem, 76
Boundary arc, 168, 238
Bounded formal sentence, 33, 42
Bounded point, 274
Bounded weak infinitesimals, 287
Bracket operation on vector fields, 134

C

Cauchy net or filter, 212
Cauchy's principle, 71, 188, 196
Čech–Stone compactification, 197, 199, 214, 228
Cesaro limits, 153
Chain rule, 84, 96, 99
Change of variables theorem, 116
Chromatic set, 198
Cluster set, 239
Coarser monad, 179
Compact enlargement, 181
Compact operator, 289
Compact point, 202
Completeness, of the hull, 273
Comprehensive model, 180, 187
Concurrent binary relation, 177, 180
Constellation, 176
Content, Jordan, 114
Continuity principles, *see* Cauchy's principle and Robinson's sequential lemma
Convergence
  pointwise, precompact, compact, uniform, 216
Corona problem, 244

D

Darboux's theorem, 87
de Rham's theorem, 135
Delta function, 300
Delta-incomplete ultrafilter, 37
Delta-stable ultrafilter, 175
Derivative, 12, 83, 94, 95
Differential form, 121
Diffusion, order of, 223
Discrete monad, 197
Discrete standard part, 181
Distribution, 299
Divergence, 139
Duality, 275

E

Element mod ultrafilter, 38
Elementarily equivalent, 191
Elementary morphism, 191
Elementary substructure, 191
Embedded standard copy, 9, 26, 44, 46
Enlargement, 178
Equal mod ultrafilter, 38
Equicontinuity, 217
Euler's product formula, 147
Exterior algebra, 118
Exterior derivative, 121
External entity, 28

F

Factorization into set theory, 43
Fiber, 250
Filter of sections, 180
Finer monad, 179
*–Finite, 48, 177, 178
*–Finite family lemma, 188
Finite linear map, 93, 276
Finite number
  rational, 9
  real, 53
  complex, 60
Finite points, locally convex, 270
*–Finite power, 13
*–Finite product, 59
*–Finite sum, 13, 59
Flat function, 106
Flow, 155
Formal language, 31
Full galaxy, 245
Fundamental theorem of calculus, 89, 123

G

Galaxy, 207
Galaxy of a hyperreal, 57
Galaxy, hyperbolic, 163
Gauss' divergence theorem, 140
Generalized limit, 152, 154
Gleason part, 245
Grothendieck's theorem, 280

H

Harnack's inequality, 171
Hausdorff property, weak, 199
Henson's lemma, 192, 194
Hilbert cube, 203
Hilbert space, 285
(HM)-space, 313
Hyperbolic geometry, 161
Hyperbolic limit, 240
Hyperbolic motions, 160, 163, 166
Hyperbolic range, 240
Hypercomplex number, 60
Hyperconvex envelope, 273
Hyperfine partition, 232
Hyperfinite dimensional space, 271, 288
Hyperrational number, 8
Hyperreal number, 49

I

Implicit function theorem, 110
Infinite number
  rational, 9
  real, 53
  complex, 60
Infinitesimal hull, 206
Infinitesimal hull
  locally convex, 270
  mixed, 306
Infinitesimal linear map, 93
Infinitesimal neighborhood, 206, 207, 212, 221
Infinitesimal number
  rational, 9
  real, 53
  complex 60
Infinitesimal partition, 81
Infinitesimal relation, 10, 55, 60, 205, 221, 228, 268, 269
Infinitesimal transformation, 127
Infinitesimal vectors, 92
Infinitesimal, hyperbolic, 163
Infinitesimal, measurable, 232
Infinitesimally close, 10, 55, 60, 205, 221, 228, 268, 269
Infinity, standard, 14, 61
Internal definition principle, 30
Internal entity, 28
Interpolating sequence, 246
Interpretation map, 35, 37
Inverse function theorem, 107

K

Kittygory, 46
Krein's theorem, 283
Kronecker approximation, 267

L

Leibniz' principle, 28
Lie derivative, 130
Littlewood's principles, 231
Los' theorem, 39
Luxemburg's *–finite lemma, 188
Lyapunov stable, 157

M

Mackey topology, 277, 309
Mackey-Arens' theorem, 279
Macles, 195
Magnitude, order of, 224
Manifold, 126
*–Map, 26, 44, 46
Maximal ideal space, 229, 244
Mazur's theorem, 282
Mean value theorem, 85, 97, 98
Metalanguage, 34
Monad, 177, 195
Monad of a uniformity, 211
Mostowski collapsing, 43

N

Near-standard point, 200
Noncompact near-standard point, 202, 217
Nonmeasurable function, 218
Nonstandard model, 25, 36
Normal function, 238

O

Oh calculus, Landau's, 90
Open mapping, infinitesimal, 240
Order ideal, 66
Ordered integral domain, 65
Ordered ring, 64

P

P-point, 175
Paracompact space, 226
Partial *L*-isomorphism, 193
Paving of Euclidean space, 111
Peano's theorem, 142
Periodic point of a flow, 156
Polars, 275
Polysaturated model, 182, 186
Pre-near-standard point, 213
Precompact point, 215
Product formula, 84, 96

R

Random pair, 176, 188, 202, 212
Real compact space, 227
Recurrent point of a flow, 156
Regular element, 63
Regular linear map, 117
Remote point, 200
Retraction theorem, 293
Riemann integral, 79, 113, 117, 139
Riemannian metric, 138
Ring of sets, 195
Robinson's sequential lemma, 150

S

S-continuous, 48, 71, 206, 209, 210, 221, 270, 276
S-regular, 115
Saturated model, 182
Saturated set, 206
Shell method, 80
Skies, 176
Small cardinality, 182, 186, 187
Small first-order theory, 188, 190
Sobolev's lemma, 304
Standard copy, embedded, 9, 26, 44, 46
Standard definition principle, 29
Standard entity, 28
Standard part, 9, 28, 55, 58, 60, 200
Stereographic projection, 61
Stokes' theorem, 141
Stone space, 231, 233, 234
Strict topology, 309
Strong topology, 277
Subinterpretation, 190
Superreflexive, 295
Superstructure monomorphism, 26
Superstructure, definition, 23

T

Tangent bundle, 130
Tangential point, 168, 251
Taylor's formula, 100
Tesselation, 236
*–Transform, 27
Tychonoff space, 199
Tychonoff's theorem, 202

U

Ultrafilter, 7, 37
Ultralimit model, 184, 186
Ultrapower models, 8, 37
Uniform differentiability, 94, 96, 97, 100
Uniform *S*-continuity, 77, 210

V

Very remote galaxies, 243
Virtual motion, 166
Volume density, 137
Volume element, 119

W

Wandering point of a flow, 157
Wattenberg's infinitesimals, 222
Weak topology, 277
Wolff's lemma, 196

A
B 6
C 7
D 8
E 9
F 0
G 1
H 2
I 3
J 4